THE TOS HANDBOOK OF TEXAS BIRDS

Number Forty-seven:
Louise Lindsey Merrick Natural Environment Series

The TOS Handbook of Texas Birds

Second Edition, Revised

MARK W. LOCKWOOD

BRUSH FREEMAN

Texas A&M University Press   College Station

The paper used in this book meets the minimum requirements
of the American National Standard for Permanence
of Paper for Printed Library Materials, Z39.48-1984.
Binding materials have been chosen for durability.

LIBRARY OF CONGRESS CATALOGING-IN-PUBLICATION DATA
Lockwood, Mark, author.
    The TOS handbook of Texas birds / Mark W. Lockwood and
Brush Freeman. — Second edition, revised.
        pages cm — (Louise Lindsey Merrick natural environment
series ; number forty-seven)
    Includes bibliographical references and index.
    ISBN 978-1-62349-176-5 (cloth : alk. paper) —
ISBN 978-1-62349-120-8 (paper, with flaps : alk. paper) —
ISBN 978-1-62349-143-7 (e-book) 1. Birds—Texas. I. Freeman,
Brush, 1951– author. II. Title. III. Title: Texas Ornithological
Society handbook of Texas birds. IV. Series: Louise Lindsey
Merrick natural environment series ; no. 47.
    QL684.T4L633 2014
    598.09764—dc23
                                                2013030346

TO GREG LASLEY—

His contribution to the documentation

of the status and distribution of Texas birds

continues to guide us.

CONTENTS

PHOTOGRAPHS

ACKNOWLEDGMENTS

THIS BOOK reflects the efforts of ornithologists and birders alike who have reported their observations over the years. We thank everyone who has taken the time to submit sightings to the Texas Bird Records Committee (TBRC) and to the journal now known as *North American Birds,* which has played a prominent role over the past 50 years in collecting and disseminating that information. Birders who formally contribute their sightings to the journal and to the committee not only play an important role in documenting the ornithological history of Texas but also generate the information so critical to the study of the status and distribution of birds on a larger scale.

The first edition of the *TOS Handbook of Texas Birds* (2004) was reviewed by many past and present members of the TBRC, as well as by regional experts on the avifauna of Texas. Their critical review was essential to developing the basis for this revision. We again thank Keith Arnold, John Arvin, Kelly Bryan, Mel Cooksey, Bert Frenz, Tony Gallucci, John Gee, Petra Hockey, Greg Lasley, Guy Luneau, Terry Maxwell, Brad McKinney, Martin Reid, David Sarkozi, Willie Sekula, Ken Seyffert, Cliff Shackelford, Matt White, and Barry Zimmer for their assistance. The species accounts in this edition were reviewed by Mary Gustafson, Randy Pinkston, Martin Reid, and Cliff Shackelford. Martin Reid and Cynthia Lindlof meticulously copyedited the manuscript as part of their review. We thank them for all their thoughtful and valued comments, which greatly enhanced the quality of this edition. In addition, we would be remiss not to thank the regional compilers of *North American Birds,* whose volunteer efforts are the cornerstone of the indispensable data collection that greatly enhances this update. These compilers include Eric Carpenter, D. D. Currie, Bert Frenz, Anthony Hewetson, Jim Paton, Willie Sekula, John Tharp, and Ron Weeks. We also thank the people who commented on the previous edition, in particular, Peter Barnes, Kelly Bryan, Mike Dillon, Bob Doe, Tim Fennell, Tony Gallucci, Anthony Hewetson, Rich Kostecke, and David Wolf, for providing often lengthy notes and input for this edition. Fred Collins graciously provided information about the occurrence of exotic species within the state.

We were very fortunate to have many photographers graciously allow us the use of their photographs to illustrate this edition. Their contributions toward the final product are certainly greatly appreciated. These kind contributors include Lynn Barber, Trey Barrow, Steve Bentsen, Brandon Best, Don Bleitz, Erik Breden, Jan Dauphin, Larry Ditto, Marc Eastman, Maryann M. Eastman, Gil Eckrich, Robert Epstein, Terry Ferguson, Tony Frank, Andy Garcia, Jay Gardner, Brian Gibbons, Karen Gleason, Michael L. Gray, Mary Gustafson, Dave Hanson, Garrett Hodne, Ken Hunt, Joseph C. Kennedy, Michael Lindsey, Scotty Lofland, Stephan Lorenz, Geoff Malosh, Matthew Matthiessen, David McDonald, Rick Nirschl, Carolyn Ohl-Johnson, Lee Pasquali, Jim Paton, Linda Gail Price, Martin Reid, Dan Roberts, Billy Sandifer, Lawrence Semo, Bruce Strange, Christopher Taylor, Kerry Taylor, Rob Tizard, and Alan Wormington. We would particularly like to thank Greg W. Lasley, Martin Reid, and Robert Epstein for the generous contribution of so many of their wonderful photographs.

Since the TOS checklist was first published in 1974, great changes to our knowledge of Texas birds have occurred. Everything we have learned over the last four decades is a direct reflection of observer participation in providing detailed data for the record. Much is still to be learned about the avifauna in Texas, and many counties remain to be explored in greater depth. It is our hope, and the hope of all those interested in preserving and building upon the record presented here, that observers will continue to gather and share their information about this state's diverse bird life.

THE TOS HANDBOOK OF TEXAS BIRDS

INTRODUCTION

T HE SECOND EDITION of *The TOS Handbook of Texas Birds* continues an effort to accurately reflect the bird life of Texas that began with the first edition of the *Check-list of the Birds of Texas* published by the Texas Ornithological Society (TOS) in 1974. As were previous editions, it is based on the observations of past and present ornithologists and naturalists.

The Texas Bird Records Committee (TBRC) has the responsibility to maintain a state checklist and as part of that responsibility has published updates on the average of every 10 years. The second edition of *The TOS Handbook of Texas Birds* is in effect the fifth edition of the society's *Check-list of the Birds of Texas*. Changes in terms of additions or deletions to the TOS state list since 2004 have appeared in the annual reports of the TBRC published in the *Bulletin of the Texas Ornithological Society* (Carpenter 2012; Lockwood 2005a, 2006, 2007a, 2008b, 2009, 2010, 2011).

This edition reflects the recent additions and deletions to the state list as a result of taxonomic revisions as well as changes in distribution. We have also attempted to capture changes in range and relative abundance. This edition has been substantially enlarged to incorporate discussion of subspecies known to occur in the state and of their distribution.

The Texas List and Its Nomenclature

This fifth edition of the TOS "checklist" includes 639 species of birds accepted for the state by the TBRC. It represents an increase of 17 species since 2004. Most of these gains are the result of new discoveries or documented occurrences; however, a number of changes reflect recent taxonomic decisions by the American Ornithologists' Union (AOU) Committee on Classification and Nomenclature.

The taxonomic treatment and species sequence in this checklist follow the *Check-list of North American Birds, Seventh Edition* (AOU 1998), as currently supplemented.

The return to inclusion of subspecies in this volume is complicated by the fact that the AOU has not treated subspecies since 1957. Primary sources for the subspecies listed here include the online Clements et al. (2012) and Gill and Donsker (2012) publications. This base list was also compared against the taxonomy used by Pyle in the *Identification Guide to North American Birds, Parts I and II* (1997, 2008) and the various volumes of the *Handbook of the Birds of the World*. Even though we have used these excellent resources, the subspecific taxonomy included here should be considered provisional. In addition, many species have occurred in Texas for which there are no specimens known and therefore no opportunity for a subspecific determination. These are primarily vagrants, and the subspecies listed are based on probability of occurrence. These cases are noted in the species account.

Documentation

For a species to be considered for inclusion on the official Texas state list of birds of the TOS, there must exist a known specimen, a recognizable and confirmed photograph or video, or a recognizable and confirmed audio recording. Any potential new state record must meet these criteria and be reviewed and accepted by the TBRC. The TBRC requests and reviews documentation on any report of a species that appears on its Review List or that has never been recorded in the state. Review Species occur, on average, fewer than four times per year for a 10-year period. Those on the current list are indicated in the species accounts, and a complete listing is found in appendix D. The TBRC reviews reports of Review List species. If the TBRC votes in favor of a report, it is considered to be an "accepted" record. The review of such reports is a continuing process, and information about the number of accepted records of any particular Review Species is likely to have changed as of this book's publication date. The files of the TBRC are housed at the Texas Cooperative Wildlife Collection at Texas A&M University in College Station. For more information about TOS and the TBRC, please visit http://www.texasbirds.org/.

Species Accounts

The purpose of this book is to define and update the known status and distribution of the birds of Texas, as defined by the TBRC. It is not intended as an identification guide, and the authors make no effort toward describing physiological features, vocalizations, other identifying traits, or a species' natural history. A broad range of excellent books in print already serves those purposes. Abundance and distribution are described for each season and for ecological regions if they differ; this information is augmented by a map for each species. Exceptional out-of-range records of nonreview species are not presented on the associated maps but are instead covered within the text (see the following "Maps" section). For each species, where appropriate, we provide the typical migration periods with extreme arrival and departure dates for species that do not typically winter or summer within the state. The issue of extreme migration dates is complex in Texas, where a lingering bird's presence is often overlapped by the arrival and departure dates of spring and fall migrants. In Texas, the lapse between late-spring and early-fall migration for some species can be as short as three weeks. Please note that the county designations of offshore records are used only as a reference to the origin point of a pelagic trip or the nearest point of land. This is particularly true for records from the central coast, where many sightings made from trips originating from Port Aransas, Nueces County, or Port O'Connor, Calhoun County, may in fact be closer to other adjacent counties.

Each TBRC Review Species is noted as such in the species account. For Review Species with five or fewer accepted records, each record is listed below the species account with date(s) of occurrence, location, the TBRC record number, and, if applicable, the Texas Photo Record File (TPRF) number. The photo file is housed at the Texas Cooperative Wildlife Collection at Texas A&M University in College Station. In a few cases, a specimen (marked with *) catalog number or the catalog number of an audio recording, archived at the Texas Bird Sounds Library (TBSL) located at Sam Houston State University in Huntsville, is also listed. The accounts in this book are current as of 1 July 2013.

Maps

Each species account includes a map that is intended to reflect the typical expected range for that species, giving an overall picture of the species' distribution in Texas. The nature of bird distribution introduces some arbitrariness to the definition of the ranges. Many species are rare to very rare in areas outside the mapped range, and we did not attempt to map all the locations of the thousands of extralimital records that have occurred within the state. However, we do mention some of the more significant out-of-range records within the species account.

Status and Abundance Definitions

This book adheres roughly to the status and abundance codes as defined in the third edition of the *Check-list of the Birds of Texas* (1995) and has been further refined from the definitions provided in the first edition of *The TOS Handbook of Texas Birds*.

permanent resident: Occurs regularly within the defined range throughout the year and implies a stable breeding population

summer resident: Implies a breeding population, although in some cases this population may be small

summer visitor: Implies a nonbreeding population or lingering migrants or winter residents

winter resident: Occurs regularly within the described range generally between December and February

winter visitor: Does not occur with enough regularity or in large enough numbers to be considered a winter resident

migrant: Occurs as a transient passing through the state in spring and/or fall (certain species may be migrants in some regions and residents in others)

local: May be found only in specific habitats or geographic area within any region, possibly in small numbers (e.g., Red-cockaded Woodpecker, Brown Jay, Colima Warbler)

abundant: Always present and in such numbers and with such general distribution in proper habitat that many may be found in a given day

common: Normally present and in such numbers that one may expect to find several in a day uncommon: Normally present in proper habitat, but one cannot be sure of finding one in a day

rare: On average, occurs only a few times a year in a given area or not at all

very rare: Not expected every year but does occur regularly

casual: One or a few records for any given area but reasonably expected to occur again. For Review Species, generally refers to species with between 6 and 15 documented records.

accidental: Species that are considered well out of the expected range with either extended intervals between occurrences or that might not be expected to occur again. For some species, changes in distribution may make additional records more likely, but currently available records do not allow for the status to be upgraded. For Review Species, generally refers to species with five or fewer records statewide.

irregular: Unpredictable and may include scattered and unrelated occurrences of individuals or small flocks outside their expected range

irruptive: Normally absent from the state or a given area but subject to large-scale nonannual incursions, typically with a variable number of years between such incursions that are linked to abnormal circumstances in their normal range

Appendixes

Appendix A: Presumptive Species List

The TBRC has accepted sight records for these species, but no specimen, photograph, video, or audio recording has been submitted, resulting in the species not yet meeting the requirements for full acceptance on the Texas list.

Appendix B: Non-accepted Species

The TBRC has not accepted a number of species for inclusion on the Texas list or Presumptive Species List. These are species about which a legitimate debate may exist and/ or a species for which published reports can be found in the literature but are open to question.

Appendix C: Exotics and Birds of Uncertain Provenance

This list includes species that are known exotics or birds of uncertain origin. Some of these species may become established in small localized areas or are obvious escaped caged birds for which geographic range and species argue against a natural occurrence in Texas. The status of a very few of these may be subject to change in the future dependent upon decisions of the AOU or as a result of large expansions of introduced breeding populations. Examples of species formerly on this list that are now accepted by the TBRC include Red-crowned Parrot, Green Parakeet, Monk Parakeet, and Eurasian Collared-Dove.

Appendix D: List of Review Species

The TBRC requests documentation of all species that appear on the Review Species List. In general, this list includes birds that have occurred four or fewer times per year anywhere in Texas over a 10-year period. Guidelines for preparing rare-bird documentation can be found in Dittmann and Lasley (1992).

NATURAL AREAS OF TEXAS

Texas can be divided into natural or ecological regions that are defined by geology and vegetation. For the purposes of this book, we have chosen to follow the natural area boundaries developed by the Lyndon B. Johnson School of Public Affairs of the University of Texas at Austin (1978). The Lyndon B. Johnson School divided the state into 11 regions with some additional subregions. Although we are generally using these boundaries, some of the regions have been combined so that only eight natural areas are used in this book. Some important subregions are specifically mentioned in the species accounts, however, and those are defined here as well.

Texas becomes more arid as one moves from east to west, and this, of course, has a definite impact on the vegetation present. Between the mesic Pineywoods in the east, which receives up to 50 inches of precipitation annually, and the deserts of far West Texas, where as little as eight inches of precipitation occurs each year, lies a complex assemblage of habitats that supports the diverse avifauna of the state. Riparian corridors along major river systems provide important habitats that are often very different from the surrounding vegetation.

The **Pineywoods** area is the easternmost ecological region in Texas. It covers 15.8 million acres or about 9.4 percent of the state. In general, the Pineywoods region is nearly level, with some gently rolling to hilly country and elevations ranging from 200 to 500 feet above mean sea level. As the name implies, mixed pine-hardwood forests dominate the vegetational communities within the Pineywoods. Native pines common to the region are loblolly (*Pinus taeda*), shortleaf (*P. echinata*), and longleaf (*P. palustris*). Slash pine (*P. elliottii*), from the southeastern United States, has been widely introduced. Throughout the uplands, hardwoods are found in mixed stands with pines. Common hardwoods in the region include sweetgum (*Liquidambar styraciflua*), various oaks (*Quercus* spp.), elms (*Ulmus* spp.), cottonwoods (*Populus* sp.), and hickories (*Carya* spp.), as well as water tupelo (*Nyssa aquatica*),

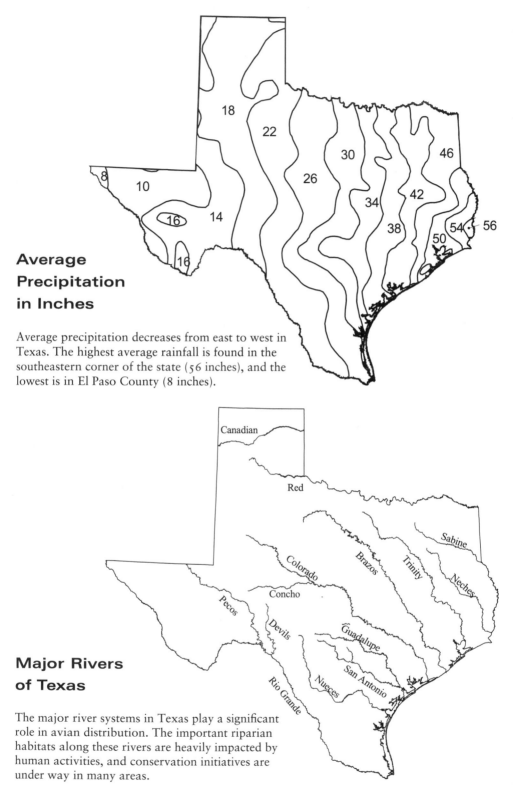

Average Precipitation in Inches

Average precipitation decreases from east to west in Texas. The highest average rainfall is found in the southeastern corner of the state (56 inches), and the lowest is in El Paso County (8 inches).

Major Rivers of Texas

The major river systems in Texas play a significant role in avian distribution. The important riparian habitats along these rivers are heavily impacted by human activities, and conservation initiatives are under way in many areas.

blackgum (*N. sylvatica*), and bald cypress (*Taxodium distichum*).

The **Coastal Prairies** is the smallest natural region in Texas and covers 10.3 million acres or about 6.1 percent of the state. This region includes prairies, marshes, estuaries, and dunes on a nearly level plain that extends along the Gulf Coast from Mexico to Louisiana and includes the barrier islands. The Coastal Prairies reaches up to 80 miles inland from the Gulf, with elevations ranging from sea level to 150 feet. The transition from Coastal Prairies to the Pineywoods and Post Oak Savannah is abruptly delineated except along drainages where riparian corridors extend into the prairies. The region grades more evenly into the South Texas Brush Country, narrowing to include the barrier islands and a thin corridor along the Laguna Madre south of Baffin Bay. The natural vegetation of the Coastal Prairies is tallgrass prairie and oak savannah. However, many of these grasslands have been invaded by trees and shrubs such as the exotic Chinese tallow (*Sapium sebiferum*), native honey mesquite (*Prosopis glandulosa*), and various acacias, in particular. Gulf cordgrass (*Spartina spartinae*), big bluestem (*Andropogon gerardii*), little bluestem (*Schizachyrium scoparium*), indiangrass (*Sorghastrum nutans*), and gulf muhly (*Muhlenbergia capillaris*) are common native grasses in these prairies.

An important natural region is the **Coastal Sand Plain**. This region is generally included as a subregion of the South Texas Brush Country. However, for ease in discussing bird distribution, this subregion is included here in the Coastal Prairies (e.g., along the Coastal Prairies north to the central coast). The Coastal Sand Plain is an area of deep sands found south of Baffin Bay that borders the narrow extension of the Coastal Prairies inland to eastern Jim Hogg County. The sand plain is stabilized by vegetation, for the most part, and is largely dominated by live oak mottes.

West of the Pineywoods a large area with belts of forest, savannah, and grassland makes up the **Post Oak Savannah** and **Blackland Prairies** regions, often referred to as the **Oaks and Prairies** region. These combined regions cover about 32.2 million acres or roughly 19.2 percent of the state. The Post Oak Savannah is found just to the west of the Pineywoods and grades into the Blackland Prairies to the south and west. This area is a gently rolling wooded plain with a distinctive pattern of post oak (*Quercus* *stellata*) and blackjack oak (*Q. marilandica*) in association with tall grasses. This basic vegetation type also characterizes the Cross Timbers subregion, which we also include in this region. The southwestern boundary with the South Texas Brush Country is indistinct. The Blackland Prairies intermingles with the Post Oak Savannah in the southeast and has divisions known as the Grand, San Antonio, and Fayette Prairies. This region was once an expansive tallgrass prairie dominated by little bluestem, big bluestem, indiangrass, tall dropseed (*Sporobolus asper*), and Silveus dropseed (*S. silveanus*). About 98 percent of the Blackland Prairies has been under cultivation for the past century. Many areas have been invaded by woody plants.

The **South Texas Brush Country** is an area of brushlands primarily found south of the Balcones Escarpment. The region covers 20.6 million acres or 12.2 percent of the state and is a moderately dissected, nearly level to rolling plain. Formerly, areas of dense brush were found only on ridges. Grazing and suppression of fires have altered the vegetation so that the region is now dominated by brushy species that include mesquite, live oak, several acacias, lotebush (*Zizyphus obtusifolia*), spiny hackberry (*Celtis pallida*), whitebrush (*Aloysia gratissima*), and Texas persimmon (*Diospyros texana*). The **Lower Rio Grande Valley** is found at the southern tip of this region. This distinctive subregion lies in the subtropical zone and is located within the delta of the Rio Grande and its alluvial terraces. Many species of plants reach their northern distribution in the Lower Valley. Historically, the floodplain of the Rio Grande supported a more diverse hardwood woodland that included sugarberry (*Celtis laevigata*), cedar elm (*Ulmus crassifolia*), ebony (*Pithecellobium ebano*), and anacua (*Ehretia anacua*). The Texas sabal (*Sabal texana*) was a locally common component of that woodland. Because of the construction of numerous dams along the Rio Grande, the seasonal flooding that maintained the natural vegetation along the floodplain has ceased, and brush species from the north are invading this area.

The **Edwards Plateau** covers approximately 11.3 million acres or 16.6 percent of the state. This region also includes the Llano Uplift or Central Mineral Region. The Balcones Escarpment bounds the Edwards Plateau on the east and south. This region is deeply dissected with numerous streams and rivers. The Balcones Canyonlands

form the true Hill Country along the escarpment and are
dominated primarily by woodlands and forests, with grasslands restricted to broad divides between drainages. Protected canyons and slopes support Ashe juniper (*Juniperus
ashei*)–oak forests. The dominant oak species differ depending on the location but include Lacey oaks (*Quercus
laceyi*), Texas red oak (*Q. buckleyi*), and plateau live oak
(*Q. fusiformis*). Much of the northern and western plateau
is characterized by semi-open grasslands and shrublands
on the uplands with riparian corridors along the drainages.

The **Rolling Plains** covers 24 million acres or 14.3 percent of the state. The region is situated between the High
Plains and the Cross Timbers and Prairies in the north-
central part of the state. These plains are nearly level to
rolling and were originally covered by prairie. The Rolling
Plains area is divided from the High Plains by the steep
Caprock Escarpment. The vegetation of this region is
tall- and midgrass prairie with a wide variety of grasses
present, including little bluestem, big bluestem, sand
bluestem (*Andropogon gerardii* var. *paucipilus*), sideoats
grama (*Bouteloua curtipendula*), indiangrass, and buffalograss (*Buchloe dactyloides*). Overgrazing and fire suppression have transformed this prairie into open shrublands. Many rivers and streams have eroded away the
Caprock Escarpment to form canyons. The largest and best
known is Palo Duro Canyon. The canyons or breaks of the
Canadian River are also included in this region. The Rolling Plains and the Edwards Plateau are ecologically similar,
but a distinct geological change defines the boundary. The
Concho Valley lies along this boundary. Overgrazing and
reduction of fires have transformed much of the Rolling
Plains from a mid- and tallgrass prairie to an open shrubland dominated by mesquite and juniper.

The **High Plains** region covers 19.4 million acres or
about 11.5 percent of the state. The High Plains area is
bounded by the Caprock Escarpment and dissected by the
Canadian River. These plains are nearly level, with many
shallow playa lakes. The original vegetation of the High
Plains consisted generally of mixed and shortgrass prairie
and was free from brush. The species of grasses present
varied based on soil types. In areas with clay soils, blue
grama (*Bouteloua gracilis*) and buffalograss were common,
while on sandy soils grasses such as little bluestem, sideoats grama, and sand dropseed (*Sporobolus cryptandrus*)

dominated. Today, about 60 percent of the High Plains is in agricultural production, and much of that is used to produce row crops. The southern extension of the High Plains, south of the Canadian River, is known as the **Llano Estacado.** The area around Lubbock, including the surrounding counties, is known as the **South Plains.**

The **Trans-Pecos** of far West Texas includes the northern extension of the Chihuahuan Desert, and it coincides with the Basin and Range Physiographic Province. The region covers approximately 18 million acres or 10.7 percent of the state. Guadalupe Peak, at an elevation of 8,751 feet, is the highest point in Texas. There are many small mountain ranges within the region—the Davis, Chisos, and Guadalupe Mountains are the best known. Desert grasslands and desert scrub are found at lower elevations, although very little desert grassland persists today. The vegetation found at the mid-elevations in the mountain ranges is dom-

inated by pinyon pines (*Pinus cembroides, P. edulis,* and *P. remota*) and junipers, while the upper elevations support pines (*P. ponderosa* and *P. arizonica*). Creosotebush (*Larrea tridentata*), tarbush (*Flourensia cernua*), and various acacias are found in the lowland basins. The **Stockton Plateau** is the subregion found west of the Pecos River and includes most of Terrell and southern Pecos Counties. This region is a transitional area between the Edwards Plateau and the Chihuahuan Desert. The dominant vegetation type is mesquite and red-berry juniper (*Juniperus pinchotii*) savannah. The Stockton Plateau is sometimes included as a subregion of the Edwards Plateau.

Texas Counties

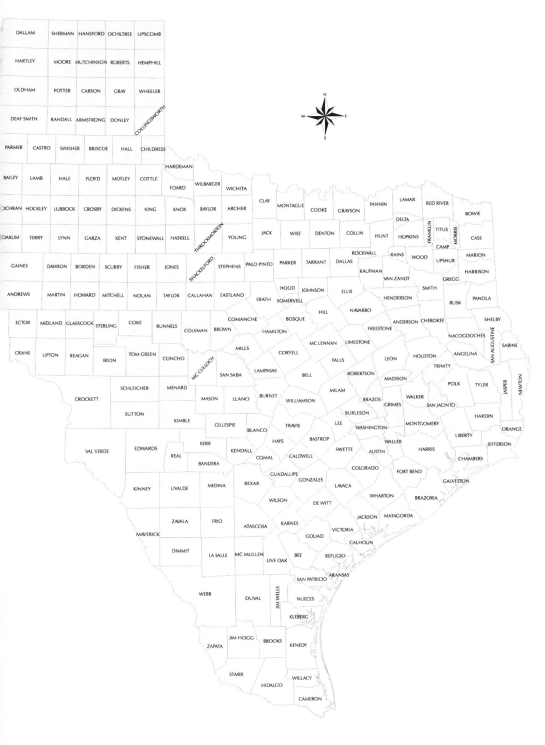

Counties Keyed to the Natural Areas of Texas

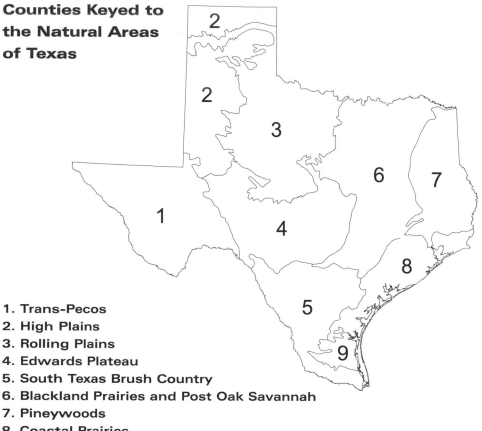

1. Trans-Pecos
2. High Plains
3. Rolling Plains
4. Edwards Plateau
5. South Texas Brush Country
6. Blackland Prairies and Post Oak Savannah
7. Pineywoods
8. Coastal Prairies
9. Coastal Sand Plain

For counties where portions of more than one Natural Area occur, all regions are listed.

Anderson 6, 7	Borden 3	Carson 2, 3	Cooke 6
Andrews 1, 2	Bosque 6	Cass 7	Coryell 4, 6
Angelina 7	Bowie 6, 7	Castro 2	Cottle 3
Aransas 8	Brazoria 8	Chambers 8	Crane 1
Archer 3	Brazos 6	Cherokee 7	Crockett 1, 4
Armstrong 2, 3	Brewster 1	Childress 3	Crosby 2, 3
Atascosa 5	Briscoe 2, 3	Clay 3	Culberson 1
Austin 6, 8	Brooks 9	Cochran 2	Dallam 2
Bailey 2	Brown 3, 4	Coke 3, 4	Dallas 6
Bandera 4	Burleson 6	Coleman 3	Dawson 2, 3
Bastrop 6	Burnet 4	Collin 6	Deaf Smith 2, 3
Baylor 3	Caldwell 6	Collingsworth 3	Delta 6
Bee 5	Calhoun 8	Colorado 6, 8	Denton 6
Bell 4, 6	Callahan 3, 4	Comal 4	DeWitt 5, 6
Bexar 4, 5, 6	Cameron 5	Comanche 4, 6	Dickens 3
Blanco 4	Camp 7	Concho 3, 4	Dimmit 5

Abbreviations

AMNH   American Museum of Natural History

ANSP   Academy of Natural Sciences of Philadelphia

BMNH   British Museum of Natural History

LSUMNS   Louisiana State University Museum of Natural
            Science

NP   National Park

NWR   National Wildlife Refuge

SFASU   Stephen F. Austin State University

SP   State Park

TBRC   Texas Bird Records Committee

TBSL   Texas Bird Sounds Library, housed at Sam Houston
            State University

TCWC   Texas Cooperative Wildlife Collection, housed at
            Texas A&M University

TPRF   Texas Photo Record File, housed at Texas A&M
            University

UKNHM   University of Kansas Natural History Museum

USNM   United States National Museum

Map Key

Year-round occurrence

Winter occurrence

Summer occurrence

Migration route

Regular but scattered summer/fall occurrence,
typically in small numbers, primarily due to
postbreeding dispersal

 Irregular winter occurrence often associated with
irruptions and may reflect an annual occurrence

 The extent of scattered, irregular occurrence (at
any time of year) beyond the mapped range. The
line in a given map is shaded to match the season as
defined above.

•   Used in the case of Review Species, a dot signifies
            a single occurrence. A much larger dot indicates
            multiple occurrences in a small geographic area.

ANNOTATED LIST OF SPECIES

ORDER ANSERIFORMES

Family Anatidae: Swans, Geese, and Ducks

Sick or injured migratory waterfowl may summer virtually anywhere in Texas. In the following accounts, only those species that have shown a pattern of summer occurrence as healthy birds are described as summering in the state.

BLACK-BELLIED WHISTLING-DUCK
Dendrocygna autumnalis (Linnaeus)

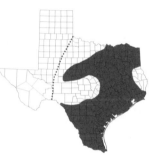

Uncommon to locally common resident in the eastern two-thirds of the state, generally becoming less common in winter. The population in Texas continues to increase, and that is also being expressed in wintering numbers. Black-bellied Whistling-Ducks primarily occur throughout the South Texas Brush Country and northward to the southern Edwards Plateau and along the Coastal Prairies. They are more localized and uncommon to locally common farther northward through the eastern third of the state to the Oklahoma border, including the Pineywoods, although they are rare in heavily forested areas. Isolated breeding populations are also established on the southern Rolling Plains west to the Concho Valley. Wintering is localized, but increasing, in north-central and northeast Texas. This species is a very rare to casual visitor to the remainder of the state, primarily between March and November. There are isolated breeding records outside the mapped range. **Taxonomy:** The subspecies that occurs in Texas is *D. a. fulgens* (Friedmann).

FULVOUS WHISTLING-DUCK
Dendrocygna bicolor (Vieillot)

Common summer resident along the Coastal Prairies. Fulvous Whistling-Ducks are local summer residents in the eastern half of the Lower Rio Grande Valley, tending to

occur in larger numbers in wet years. This species is a rare to locally uncommon winter resident north to Baffin Bay, becoming increasingly rare and irregular through the remainder of the breeding range. These ducks are also rare west and north of their breeding range and are very rare and irregular spring visitors to the Oaks and Prairies region. Fulvous Whistling-Ducks are accidental visitors to the Trans-Pecos, Panhandle, and South Plains. **Timing of occurrence:** The breeding population arrives in early to mid-March and is present through October. **Taxonomy:** Monotypic.

GREATER WHITE-FRONTED GOOSE
Anser albifrons (Scopoli)

Common to uncommon migrant through the eastern three-quarters of the state, becoming rare in the Trans-Pecos. Greater White-fronted Geese are uncommon to locally abundant winter residents on the Coastal Prairies. They can also be found in agricultural areas from Medina and Uvalde Counties southward through the South Texas Brush Country. There are also isolated wintering populations in the Rolling Plains and northeast Texas. This species is a rare to locally uncommon winter visitor in the northwestern portion of the state and irregular in occurrence elsewhere. **Timing of occurrence:** Fall migrants and winter residents normally appear in late September before other species of geese, and a few wintering birds often linger as late as early April and rarely early May. An early arrival date of 16 August 1954 may refer to a bird that was present during the summer. **Taxonomy:** The taxonomy of Greater White-fronted Goose has long been a complicated subject and has recently been reviewed by Banks (2011). The following traditional taxonomy is conservatively presented until it is clear whether Banks's revision will be widely accepted.

A. a. gambelli Hartlaub
>   Status uncertain. This subspecies has been described as common (Oberholser 1974), but it appears to be uncommon at best and far outnumbered by *A. a. frontalis*. More research is needed to determine its status in the state.

A. a. frontalis Baird
>   Common migrant and winter resident. Banks (2011) includes this subspecies under *A. a. gambelli*.

CANADA GOOSE *Branta canadensis* (Linnaeus)

Uncommon to rare migrant throughout the state. Canada Geese winter primarily in two areas of Texas: the High Plains, where they are abundant; and along the upper and central Coastal Prairies, where they are common. This species is erratic in occurrence along the lower coast but is generally very rare. Canada Geese are very rarely detected in the southern Oaks and Prairies region, southern Edwards Plateau, southern Trans-Pecos, and western South Texas Brush Country at any season. This species is a localized, but increasing, breeder in the northern third of the state. These breeding birds are largely the result of introductions in Oklahoma. Some breeding records involved pairs in which one individual was known to be injured. **Timing of occurrence**: Fall migrants begin to appear in late September, with the majority of the wintering population arriving in late October. Most have departed by March, although a few linger into early April. **Taxonomy**: Four subspecies are known to occur in Texas.

B. c. maxima Delacour
Rare to locally uncommon, but apparently increasing, resident in the northern third of the state.

B. c. parvipes (Cassin)
Abundant migrant and winter resident in the northwestern portion of the state. Rare migrant in Central Texas and rare winter resident on the Coastal Prairies and in the Trans-Pecos.

B. c. interior Todd
Uncommon migrant in eastern Texas. Uncommon winter resident along the upper coast.

B. c. moffitti Aldrich
Uncommon migrant and winter resident in the Trans-Pecos, High Plains, and Rolling Plains. Uncommon winter resident on the central and upper coasts.

TBRC Review Species   ### TRUMPETER SWAN *Cygnus buccinator* Richardson

The Trumpeter Swan was a regular winter resident in the eastern two-thirds of Texas until the early 1900s. Nehrling (1882) reported large numbers in Galveston Bay and elsewhere along the upper coast in winter. Since then there have been only nine accepted records for the state. The occurrence of Trumpeter Swan in Texas has been compli-

cated by active reintroduction projects in the midwestern United States. In the 1990s there were several instances of collared swans documented from these efforts in Texas that were not considered as state records. However, since 2001 these midwestern populations have been considered established, and the swans have not been marked. Virtually all of the Trumpeter Swans found in Texas since then have been unmarked, and therefore their breeding grounds could not be determined. It was expected that Trumpeter Swans from the reintroduction programs would be more common winter visitors to Texas, but that has not occurred. Of particular interest is an immature bird found dead near Vega, Oldham County, on 8 April 1993 (TBRC 1996-15; TPRF 1521). This individual was banded in the nest in Wyoming and is the only modern record of a bird known to have originated from the natural population. **Timing of occurrence:** All records to date have occurred between 15 November and 8 April. **Taxonomy:** Monotypic.

TUNDRA SWAN *Cygnus columbianus* (Ord)

Rare to very rare and irregular winter visitor to all regions of the state. Tundra Swans occur annually, or nearly so, only in the Panhandle and eastward along the Red River in northern Texas. Oberholser (1974) reports that this swan was a common to uncommon winter resident throughout the state prior to 1900. **Timing of occurrence:** Most of the records occur between late October and mid-March, but individuals have lingered as late as early May. Out-of-season reports include an adult in Moore County on 13 July 1985 and another in Potter County on 21 September 1995 (Seyffert 2001b). **Taxonomy:** The subspecies that is known to occur in Texas is *C. c. columbianus* (Ord).

MUSCOVY DUCK *Cairina moschata* (Linnaeus)

Rare and local resident along the Rio Grande in Starr and Zapata Counties. This tropical duck is generally more difficult to find during the summer. Muscovy Ducks were first discovered in Texas in December 1984. A small population, centered below Falcon Dam, became established and has persisted. The first reported nesting of apparently wild birds was near Bentsen–Rio Grande Valley SP, Hidalgo County, during July 1994 (Brush and Eitniear 2002). An

adult that was believed to be wild was seen along the Rio Grande in northern Maverick County on 5 June 2000, providing the northernmost report. There was some dispersal of wild-type Muscovy Ducks away from the Rio Grande following Hurricane Alex in 2010. This species is a commonly kept domestic duck and is abundant in parks and other areas in the Lower Rio Grande Valley, as well as other areas of the state. Populations originating from domesticated stock very rarely exhibit wild-type plumage and normally have large areas of white on the body and extensive red warty protuberances on the face. There is some anecdotal evidence of domesticated stock breeding with the wild-type birds in Hidalgo County. **Taxonomy:** Monotypic.

WOOD DUCK *Aix sponsa* (Linnaeus)

Locally uncommon to common resident in the eastern three-quarters of the state. Wood Ducks are summer residents as far west as the eastern Panhandle and southwestern Edwards Plateau. This species is also an uncommon resident along the Rio Grande in the western Trans-Pecos. During the winter months, Wood Ducks largely withdraw from the Panhandle but are regular during this season on the South Plains and sporadically in the central Trans-Pecos. This species becomes more common in winter in the eastern half of the state as migrants from more northerly populations arrive. Wood Duck populations have rebounded from a low point in the early 1900s. Strecker (1912) considered the species to be very rare, and Simmons (1925) thought them to be almost extinct in the Austin area during that time. **Taxonomy:** Monotypic.

GADWALL *Anas strepera* Linnaeus

Common to abundant migrant and winter resident throughout the state. This species is also a rare and local summer resident in the Panhandle, with fewer than 20 nesting records (Johnson and Lockwood 2013). In addition to a few other scattered nesting records from that area, two more are from Dallas County in 1961, which have been questioned by Pulich (1988), and one from San Antonio in 1987. Aside from nesting records, Gadwalls are known to summer irregularly in small numbers throughout the state. **Timing of occurrence:** Fall migrants and winter residents arrive during October, and spring migrants are present as late as mid-May. **Taxonomy:** The subspecies that occurs in Texas is *A. s. strepera* Linnaeus. Gadwall is sometimes considered monotypic with one other extinct subspecies, *A. s. couesi,* described from islands in the central Pacific Ocean.

TBRC Review Species

EURASIAN WIGEON *Anas penelope* Linnaeus

Very rare to casual visitor, with 53 documented records for Texas. This species occurs with greater frequency in the western half of the state, with many records from the El Paso area. As might be expected, all of the documented records are of the more conspicuous adult male. There are also a few documented occurrences of presumed Eurasian Wigeon × American Wigeon hybrids. Careful examination of the feather pattern and coloration of the head, breast, and flanks is needed to identify hybrids. There are 24 reports of Eurasian Wigeons from prior to the development of the Review List in 1988 for which there is no documentation on file. **Timing of occurrence:** All records to date have occurred between 3 October and 8 May. **Taxonomy:** Monotypic.

AMERICAN WIGEON *Anas americana* Gmelin

Common to abundant migrant and winter resident throughout the state. American Wigeons also occur as local summer visitors to the Panhandle and the El Paso area, but there is no evidence of nesting in the state. This species is occasionally encountered elsewhere in the state during the summer. **Timing of occurrence:** Fall migrants and winter residents begin arriving in early August, with the majority of the population present by early November. Spring migrants depart during March, but a few linger into late May. **Taxonomy:** Monotypic.

TBRC Review Species

AMERICAN BLACK DUCK *Anas rubripes* Brewster

Accidental. There are eight accepted records for the state, and only four since 1950. In contrast to this very small number, more than 40 American Black Ducks are reported to have been collected by hunters between 1920 and 2009 (Johnson and Garrettson 2010). Unfortunately, there is no documentation of these occurrences available for review

by the TBRC. In addition, American Black Duck appears to have been a more regular visitor prior to 1950, although detailed information about specific sightings is lacking here as well. A factor that complicates documentation of American Black Ducks is the identification challenge of separating them from the subspecies of Mottled Duck that occurs in Texas. Male Mottled Ducks of this subspecies can appear as dark as American Black Ducks, show a purple speculum (depending upon light angle), and have gleaming white wing linings. Documentation requires very careful study of the pattern, including the internal markings, of the feathers of the breast, scapulars, flanks, and back. Eliminating potential American Black Duck × Mallard hybrids only adds to the problem. In addition to the accepted TBRC records, there are 95 reports of American Black Duck from prior to the development of the Review List in 1988 for which there is no documentation on file. **Timing of occurrence:** All records to date have occurred between 7 December and 15 April. In addition, there are recoveries from November listed in Johnson and Garrettson (2010). **Taxonomy:** Monotypic.

MALLARD *Anas platyrhynchos* Linnaeus

Common to locally abundant winter resident in the northern half of the state, becoming increasingly less common farther south. The primary wintering areas in Texas appear to have shifted northward over the past 20 years. Mallards are an uncommon summer resident in the Panhandle and are rare and local breeders in many other parts of the state. Excluded are the numerous domesticated Mallards present throughout the state. The distinctive Mexican Duck is a rare to locally common resident throughout the Trans-Pecos and along the Rio Grande south to Hidalgo County. There are occurrences of Mexican Ducks out of the normal range in Texas in Crosby, Lubbock, Midland, and Swisher Counties. Mexican Ducks were formerly considered a separate species, but Hubbard (1977) concluded that most, if not all, individuals in the contact zone between the two taxa are intergrades between green-headed Mallards and the true Mexican Ducks of central Mexico. He extrapolated his results to suggest that all Mexican Ducks in the United States are intergrades. This may be correct, but there are no breeding populations of green-headed Mallards in the southern portion of the Mexican Duck range in the United States, including the entire Trans-Pecos. Recent studies using DNA techniques have suggested that the Mexican Duck is more closely related to Mottled and American Black Ducks than to Mallards (McCracken, Johnson, and Sheldon 2001; Gonzales, Düttman, and Wink 2009) and that they should not be considered conspecific. However, these studies have not examined individuals from the intergrade zone defined by Hubbard. **Timing of occurrence:** Fall migrants and winter residents begin arriving in early September, and spring migrants generally have departed by late March, although some individuals linger until late April and exceptionally to early May. **Taxonomy:** Two subspecies are known to occur in Texas.

A. p. platyrhynchos Linnaeus
>   Distribution as given in the beginning of the species account. This subspecies refers to the population with green-headed males.

A. p. diazi Ridgway
>   Distribution as given for Mexican Duck above.

MOTTLED DUCK *Anas fulvigula* Ridgway

Locally common resident along the Coastal Prairies south to the eastern Lower Rio Grande Valley. Wandering individuals, not limited to postbreeding dispersal, are regularly encountered as far inland as Travis County. The occurrences of Mottled Ducks in north-central and northeast Texas have declined over the past decade, and they are now rare. Current records may represent vagrants from the Coastal Prairies. Mottled Ducks have shown significant population declines during the past 20 years, and hybridization with feral Mallards is a conservation concern. Although only a small number of such hybrids have been found in Texas, the Florida population has been dramatically impacted by hybridization (Bielefeld et al. 2010). **Taxonomy:** The subspecies that occurs in Texas is *A. f. maculosa* Sennett.

BLUE-WINGED TEAL *Anas discors* Linnaeus

Common to abundant migrant throughout the state. This species is an uncommon summer resident throughout the Panhandle and South Plains and along the Upper Texas

Coast. It is a rare and local breeder throughout most of the remainder of the state. Blue-winged Teal are uncommon to common winter residents in the Coastal Prairies and South Texas Brush Country. They become increasingly less common northward to Travis, Burleson, and Montgomery Counties and are generally absent farther north and west. **Timing of occurrence**: Spring migration begins in mid-February and continues through April, when the breeding population is in place. Fall migration occurs between late July and late October, with a few lingering into early November. **Taxonomy**: Monotypic.

CINNAMON TEAL *Anas cyanoptera* Vieillot

Common migrant in the western half of the state, becoming uncommon to rare eastward and very rare in the Pineywoods. Cinnamon Teal are uncommon and somewhat local winter residents along the coast. They are rare to very uncommon inland during the winter in much of the western half of the state from the South Plains southward, with scattered records from farther east. This species is a local summer resident in the Panhandle and South Plains. There are also isolated breeding records from Bexar, Colorado, and El Paso Counties. **Timing of occurrence**: Fall migrants can be found beginning in mid-August, with the wintering population in place by early October. Spring migration is from late February through April. **Taxonomy**: The subspecies that occurs in Texas is *A. c. septentrionalium* Snyder & Lumsden.

NORTHERN SHOVELER *Anas clypeata* Linnaeus

Common to abundant migrant and winter resident throughout the state. This species is an uncommon to rare summer visitor to most areas of the state. Despite the regular summer occurrence in the state, nesting records are few and primarily from the Panhandle and Trans-Pecos. There are isolated breeding records from as far south and east as Bastrop and Bexar Counties. **Timing of occurrence**: Fall migrants arrive in the state in late August, with a peak in numbers in late September and October, when the majority of the winter population arrives. Spring migrants can be detected from late February through May. **Taxonomy**: Monotypic.

WHITE-CHEEKED PINTAIL *Anas bahamensis* Linnaeus

Accidental. There is one accepted record for Texas. This single individual represents the westernmost record for the United States. Because of the popularity of this species with aviculturists, the provenance of White-cheeked Pintails found in the United States, including those from Florida, has been debated (American Birding Association 2011). Free-flying, presumed escaped, White-cheeked Pintails were present on the upper coast until the passage of Hurricane Ike in 2008. Seasonal movements are known for this species throughout its range, including populations in the West Indies. **Taxonomy:** The subspecies that occurs naturally in the Caribbean is *A. b. bahamensis* Linnaeus.

20 NOV. 1978–15 APR. 1979, LAGUNA ATASCOSA NWR, CAMERON CO. (TPRF 141)

NORTHERN PINTAIL *Anas acuta* Linnaeus

Locally uncommon to abundant migrant and winter resident throughout the state. Northern Pintail is a rare to locally uncommon summer resident in the Panhandle and South Plains. There are also isolated breeding records from various other parts of the state, including Baylor, Calhoun, El Paso, Medina, and Wilbarger Counties, but most of these records date from the 1930s. **Timing of occurrence:** Migrants normally begin to arrive in Texas in early August, with most of the wintering population in place by November. Spring migration begins in late February, and lingering individuals can often be found well into May. **Taxonomy:** Monotypic.

GARGANEY *Anas querquedula* Linnaeus

Accidental spring migrant. There are four documented records for the state, all pertaining to alternate males. These sightings have occurred within a well-established pattern of spring Garganey records from the central United States. These individuals are believed to have wintered much farther south and may have been overlooked during fall migration, when they are in basic plumage. **Timing of occurrence:** The records for Texas are all between 4 April and 17 May. **Taxonomy:** Monotypic.

11 APR.–17 MAY 1985, NEAR RIVIERA, KLEBERG CO. (TPRF 354)
29 APR.–6 MAY 1994, PRESIDIO, PRESIDIO CO.
  (TBRC 1994-77; TPRF 1296)
17 APR. 1998, BOLIVAR PENINSULA, GALVESTON CO.
  (TBRC 1998-93)
4–7 APR. 2001, KING RANCH, KLEBERG CO.
  (TBRC 2001-88; TPRF 1968)

GREEN-WINGED TEAL *Anas crecca* Linnaeus

Common to abundant migrant and uncommon to locally common winter resident throughout the state. This species is particularly abundant on the upper coast during winter. Paired Green-winged Teal are frequently encountered during early summer in the Panhandle and Trans-Pecos, but there are only two reports of breeding in Texas: a pair with young was found in Hutchinson County in 1975, and an unsuccessful nesting attempt was reported from Crosby County in 1982. Oberholser (1974) discounted a report of several nesting pairs in Dallam County from 1936, but Seyffert (2001b) gave this report more weight based on the regular occurrence of this species in the area. **Timing of occurrence**: Fall migrants arrive in the state in late August, and most of the wintering population has departed by late April. Lingering migrants are frequently seen well into June. **Taxonomy**: Two subspecies are known to have occurred in Texas.

A. c. carolinensis Gmelin
  Distribution as given above.

A. c. crecca Linnaeus
  There is a one documented record from the Village Creek Drying Beds, Tarrant County, from 30 January to 1 February 1994 (TBRC 1994-135; TPRF 1279). This taxon is frequently recognized as a separate species, Eurasian (or Common) Teal. Documentation is requested by the TBRC for all reports of this taxon.

CANVASBACK *Aythya valisineria* (Wilson)

Uncommon to locally common migrant and winter resident throughout the state. This species occasionally remains through the summer, but the only known nesting records are four broods from the southwestern Panhandle in 1975 (Seyffert 2001b) and a female with young in Jeff

Davis County on 13 May 1986. **Timing of occurrence:** Migrants begin arriving in mid-October and peak in November, with the wintering population present by early December. Spring migration is from mid-February through mid-March. Most depart the state by the end of March, although a few may linger as late as early May. **Taxonomy:** Monotypic.

REDHEAD *Aythya americana* (Eyton)

Uncommon to common migrant throughout the state. The Laguna Madre is the winter home to the largest concentration of Redheads in the world, and the species is considered a common to abundant winter resident along the central and lower coasts. They are also rare to locally common winter residents throughout the remainder of the state. Redheads routinely linger through the summer in the Laguna Madre. In the Panhandle they are uncommon to locally common summer visitors, but despite this abundance, nesting has rarely been observed (Seyffert 2001b). In the Trans-Pecos, Redheads are rare to very rare nesters in El Paso and Hudspeth Counties, and nesting has been documented as far east as Pecos County. They are casual in summer throughout the remainder of the state. **Timing of occurrence:** This species arrives in the state in late September, and the majority of the population departs in March, with a few lingering into early April. **Taxonomy:** Monotypic.

RING-NECKED DUCK *Aythya collaris* (Donovan)

Uncommon to locally common migrant and winter resident throughout the state. This species regularly lingers into the summer, particularly from the Panhandle southward through the Trans-Pecos. Despite the regular occurrence of this species in summer, there is no evidence of nesting in the state. **Timing of occurrence:** Migrants and winter residents arrive in the state in early October, and spring migration occurs from late February through March, with some lingering until mid-April. **Taxonomy:** Monotypic.

GREATER SCAUP *Aythya marila* (Linnaeus)

Rare to uncommon migrant in all parts of the state. Greater Scaup are rare to locally uncommon winter residents along

the upper and central coasts and on reservoirs in the eastern half of the state. They are rare to very rare along the lower coast and elsewhere in the state. **Timing of occurrence:** Greater Scaup arrive in late October, with the peak of migration occurring during November, when the majority of the winter population is present. Spring migrants depart the state by March. Most years a few apparently healthy individuals linger through May and occasionally into June. **Taxonomy:** The subspecies that occurs in Texas is *A. m. nearctica* Stejneger.

LESSER SCAUP *Aythya affinis* (Eyton)

Common to abundant migrant and winter resident east of the Pecos River. In the Trans-Pecos Lesser Scaup are common migrants and uncommon to locally rare winter residents. They are rare summer visitors primarily to the Panhandle, and there are only two reports of breeding for the state: Muleshoe NWR, Bailey County, in 1942 (Hawkins 1945) and a female with ducklings in southern Swisher County in early July 1977 (Seyffert 2001b). This species is a rare summer visitor to virtually all regions of the state. Although Lesser Scaup are considered abundant, population surveys show that the species is in sharp decline. **Timing of occurrence:** Lesser Scaup begin arriving in October, but winter residents are not common until November. Most of the wintering population departs by early April, but some birds linger into mid-May. **Taxonomy:** Monotypic.

TBRC Review Species   ### KING EIDER *Somateria spectabilis* (Linnaeus)

Accidental. Texas has two documented records from the upper coast. Both records refer to first-year males in worn plumage. The plumage of the first was in particularly bad condition, so it was captured and transported to Ohio for rehabilitation, where it died. Typically, King Eiders are not normally found farther south than New England during the winter, but there are numerous extralimital records south to Florida. There is also a sight record of a male at Rockport on 23 October 1968 for which there is no documentation on file. **Timing of occurrence:** The two documented Texas records and three from Louisiana all date from the late spring and early summer. **Taxonomy:** Monotypic.

30 APR.–7 MAY 1998, QUINTANA, BRAZORIA CO.
  (TBRC 1998-59; TPRF 1729)
10 APR.–10 MAY 2005, PORT BOLIVAR, GALVESTON CO.
  (TBRC 2005–60; TPRF 2305)

TBRC Review Species

COMMON EIDER *Somateria mollissima* (Linnaeus)

Accidental. There is one record for this species for the state. The lone record refers to an adult male collected by a hunter and is the first for the western Gulf. Common Eiders are rare winter visitors to Florida, with two documented records from the Gulf Coast (Petrovic and King 1972; B. Pranty, pers. comm.). **Taxonomy:** The lone Texas specimen belongs to *S. m. dresseri* Sharpe.

8 JAN. 2007, NORTH LAGUNA MADRE, NUECES CO.
  (TBRC 2007–11; TPRF 2480)

TBRC Review Species

HARLEQUIN DUCK *Histrionicus histrionicus* (Linnaeus)

Accidental. Texas has two documented records of Harlequin Duck: the first a male; the second a pair. The Harlequin Ducks found in Texas may have originated from either of the disjunct North American populations. Perhaps the more likely point of origin is from the western population, which occurs as close as northwestern Wyoming; however, these birds migrate to the Pacific for the winter. Winter records from the intermountain west are rare, and there is only one recent record for Colorado. Eastern birds breed east of Hudson Bay and in Greenland and winter along the Atlantic Coast as far south as Virginia. There are three sight records from Aransas County, two in January 1945 (Hagar 1945) and one in November 1964; however, there is no documentation of these sightings on file. **Timing of occurrence:** Both Texas occurrences are from midwinter. **Taxonomy:** Monotypic.

30 JAN.–4 FEB. 1990, SOUTH PADRE ISLAND, CAMERON CO.
  (TBRC 1990-21; TPRF 858)
5 JAN. 1995, LAKE TAWAKONI, VAN ZANDT CO. (TBRC 1995-12)

SURF SCOTER *Melanitta perspicillata* (Linnaeus)

Uncommon and local winter resident along the immediate coast, most commonly north of Baffin Bay. Surf Scoters are also rare migrants and winter visitors over the remainder of the state. This species is the most frequently encountered scoter in Texas. The wintering population along the coast is mostly confined to saltwater habitats, such as bays and

nearshore Gulf waters. The majority of Surf Scoters found in Texas are in first-winter plumage, although second-year and full-adult males are also present. **Timing of occurrence:** Fall migrants are found in Texas between late October and early December. Spring migrants are present inland between late February and early April. There are many instances of birds lingering into late May, and on rare occasions birds have remained well into June and once into July. **Taxonomy:** Monotypic.

WHITE-WINGED SCOTER *Melanitta fusca* (Linnaeus)

Very rare winter resident along the coast, most commonly along the upper and central portions. It is also a very rare migrant throughout the state. Wintering White-winged Scoters are more commonly encountered in saltwater habitats than elsewhere. The majority of White-winged Scoters found in Texas are first-winter birds; adults are rarely encountered. **Timing of occurrence:** Fall migrants are found in Texas between early November and mid-December. Spring migrants are present inland from mid-February to mid-April. The earliest a migrant has been reported is 22 September, and the latest is 17 June. **Taxonomy:** The subspecies that occurs in Texas is *M. f. deglandi* (Ridgway).

BLACK SCOTER *Melanitta americana* (Swainson)

Very rare winter resident along the coast, primarily along the upper and central portions. Black Scoters are very rare to casual migrants, usually on inland reservoirs. As are other scoters, wintering Black Scoters are found primarily in saltwater habitats and are casual inland. **Timing of occurrence:** Fall migrants have been found in Texas between late October and mid-December, with spring migrants present inland from mid-February to early April. An exceptionally late departing bird was at Aransas Bay, Aransas County, on 17 May 2005. **Taxonomy:** Monotypic. Although this species retained the same North American common name, it was split from the *M. nigra* (Linnaeus) in 2010 (Chesser et al. 2010).

LONG-TAILED DUCK *Clangula hyemalis* (Linnaeus)

Rare to very rare winter visitor along the entire coast, and a very rare to casual migrant and winter visitor in nearly all

other parts of the state. As in the case of the scoters, most individuals are in first-winter plumage, and adult males are rarely observed. **Timing of occurrence:** Fall migrants have been found in Texas between early November and mid-December. Based on inland sightings, wintering birds depart between mid-February and early April. Exceptional was one that stayed at Amarillo, Potter County, until 12 June 2004. **Taxonomy:** Monotypic.

BUFFLEHEAD *Bucephala albeola* (Linnaeus)

Common migrant and winter resident throughout the state. The population of Buffleheads has increased markedly since the 1950s; however, there also seems to be a northward shift in their wintering range during this period. They are now much more common in the northern portions of that range. **Timing of occurrence:** Buffleheads begin arriving in Texas in late September, and most have departed by early May. In most years, wintering populations along the coast and in the southern third of the state have migrated north by early April. Buffleheads very rarely linger later in the spring, but a few records exist from early June. **Taxonomy:** Monotypic.

COMMON GOLDENEYE *Bucephala clangula* (Linnaeus)

Uncommon to locally common winter resident along the upper and central coasts, becoming rare to very rare on the lower coast. Uncommon to rare migrant and winter resident in the rest of the state. Inland, Common Goldeneyes are most frequently encountered on larger reservoirs; however, they are sometimes found on small lakes and even on rivers. **Timing of occurrence:** This species is present primarily between early November and early March. A few regularly linger into late April, particularly along the coast. The earliest a migrant has been reported is 12 October, and the latest is 6 June. **Taxonomy:** The subspecies that occurs in Texas is *B. c. americana* (Bonaparte).

TBRC Review Species ### BARROW'S GOLDENEYE *Bucephala islandica* (Gmelin)

Casual to accidental. There are 10 documented records in the state. All but one have involved the more easily identified adult male. Barrow's Goldeneyes are regular winter visitors to northern New Mexico and Arizona and could

be expected to occur in Texas more frequently than currently reported. The first record for the state was a male taken by a hunter near Greenville, Hunt County, on 6 November 1958 (TBRC 1989-202; TPRF 787). Reportedly, this bird was with two others of its kind. One other Texas record included more than a single bird: a pair at Loy Lake, Grayson County, from 5 January to 22 February 2002 (TBRC 2002-12; TPRF 1981). There are six reports of Barrow's Goldeneyes from prior to the development of the Review List in 1988 for which there is no documentation on file. **Timing of occurrence:** All of the Texas records have occurred between 6 November and 22 February. **Taxonomy:** Monotypic.

HOODED MERGANSER
Lophodytes cucullatus (Linnaeus)

Uncommon to common migrant and winter resident throughout most of the state. Hooded Mergansers are uncommon to rare and local winter visitors on the Edwards Plateau and west throughout the Trans-Pecos. They are rare but increasing summer residents in northeast Texas, south to Nacogdoches County. Since the early 1990s the number of reported nesting attempts has increased, primarily due to use of nest boxes intended for Wood Ducks. There are cases of Hooded Mergansers laying eggs in active Wood Duck nests, with all young from the joint clutches reared by the Wood Ducks. Hooded Mergansers have successfully nested in the eastern Panhandle in Hemphill County and on the upper coast in Brazoria County. **Timing of occurrence:** Fall migrants begin to arrive in early November, and most have departed by mid-March. **Taxonomy:** Monotypic.

COMMON MERGANSER *Mergus merganser* Linnaeus

Locally common to occasionally abundant winter resident in the Panhandle and the El Paso area in the western Trans-Pecos. Common Mergansers are uncommon to rare on the South Plains south to the San Angelo area and in the northeastern portion of the Trans-Pecos. This species is rare to very rare eastward, including the northern half of the Pineywoods, and generally absent throughout the remainder of the state. Common Mergansers are very rarely found in saltwater habitats, although there are a small number of coastal records. This species is a very rare summer visitor to the Panhandle. **Timing of occurrence:** Fall migrants arrive in mid-November, and most of the population departs by early April, although lingering individuals have been noted into late May. **Taxonomy:** The subspecies that occurs in Texas is *M. m. americanus* Cassin.

RED-BREASTED MERGANSER
Mergus serrator Linnaeus

Common winter resident along the coast and uncommon to rare migrant and winter resident on inland reservoirs in the eastern half of the state. Red-breasted Mergansers are generally casual to rare migrants and winter residents in the western half of Texas, except in the El Paso area, where they are locally uncommon. This species is a rare to very rare summer visitor along the entire coast. There is one breeding record from the state: two females with young were discovered at Laguna Atascosa NWR, Cameron County, during May and June 1995 (Rupert and Brush 1996). One fledgling from this discovery was found dead and preserved as a specimen in the collection at the University of Texas–Pan American. **Timing of occurrence:** Fall migrants arrive in mid-October, and most of the population departs by early April. Fall migrants have been noted as early as 5 August, while lingering birds have been noted into early June. Summer records away from the coast include a bird at Lavon Lake, Collin County, on 6 July 2012. **Taxonomy:** Monotypic.

TBRC Review Species   **MASKED DUCK** *Nomonyx dominicus* (Linnaeus)

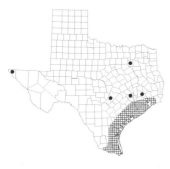

Rare and irregular visitor along the Coastal Prairies and in the Lower Rio Grande Valley. Masked Ducks have been nearly, if not actual, annual visitors to Texas over the past 20 years. Very small numbers have been discovered since the last large influx of birds in the mid-1990s. This tropical duck can be locally uncommon during such invasion periods (Lockwood 1997), as was demonstrated by a total of 37 individuals on a lake in San Patricio County in February 1993 (Blankenship and Anderson 1993). Of the 94 documented records for the state, only five are away from the Coastal Prairies: single birds in El Paso County on 11 July 1976 (TBRC 1999-42); Hays County, from 20 September to early October 1980 (TPRF 211); Huntsville, Walker County, on 10 June 1993 (TBRC 1993–103; TPRF 1169); near Seagoville, Kaufman County, from 14 to 25 June 2008 (TBRC 2008–43; TPRF 2585); and near Zionsville, Washington County, from 16 August to 3 September 2011 (TBRC 2011–079; TPRF 2975). There are two well-documented nesting records for the state: the first at Anahuac NWR, Chambers County, in 1967; and the second in Live Oak County in 2007 (TBRC 2007–81; TPRF 2516). Other reports of nesting have come from Brooks, Brazoria, Cameron, Colorado, Jefferson, Hidalgo, and San Patricio Counties. This species has been widely reported as a rare permanent resident in Texas; however, there is very little evidence to support this contention. There are 76 reports of Masked Ducks from prior to the development of the Review List in 1988 for which there is no documentation on file. **Timing of occurrence**: Masked Ducks have been found in Texas during every month of the year, although there are more records in April and December than other months. **Taxonomy**: Monotypic.

RUDDY DUCK *Oxyura jamaicensis* (Gmelin)

Common migrant and winter resident throughout most of the state but rare in winter in the Panhandle. Ruddy Ducks are opportunistic breeders in Texas. There are nesting records from most regions of the state, but this species is generally absent or is present in very low numbers during the summer. In the Panhandle and El Paso areas Ruddy Ducks can be uncommon to locally common during the summer, yet even with this higher relative abundance, nesting is not found every year. **Timing of occurrence**: Ruddy Ducks begin arriving in Texas in early September, with the bulk of the population in place by the end of October. Spring migration begins in late March, and most have departed the state by mid-May. **Taxonomy**: The subspecies that occurs in Texas is *O. j. rubida* (Wilson).

ORDER GALLIFORMES

Family Cracidae: Chachalacas, Guans, and Curassows

PLAIN CHACHALACA *Ortalis vetula* (Wagler)

Uncommon to common resident in the Lower Rio Grande Valley north to Kenedy and Zapata Counties. Plain Chachalacas appear to be colonizing areas along the Rio Grande in southern Webb County, where the species had been extirpated previously. A small population in Kleberg and Nueces Counties originated from a release program on private ranches in the 1970s and 1980s. Similar releases occurred until the mid- to late 1980s in Brooks, Dimmitt, Jim Wells, Kenedy, La Salle, and San Patricio Counties, which likely account for the occasional reports from those counties. **Taxonomy:** The subspecies that occurs in Texas is *O. v. mccalli* Baird.

Family Odontophoridae: New World Quail

SCALED QUAIL *Callipepla squamata* (Vigors)

Uncommon to locally common resident in the Trans-Pecos and from the Panhandle south through the western Rolling Plains (east to Wichita and Coleman Counties) to the western South Texas Brush Country. Scaled Quail populations have experienced sharp declines in recent years, in some areas as much as a 75 percent drop in the population. This is particularly evident on the South Plains and western Edwards Plateau, but the reasons behind this decline are unclear. Habitat loss or degradation seems to be a primary factor overall. Scaled Quail formerly ranged eastward to the eastern Rolling Plains and eastern Edwards Plateau (Strecker 1912; Simmons 1925). **Taxonomy:** Three subspecies are known to occur in Texas.

C. s. pallida Brewster

Common resident in the Trans-Pecos and from the southern Panhandle to the Balcones Escarpment. Intergrades with *C. s. castanogastris* are reported from Val Verde and Kinney Counties.

C. s. castanogastris Brewster
> Common to abundant resident in the western South Texas Brush Country from Val Verde and Kinney Counties south to Starr and western Hidalgo Counties.

C. s. hargravei Rea
> Common resident in the northern third of the Panhandle.

GAMBEL'S QUAIL *Callipepla gambelii* (Gambel)

Common resident in El Paso and western Hudspeth Counties southward along the Rio Grande to southern Presidio County. Gambel's Quail are locally uncommon to rare residents through the remainder of Hudspeth County and in extreme western Culberson County. They have also been found irregularly farther east along the Rio Grande to western Brewster County, although no resident population is known to occur in these areas. **Taxonomy:** The subspecies that occurs in Texas is *C. g. gambelii* (Gambel).

NORTHERN BOBWHITE *Colinus virginianus* (Linnaeus)

Occurs throughout most of the state, including as far west as the eastern Trans-Pecos. The overall population of the species in Texas is in sharp decline. Northern Bobwhites are locally uncommon in the Panhandle south through the South Texas Brush Country. Farther east the species is now rare and local or absent in most locations, although there are still areas where it can be characterized as uncommon. Studies on the population of Northern Bobwhites in the state show declines of up to 90 percent between 1967 and 2011 in some areas and sharp declines virtually statewide. The last stronghold for the species in Texas appears to be the South Texas Brush Country and the Rolling Plains. Northern Bobwhites are rare and local residents in the eastern Trans-Pecos from eastern Reeves County southeast to Terrell County. **Taxonomy:** Three subspecies of Northern Bobwhites are known to occur in Texas. However, the innumerable releases of pen-raised birds, some originating from populations in Mexico, have confused distributional ranges of the subspecies.

C. v. virginianus (Linnaeus)
> Rare to locally uncommon and declining resident in the eastern third of the state.

C. v. taylori Lincoln

Uncommon resident in the Panhandle and Rolling Plains south to the southern Llano Estacado (Andrews and Howard Counties).

C. v. texanus (Lawrence)

Uncommon resident from the eastern Trans-Pecos eastward through the Edwards Plateau and south through the western South Texas Brush Country.

MONTEZUMA QUAIL *Cyrtonyx montezumae* (Vigors)

Uncommon and local resident in the Davis and Del Norte Mountains of the Trans-Pecos. This quail is rare to locally uncommon in the Chinati and Glass Mountains, as well as the Sierra Diablo and Sierra Vieja. A relict population still exists on the western Edwards Plateau and is centered in Edwards County, where the birds are rare to locally uncommon. There have been unsuccessful attempts at reintroduction into the Guadalupe and Chisos Mountains. Montezuma Quail were found on several occasions in the Chisos Mountains from 2005 to 2007 (Holderman, Sorola, and Skiles 2007), but whether these birds are a remnant of the natural population, part of the reintroduction attempts, or colonizers is unknown. There have been no reported sightings since that period. Montezuma Quail have been shown to be sensitive to changes in the quality of the grasslands and savannahs they inhabit. Albers and Gehlbach (1990) showed that when 40–50 percent of tallgrass cover is removed, the quail are extirpated. Montezuma Quail formerly occurred over much of the Edwards Plateau and southward into Maverick County. **Taxonomy:** The subspecies that occurs in Texas is *C. m. mearnsi* Nelson.

Family Phasianidae: Pheasants, Grouse, and Allies

RING-NECKED PHEASANT

Phasianus colchicus Linnaeus

Common resident in the northern and western parts of the Panhandle, becoming increasingly less common southward onto the northern South Plains and farther east. Ring-necked Pheasants were formerly more widespread but have

declined in many areas, including most of the South Plains. Small populations were once found in Chambers County on the upper coast and in the Trans-Pecos, but there are no recent sightings to suggest that any remain in those areas. Originally a species of eastern Asia, the Ring-necked Pheasant was introduced into Texas for hunting starting in the 1930s. Releases by the Texas Parks and Wildlife Department and private entities have occasionally supplemented these populations, producing temporary, nonviable populations elsewhere in the state. **Taxonomy:** The population in Texas cannot be identified to a subspecies. The initial introduced populations in Texas (and North America in general) were primarily from populations in eastern China of *P. c. torquatus* Gmelin. However, other subspecies have been released, as well as stock that originated from Great Britain (where they were introduced from Asia more than 2,000 years ago), resulting in intergrades of various subspecies.

GREATER PRAIRIE-CHICKEN

Tympanuchus cupido (Linnaeus)

Extremely rare on remnant prairies in the upper and central Coastal Prairies. There are now three tiny populations, and all of the birds remaining in Texas have originated from a captive breeding program and are dependent on human intervention for survival. The stronghold of the introduced population is at Attwater Prairie Chicken NWR in Colorado County. There is a remnant population at The Nature Conservancy's Texas City Prairie Preserve, and efforts are now under way to establish another population on private lands in Refugio County. The Coastal Prairies of Texas and Louisiana once hosted more than 1 million birds (Lehmann 1941), but currently there are fewer than 120 individuals within these three populations. **Taxonomy:** Two subspecies have been documented in Texas.

T. c. pinnatus (Brewster)

Extirpated. Once common in the tallgrass prairies of north-central and northeast Texas as far south as Travis County (Simmons 1925). By 1920 they were largely extirpated, with the last small flocks reported in Harrison and Smith Counties. The subspecies was last reported in 1956 from Smith County (Oberholser 1974).

T. c. attwateri Bendire

As given above. This population is popularly known as the Attwater's Prairie-Chicken. This subspecies formerly ranged from the Louisiana border southwest to Refugio County. The last remnants of the wild Refugio County population were last reported in 1999.

LESSER PRAIRIE-CHICKEN
Tympanuchus pallidicinctus (Ridgway)

Rare to uncommon and local resident in the eastern Panhandle and western South Plains. There are currently two disjunct populations in Texas remaining from the much more widespread occurrence of the species. The population on the western South Plains extends from Bailey County southward to Gaines and northern Andrews Counties. The population in the eastern Panhandle is found from Lipscomb County south to Collingsworth County. Lesser Prairie-Chickens formerly ranged south to Menard and Jeff Davis Counties (Strecker 1912). They have been reported to have formerly ranged as far east as Cooke and Tarrant Counties. Pulich (1988) questioned the validity of the easternmost reports and postulated that the eastern boundary was probably several counties farther west. Pulich also noted that G. H. Ragsdale is credited with collecting both Greater and Lesser Prairie-Chickens from Cooke County in January 1878, which he considered unlikely. Lesser Prairie-Chickens formerly wandered southward during the winter months. The last incursion into the western Edwards Plateau was during the winter of 1885–86 (Lacey 1911). This species was last documented in the San Angelo area in 1912 and was present in Howard County until at least 1945 (Maxwell 2013). **Taxonomy:** Monotypic.

WILD TURKEY *Meleagris gallopavo* Linnaeus

Common to uncommon resident from the eastern Panhandle southward through the Rolling Plains to the Edwards Plateau. This species is also common to uncommon in the South Texas Brush Country north of the Lower Rio Grande Valley, on the Coastal Prairies to Jackson County, and in the mountains of the central Trans-Pecos. The Wild Turkey was formerly a common resident in the eastern half of the state and is now rare and local in many areas. The Texas Parks and Wildlife Department has reintroduced

Wild Turkeys to many areas of the state to bolster or re-store local populations. Since 1979 more than 7,000 Wild Turkeys from 16 states have been transplanted to East Texas (Seidel 2010). **Taxonomy:** Three subspecies are known to occur in the state.

M. g. silvestris Vieillot

Rare and local resident in eastern Texas. Released extensively during the 1980s and early 1990s in attempts to reestablish local populations.

M. g. intermedia Sennett

Common to locally uncommon in the eastern Panhandle south through the Edwards Plateau to the South Texas Brush Country.

M. g. merriami Nelson

Formerly a rare resident in the Guadalupe Mountains but ex-tirpated around 1907. Reintroduction releases began in 1928, and this subspecies is now locally common in the Guadalupe and Davis Mountains.

ORDER GAVIIFORMES

Family Gaviidae: Loons

RED-THROATED LOON *Gavia stellata* (Pontoppidan)

Rare winter resident along the upper coast and on reser-voirs in the northeastern part of the state. Red-throated Loons are casual winter visitors to most other areas of Texas. This species was on the Review List from 1987 until 2002, and 48 records were accepted, with the majority oc-curring since 1990. Since 2002 there have been more than 40 reports of Red-throated Loons, many of which were of multiple individuals. This would suggest the species is oc-curring in greater numbers, although increasing observer awareness may partially account for this change. **Timing of occurrence:** Records exist between 26 October and 3 May. **Taxonomy:** Monotypic.

PACIFIC LOON *Gavia pacifica* (Lawrence)

Rare winter resident along the upper coast and on reservoirs in the northeastern part of the state. Pacific Loons occur with greater frequency, and are more evenly distributed statewide, than Red-throated Loons. This species has been an annual visitor to the state since at least 1980 and was removed from the Review List in 1996. **Timing of occurrence:** Fall migrants arrive in the state as early as late October, but most are found from mid-November to mid-December. Spring migrants leave the state from late February through early April, but individuals regularly linger well into May. There are several occurrences of Pacific Loons remaining along the coast well into June and, on rare occasions, early July. **Taxonomy:** Monotypic. This taxon was previously included under Arctic Loon (*G. arctica*). No accepted records of *G. arctica* exist for Texas, but there is a record from Colorado, and the potential exists for one to be found in the state.

COMMON LOON *Gavia immer* (Brünnich)

Uncommon to rare migrant throughout the state. This species is a common winter resident on lakes and in bays and estuaries along the coast, where loose groups exceeding 60 individuals are occasionally found in late winter and early spring. Common Loons are locally common winterers, particularly on large reservoirs, in the eastern half of the state and become increasingly uncommon through the west. They are rare in winter in the western Trans-Pecos and the High Plains. **Timing of occurrence:** Fall migrants begin arriving in the state in late September, with the bulk of the wintering population arriving by early November. Most Common Loons depart in late March or early April, although some birds routinely linger until late May. Annually a few birds remain throughout the summer, most often along the coast. These summering individuals are non-breeders and typically retain basic plumage. **Taxonomy:** Monotypic.

TBRC Review Species   ### YELLOW-BILLED LOON *Gavia adamsii* (Gray)

Casual. Texas has six documented records of Yellow-billed Loons from locations scattered across the state. A bird of the high Arctic, it has occurred with greater frequency in

the interior United States since 1980 (Patten 2000). To date, all individuals discovered in Texas have been in first-winter plumage. An individual at South Padre Island, Cameron County, from 22 December 2000 to 26 May 2001 provided a very late departure date and appears to be the southernmost ever documented in the New World. **Timing of occurrence**: In addition to the Cameron County record, there are four winter records between 25 November and 9 January. One was at Flour Bluff, Nueces County, from 25 to 27 March 2006. **Taxonomy**: Monotypic.

ORDER PODICIPEDIFORMES

Family Podicipedidae: Grebes

LEAST GREBE *Tachybaptus dominicus* (Linnaeus)

Uncommon to locally common resident of the Lower Rio Grande Valley north through the South Texas Brush Country and along the Coastal Prairies to the central coast. Least Grebes are uncommon and local farther up the coast to Galveston Bay and inland north to Bastrop and Travis Counties. They are rare visitors at all seasons farther up the coast to Jefferson County and inland through the eastern half of the state as far north as Tarrant, Kaufman, and Harrison Counties. The northernmost breeding record is from Richland Creek WMA, Freestone County, in 2004. They also occur irregularly farther west on the Edwards Plateau and have been noted with increasing frequency up-stream along the Rio Grande to southern Brewster County. More far-flung individuals include two records from El Paso County. The greatest population densities are found from Brooks, Kenedy, and Kleberg Counties southward to Cameron and Hidalgo Counties. The number of individuals present fluctuates greatly from year to year depending on rainfall and, consequently, habitat availability. Populations may also decline temporarily in response to stress resulting from unusually cold weather (James 1963b). Some individuals retreat southward during the winter, particularly those from the northernmost portions of their range. Breeding may occur during any month in the Lower Rio

Grande Valley. **Taxonomy:** The subspecies found in Texas is *T. d. brachypterus* (Chapman).

PIED-BILLED GREBE *Podilymbus podiceps* (Linnaeus)

Uncommon to common migrant and winter resident throughout the state. Pied-billed Grebes are uncommon, local breeders through the eastern half of the state and can be locally common during a given year when habitat conditions are favorable. This species is a rare and irregular breeder in the Trans-Pecos. The southern Rolling Plains, western Hill Country, and western South Texas Brush Country have limited nesting habitat available, so there are very few reported breeding records from this large region. Nonbreeding Pied-billed Grebes can be found during the summer in all areas of the state. **Timing of occurrence:** Fall migrants arrive in mid-August through early September. In spring, wintering birds have generally left the state by early May. **Taxonomy:** The subspecies found in Texas is *P. p. podiceps* (Linnaeus).

HORNED GREBE *Podiceps auritus* (Linnaeus)

Uncommon to locally common winter resident on reservoirs in the northeastern quarter of the state, becoming uncommon to rare along the upper and central coasts. Horned Grebes are generally rare across the remainder of the state, becoming casual in the South Texas Brush Country. **Timing of occurrence:** This species generally occurs between mid-October and mid-March. There are reports as early as mid-September, and birds have lingered into mid-May. **Taxonomy:** The subspecies found in Texas is *P. a. cornutus* (Gmelin).

TBRC Review Species

RED-NECKED GREBE *Podiceps grisegena* (Boddaert)

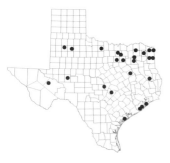

Very rare winter visitor for which there are 25 documented records. The first to be documented was found at Big Lake, Reagan County, from 23 to 30 November 1975 (TBRC 1992-129). Nearly half of those documented have been from north-central and northeast Texas. The greatest number of individuals present for a single accepted record was three at White River Lake, Crosby County, from 21 to 22 January 1978 (TBRC 1990–11). There are 35 reports of Red-necked Grebes from prior to the development of the

Review List in 1988 for which there is no documentation on file. **Timing of occurrence:** Extreme dates of occurrence are 5 November and 30 March. **Taxonomy:** The subspecies found in Texas is *P. g. holbollii* Reinhardt.

EARED GREBE *Podiceps nigricollis* Brehm

Common to uncommon migrant in the western three-quarters of the state and rare to very rare in the Pineywoods. Eared Grebes are uncommon to common winter residents in most areas, except the High Plains and Pineywoods, where they are rare to very rare. They are common to locally abundant along the coast, where flocks of several hundred are occasionally encountered. This species is an uncommon to rare summer resident in the Panhandle and the westernmost Trans-Pecos. Large concentrations of breeding birds have occasionally been reported from the Panhandle, including a total of 61 nests on a single playa in Randall County in 1999. Breeding records are rare in other areas of the state but exist from as far south as Bee County. Nonbreeding Eared Grebes are frequently found summering throughout the state, including in coastal bays. **Timing of occurrence:** Fall migrants begin to arrive in Texas in late July, but the main population arrives between late September and the end of October. In spring, birds depart the state primarily between early April and early May, and some linger until early June. **Taxonomy:** The subspecies found in Texas is *P. n. californicus* Heermann.

WESTERN GREBE

Aechmophorus occidentalis (Lawrence)

Uncommon to locally common resident in the western Trans-Pecos. Western Grebes are rare migrants and winter visitors through the remainder of the Trans-Pecos as well as the northwestern portion of the state. They are rare to very rare east to the eastern Edwards Plateau and north-central Texas and casual farther east through the Pineywoods. The status of this species has changed dramatically since the 1990s. The first nesting record of *Aechmophorus* grebes in Texas consisted of a mixed pair of Western × Clark's found at Balmorhea Lake, Reeves County, in 1991 (Lockwood 1992). The first breeding record involving only Western Grebes was discovered at McNary Reservoir, Hudspeth County, in December 2001, where the species is now a consistent breeder. Small breeding populations can be found on reservoirs in the El Paso area and at Balmorhea Lake, Reeves County. Nesting can occur at all seasons but is dependent on water levels in these impoundments. The number of wintering birds has steadily increased in the western Trans-Pecos. **Timing of occurrence**: Fall migrants begin to arrive in Texas in late September, with most of the population arriving by late October. Spring migrants can be found between early April and mid-May. **Taxonomy**: The subspecies found in Texas is *A. o. occidentalis* (Lawrence).

CLARK'S GREBE *Aechmophorus clarkii* (Lawrence)

Common to locally abundant resident in the western Trans-Pecos. Clark's Grebes are rare migrants and winter visitors through the remainder of the Trans-Pecos as well as the northwestern portion of the state south to the Concho Valley. They are winter vagrants farther east. As in the case of the Western Grebe, the population of Clark's Grebe in the Trans-Pecos has increased significantly since 1997. This species is now the most common of the *Aechmophorus* grebes in the El Paso area as well as at Balmorhea Lake, Reeves County. Nesting can occur in all seasons and is dependent on water levels. The number of birds present during a given season varies considerably from high counts of more than 150 individuals to just a few birds. **Timing of occurrence**: Fall migrants begin to arrive in Texas in late September, with most of the population arriving by mid-October. Spring migrants can be found between early April and mid-May. **Taxonomy**: The subspecies found in Texas is *A. c. transitionalis* Dickerman.

Family Phoenicopteridae: Flamingo

TBRC Review Species   **AMERICAN FLAMINGO** *Phoenicopterus ruber* Linnaeus

Casual with eight accepted records from Texas, although there have been numerous additional undocumented reports. All Texas records have been from the immediate coast. The first accepted record for the state was a bird discovered in spring 1978 at Bird Island, Kleberg County; it remained through the summer, returning each spring from 1979 through 1982. However, the most significant

record for the state was a bird banded as a nestling at the Rio Lagartos Biosphere Reserve in the state of Yucatán, Mexico, that was discovered at Shoalwater Bay, Calhoun County, on 14 October 2005. This provided the first solid evidence that birds from the large colony on the northern tip of the Yucatán Peninsula could wander to the Texas coast. As of spring 2012, this bird occasionally was still being seen in Texas, with no evidence that it has returned to the Yucatán, although it has wandered east to southern Louisiana and south to Aransas County. During much of its stay it has been in the company of a Greater Flamingo that escaped from the zoo in Wichita, Kansas. Escaped Greater Flamingos have been documented on the Texas coast previously, which underscores the need for thorough documentation of all flamingos. **Timing of occurrence**: With the exception of the long-staying bird described above, the American Flamingos found in Texas have occurred in the spring and summer, with dates ranging from 5 April to 29 July. **Taxonomy**: Monotypic. This species was formerly considered a subspecies of *P. roseus* Pallas, under the name Greater Flamingo.

ORDER PROCELLARIIFORMES

Family Diomedeidae: Albatrosses

TBRC Review Species

YELLOW-NOSED ALBATROSS
Thalassarche chlororhynchos (Gmelin)

Accidental. There are four documented records for the state. The first three were found on beaches: two on the lower coast and one on the central coast. Two of the individuals were found dead or dying and were preserved as specimens. **Taxonomy**: The adult found on San Jose Island is *T. c. chlororhynchos* (Gmelin). The other records are presumed to be this subspecies, but the most recent two records show some features associated with *T. c. carteri* (Rothschild), which is sometimes considered a separate taxon.

14 MAY 1972, PORT ISABEL, CAMERON CO. (TBRC 1988-50)
28 OCT. 1976, SOUTH PADRE ISLAND, WILLACY CO.
  (*U.T. PAN AMERICAN; TPRF 118)

11 JULY 1997, SAN JOSE ISLAND, ARANSAS CO.
  (*TCWC 13,338; TBRC 1997-129; TPRF 1651)
26 SEPT. 2003, OFF SOUTH PADRE ISLAND, CAMERON CO.
  (TBRC 2003–74; TPRF 2119)

Family Procellariidae: Shearwaters and Petrels

TBRC Review Species

BLACK-CAPPED PETREL *Pterodroma hasitata* (Kuhl)

Accidental. There are two records from the offshore waters of Texas. Both records involve single individuals found in different years 60+ miles offshore Port O'Connor, Calhoun County. Black-capped Petrels have rarely been detected in the Gulf of Mexico, and the species has not been found on Louisiana pelagic trips. The eastern extreme of the normal foraging range is southern Florida. **Taxonomy**: Monotypic.

28 MAY 1994, OFF PORT O'CONNOR, CALHOUN CO.
  (TBRC 1996-129; TPRF 1631)
26 JULY 1997, OFF PORT O'CONNOR, CALHOUN CO.
  (TBRC 1997-118; TPRF 1633)

TBRC Review Species

STEJNEGER'S PETREL

Pterodroma longirostris (Stejneger)

Accidental. There is a single record of this species for Texas. A decomposing carcass was retrieved from a beach near Port Aransas, Nueces County. Specific identification was made by comparing the skeleton to that of other similar-sized *Pterodroma* species (S. Olson and D. Lee, pers. comm.). Stejneger's Petrel is found in the southern Pacific, and its occurrence in the Gulf of Mexico was quite unexpected. **Taxonomy**: Monotypic.

15 SEPT. 1995, PORT ARANSAS, NUECES CO.
  (TBRC 1997-59; TPRF 1746)

TBRC Review Species

WHITE-CHINNED PETREL

Procellaria aequinoctialis (Linnaeus)

Accidental. There is a single record of this species for Texas. The bird was discovered floundering in the surf and was taken to a wildlife rehabilitator. Unfortunately, the specimen was not saved, but it was photographed. This individual was the first documented occurrence for North America. The AOU (1998) and ABA (1994) originally

1. The distribution of Ross's Geese (*Chen rossii*) within Texas has changed significantly since 1990 as the overall population has increased. This species is now a common sight on the High Plains and Coastal Plains. In areas with high concentrations of Ross's and Snow Geese, the two species segregate even within a single flock. This adult was in Andrews, Andrews County, on 28 November 2009. *Photograph by Mark W. Lockwood.*

2. Although Brant is a TBRC Review Species, reports from hunters suggest that this dark goose occurs with greater frequency than the accepted records would indicate. This (Black) Brant (*Branta bernicla nigricans*) was in Andrews, Andrews County, on 10 November 2007. *Photograph by Mark W. Lockwood.*

3. The separation of Canada Goose into two species has provided an identification challenge when comparing the Lesser Canada Goose to Cackling Goose (*Branta hutchinsii*). This Cackling Goose was in Lubbock, Lubbock County, on 24 December 2007. *Photograph by Mark W. Lockwood.*

4. Tundra Swan (*Cygnus columbianus*) is a rare visitor to Texas, and all discoveries are noteworthy. The majority of sightings are from the Panhandle and across northern Texas, but there are scattered records from almost all regions of the state. These two were at Fort Hood, Bell County, on 4 January 2011. *Photograph by Gil Eckrich.*

5. The population of Mallards found in the Trans-Pecos and southward along the Rio Grande belongs to the distinctive Mexican Duck (*Anas platyrhynchos diazi*). Additional study of the amount of introgression with green-headed Mallards within populations in Texas and northern Mexico is needed. This male Mexican Duck was at Midland, Midland County, on 21 December 2011. *Photograph by Mark W. Lockwood.*

6. Surf Scoter (*Melanitta perspicillata*) is the most commonly occurring of the three scoter species in Texas. Although fully adult males are regularly seen, first- and second-year birds make up the majority of the wintering population in the state. This second-year male was at Port Bolivar, Galveston County, on 26 May 2010. *Photograph by Joseph C. Kennedy.*

7. Long-tailed Duck (*Clangula hyemalis*) was long known under the name Oldsquaw. These ducks are rare to very rare in winter in the state, with most sightings coming from along the coast. This female was photographed in Austin, Travis County, on 22 January 2010. *Photograph by Lawrence Semo.*

8. Hooded Mergansers (*Lophodytes cucullatus*) have been increasing as a nesting species in the northeastern portion of the state. They are often found using nest boxes designed for Wood Ducks and laying eggs in active Wood Duck nests. This male was at Balmorhea SP, Reeves County, on 28 February 2012. *Photograph by Mark W. Lockwood.*

9. Gambel's Quail (*Callipepla gambelii*) are restricted to the westernmost counties in the state. The easternmost edge of the range is in extreme western Culberson County and along the Rio Grande to eastern Presidio County. This male was at Fort Leaton State Historic Site, Presidio County, on 5 April 2007. *Photograph by Mark W. Lockwood.*

10. Northern Bobwhite (*Colinus virginianus*) populations have declined precipitously in recent decades, and this once-abundant species is now very rare or even absent in many areas of its historic range. This pair was near Rio Grande City, Starr County, on 2 December 2007. *Photograph by Greg W. Lasley.*

11. Montezuma Quail (*Cyrtonyx montezumae*) were once more widespread on the Edwards Plateau but are now restricted to the southwestern portion of that region. Studies have shown that they are very sensitive to changes in the tallgrass cover of their preferred savannah-like habitat. This male was in northern Jeff Davis County on 11 January 2010. *Photograph by Mark W. Lockwood.*

12. Lesser Prairie-Chicken (*Tympanuchus pallidicinctus*) populations have continued to decline in Texas, and the species appears imperiled. The remaining populations are being intensely studied in attempts to develop more successful conservation strategies. This displaying male was in northern Yoakum County on 30 March 2010. *Photograph by Mark W. Lockwood.*

13. Red-throated Loons (*Gavia stellata*) are a rare visitor to Texas, although they do appear to occur with much higher frequency in the northeastern quarter of the state. This winter-plumaged adult was at Balmorhea Lake, Reeves County, on 26 November 2010.
Photograph by Mark W. Lockwood.

14. Although still considered rare visitors to Texas, Pacific Loons (*Gavia pacifica*) are found much more frequently than are Red-throated Loons, and there are records from most regions of the state. This one was at Balmorhea Lake, Reeves County, on 14 January 2009.
Photograph by Greg W. Lasley.

15. Red-necked Grebe (*Podiceps grisegena*) is a rare visitor to the state, with only about 25 records. These occurrences have primarily been in the northern half of the state, with records from the northeastern corner westward to the eastern Trans-Pecos. This adult was at Lake Kickapoo, Archer County, on 4 March 2007. *Photograph by Scotty Lofland.*

16. The population of Clark's Grebes (*Aechmophorus clarkii*) in the western Trans-Pecos has grown at an unexpected rate since 1995. This species was once far outnumbered by Western Grebe but now is the much more common of the *Aechmophorus* grebes and is a regular breeder. This adult was at Balmorhea Lake, Reeves County, on 11 November 2010. *Photograph by Mark W. Lockwood.*

17. This banded American Flamingo (*Phoenicopterus ruber*) provided important evidence that the individuals found in Texas could have originated on the Yucatán Peninsula. It was banded in the nest in 2005 at Rio Lagartos on the northern tip of the Yucatán and discovered on the Texas coast in October 2005. It wandered widely after its discovery, including to Goose Island SP, Aransas County, on 13 November 2005. *Photograph by Lynn Barber.*

18. Yellow-nosed Albatross (*Thalassarche chlororhynchos*) has been documented in Texas on only four occasions. This near-adult-plumaged bird was off South Padre Island, Cameron County, on 26 September 2003 and represents one of the more spectacular finds from deepwater pelagic trips in Texas waters. *Photograph by Garrett Hodne.*

19. Two subspecies of Cory's Shearwater (*Calonectris diomedea*) occur in Texas waters. This individual belongs to C. *d. diomedea*, based on the extensive white of the underwing extending into the primaries. It was off South Padre Island, Cameron County, on 19 September 2009. *Photograph by Garrett Hodne.*

20. Great Shearwater (*Puffinus gravis*) was formerly known as Greater Shearwater. There are 17 records for the state. This adult was off South Padre Island, Cameron County, on 17 July 2010. *Photograph by Mary Gustafson.*

21. The Band-rumped Storm-Petrel (*Oceanodroma castro*) is now known to be the expected storm-petrel in Texas waters; however, recent discoveries related to discrete populations of this species have brought up taxonomic questions relating to the birds found in the state. More research is needed, but the majority appear to be winter breeders from islands in the eastern Atlantic Ocean. This one was off South Padre Island, Cameron County, on 25 July 2009. *Photograph by Tony Frank.*

22. Red-billed Tropicbird (*Phaethon aethereus*) is now known to be the expected species of tropicbird found along the Texas Coast. The number of sightings for the state has increased since 2004. This may reflect an actual increase in occurrence, or they may simply be more regular off South Padre Island than farther north. This adult was off South Padre Island, Cameron County, on 17 June 2005. *Photograph by Garrett Hodne.*

23. Jabiru (*Jabiru mycteria*) has been documented in Texas on 10 occasions. All but one have been postbreeding wanderers that reached the state between late July and late October. This individual was the longest staying of these records and was present for almost two weeks near Raymondville, Willacy County, from 10 to 22 August 2008. *Photograph by Jan Dauphin.*

24. The number of documented records of Brown Booby (*Sula leucogaster*) in Texas from 2004–8 was substantially above average, suggesting an increase in abundance; however, since then the number of sightings has dropped, raising questions about the reasons for this four-year increase. This immature bird was at the Packery Channel Jetties, Nueces County, on 16 July 2008. *Photograph by Jay Gardner.*

25. An outstanding find for Texas was a Bare-throated Tiger-Heron (*Tigrisoma mexicanum*) present at Bentsen–Rio Grande Valley SP, Hidalgo County, during the winter of 2009–10. Although often difficult to find, it did frequent an area of tall grass, where it hunted for prey, including small mammals. This image was obtained on 1 January 2010. *Photograph by Rick Nirschl.*

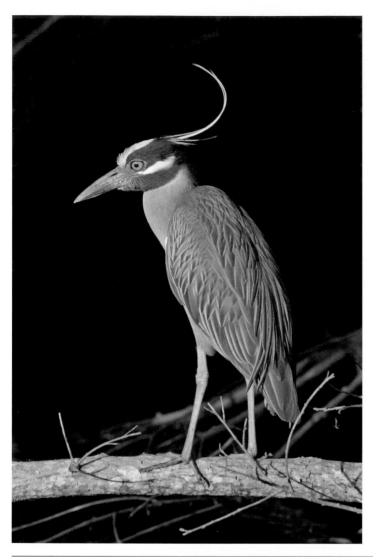

26. Yellow-crowned Night-Heron (*Nyctanassa violacea*) is a locally common summer resident of wooded wetlands in the eastern half of the state. As their name implies, these striking herons often forage at night, when one of their favored prey items, crayfish, are most active. This adult was near Baytown, Chambers County, on 19 April 2012. *Photograph by Mark W. Lockwood.*

27. With only four records, the occurrence of any Snail Kite (*Rostrhamus sociabilis*) in Texas is very noteworthy. This adult male was at El Franco Lee Park in Houston, Harris County, on 17 June 2011. The other state records are from the South Texas Brush Country. *Photograph by Stephan Lorenz.*

28. The discovery of a near-adult-plumaged Double-toothed Kite (*Harpagus bidentatus*) at High Island, Galveston County, on 3 May 2011 has to be one of the most unexpected records, ever for Texas. Fortunately, it was cooperative and allowed excellent photographs to be obtained, but it was never relocated after the initial observation. *Photograph by David Hanson.*

29. Northern Goshawk (*Accipiter gentilis*) is a frequently reported species in the state but very seldom conclusively documented. This has led to a large number of records that have not been endorsed by the TBRC. Only nine of the 24 accepted records include a specimen or photograph. This immature bird was at Lake Meredith, Moore County, on 11 January 2010. *Photograph by Trey Barron.*

30. The occurrence of three Roadside Hawks (*Buteo magnirostris*) in the Lower Rio Grande Valley between January and March 2005 was unprecedented. This incursion almost doubled the number of records for the United States at the time. In early 2010 two more records were documented, including this adult at Falcon SP, Starr County. This image was obtained on 28 February 2010. *Photograph by Alan Wormington.*

31. Although still very local in distribution, Gray Hawks (*Buteo plagiatus*) have expanded their breeding range in Texas since 1980. They are now found from the Lower Rio Grande Valley to Big Bend NP and the Davis Mountains, with nesting areas in Val Verde County discovered in 2012. This adult was at Cottonwood Campground in Big Bend NP, Brewster County, on 23 June 2012. *Photograph by Marc Eastman.*

32. Like several species for which Texas is in the southernmost portion of their winter range, Rough-legged Hawks (*Buteo lagopus*) are becoming harder to find away from the Panhandle. The winter range appears to be shrinking northward, and this species is now regular in occurrence only in the Panhandle and northern South Plains. This adult was near Dumas, Moore County, on 22 December 2012. *Photograph by Mark W. Lockwood.*

33. Clapper Rails (*Rallus longirostris*) are most often associated with saltwater marshes in Texas, but they can also be found in freshwater habitats that are close to the immediate coast. They are very similar to King Rails in appearance, making birds in these freshwater marshes a challenge to identify. This Clapper Rail was at Sabine Pass, Jefferson County, on 31 October 2009. *Photograph by Greg W. Lasley.*

34. In 2011 Common Moorhen was split into two species. The taxa found in the Americas reverted to the common name Common Gallinule (*Gallinula galeata*). Differences in morphology, voice, and genetics led to this change in taxonomy. This adult Common Gallinule was at Anahuac NWR, Chambers County, on 22 April 2012. *Photograph by Mark W. Lockwood.*

35. The population of Whooping Cranes (*Grus americana*) that winters in Texas continues to slowly increase and now totals about 245 individuals. This has led to an expansion of the wintering range to beyond Aransas NWR and, on rare occasions, at inland sites well away from the coast. This adult was at Aransas NWR, Aransas County, on 20 February 2010. *Photograph by Greg W. Lasley.*

36. The Mountain Plover (*Charadrius montanus*) has been considered a candidate for the Endangered Species list, but a review of the species status in 2011 concluded that the population was larger than previously thought and not in immediate danger of extinction. This adult was near Pearsall, Frio County, on 3 March 2005. *Photograph by Martin Reid.*

37. Northern Jacanas (*Jacana spinosa*) were found in Texas fairly regularly in the 1980s and early 1990s, including eight documented records in 1993 and early 1994. This rate of occurrence made the fact that none were found between the spring of 1994 and the fall of 2006 perplexing. This adult was at Choke Canyon SP, McMullen County, on 6 November 2009. *Photograph by Lee Pasquali.*

questioned the provenance of this bird. However, the ABA reevaluated the record in 2007 (Pranty et al. 2008), adding it to its checklist. The AOU added this species to the North America list in 2011 (Chesser et al. 2011), which included records from California, Maine, and Texas. **Taxonomy:** Monotypic.

27 APR. 1986, ROLLOVER PASS, GALVESTON CO.
  (TBRC 1990-129; TPRF 957)

CORY'S SHEARWATER *Calonectris diomedea* (Scopoli)

Uncommon, but regular, offshore along the entire Texas coast during summer and fall. The status of Cory's Shearwater in Texas was not well understood until birders visited waters along the Continental Shelf more frequently. Unlike other tubenoses, Cory's Shearwaters were regularly encountered in waters near the Continental Shelf during the 1970s and 1980s. **Timing of occurrence:** This species occurs in Texas waters from June through early November. The numbers peak between late July and October. **Taxonomy:** There have been two subspecies of Cory's Shearwater documented in Texas. These taxa are sometimes considered to be separate species.

C. d. diomedea (Scopoli)
  Uncommon during summer and fall. Exact status of this and the next taxon in Texas requires additional research.

C. d. borealis (Cory)
  See comments on status under *C. d. diomedea*.

TBRC Review Species

GREAT SHEARWATER *Puffinus gravis* (O'Reilly)

Formerly known as Greater Shearwater, this species is a very rare visitor to offshore waters along the upper and central coasts of the Gulf. There are 20 documented records of Great Shearwater for the state. Six of these records involve dead or dying birds discovered on beaches, and five have been preserved as specimens. The first record for Texas was found on Mustang Island, Nueces County, on 27 December 1971 (TBRC 2006–02; TPRF 2361). An exceptional find was that after Hurricane Isaac at Lake Wright Patman, Cass County, on 31 August 2012. **Timing of occurrence:** The majority of Texas records are between 24 June and 13 October; however, this coincides with the period when the vast majority of pelagic trips have occurred. There are also single records for April and November and two from December. **Taxonomy**: Monotypic.

SOOTY SHEARWATER *Puffinus griseus* (Gmelin)

Very rare visitor to the offshore waters of the Gulf. There are 16 documented records from Texas. Of these, 11 refer to beached individuals, with seven preserved as specimens. The Sooty Shearwater is very similar in appearance to the Short-tailed Shearwater (*P. tenuirostris*), which is found in the northern Pacific Ocean. There is a single specimen record of the latter for the Gulf of Mexico pertaining to a bird found 25 miles west of Sanibel Island, Lee County, Florida (Kratter and Steadman 2003). This record suggests the possibility that Short-tailed Shearwaters could occur in Texas waters and must be considered when dealing with all-dark shearwaters. **Timing of occurrence**: The majority of records fall between 12 May and 19 September. There are also single records for October, November, and December and two for January. **Taxonomy**: Monotypic.

MANX SHEARWATER *Puffinus puffinus* (Brünnich)

Casual. There are eight documented records for the state, all of which involve individuals discovered on the central or lower coast. The majority of records are of birds that were either dead or dying when discovered on Mustang Island, Nueces County, or North Padre Island, Nueces or Kenedy Counties. Five of these are preserved as specimens. The first record for the state was discovered on North Padre Island on 15 February 1975. **Timing of occurrence**: The majority of records for this species have occurred between 11 August and 13 November. There are also single records from February, June, and December. **Taxonomy**: Monotypic.

AUDUBON'S SHEARWATER
Puffinus lherminieri Lesson

Uncommon in the Gulf of Mexico. Audubon's Shearwater was once considered extremely rare in Texas, but the more intensive exploration by birders of waters along the Continental Shelf since 1994 has clarified the status of this spe-

cies. A pelagic trip from Port O'Connor, Calhoun County, on 30 September 1995 tallied no fewer than 206 individuals. **Timing of occurrence**: Audubon's Shearwaters are present between mid-April and early November, with the peak of occurrence during August and September. There is one midwinter specimen of an individual picked up on Mustang Island, Nueces County, on 23 January 1989. **Taxonomy**: The records for Texas are believed to belong to *P.l. lherminieri* Lesson.

Family Hydrobatidae: Storm-Petrels

TBRC Review Species

LEACH'S STORM-PETREL
Oceanodroma leucorhoa (Vieillot)

Rare to very rare in the offshore waters of the Gulf. There are 30 documented records for the state, some involving multiple birds. Before 1991 there were only three accepted records for the state. The exploration of waters along the Continental Shelf beginning in 1994 has greatly enhanced our understanding of the occurrence of this species and other tubenoses in Texas. **Timing of occurrence**: Documented records in Texas waters range from 28 May through 30 September. **Taxonomy**: The subspecies that occurs in Texas is *o.l. leucorhoa* (Vieillot).

BAND-RUMPED STORM-PETREL
Oceanodroma castro (Harcourt)

Uncommon to rare summer visitor offshore. The Band-rumped Storm-Petrel was considered accidental in Texas prior to the mid-1990s, but it has been discovered to be the storm-petrel most likely to be encountered. The highest count was 70 individuals found on 14 July 2000 off South Padre Island, Cameron County. There are two documented inland records: three individuals were discovered near Edinburg, Hidalgo County, after Hurricane Alice on 25 June 1954, and a single bird was found near San Antonio, Bexar County, on 14 June 1984. A storm-petrel photographed on Lake Corpus Christi, Live Oak/San Patricio Counties after Hurricane Dolly on 24 July 2008 was most likely to have been this species. **Timing of occurrence**: Band-rumped Storm-Petrels occur in the western Gulf of Mexico from mid-May through September. **Taxonomy:** The taxonomy of this species is currently under investigation. Results of research on the breeding grounds suggest that at least four taxa (some or all could be distinct species) are present in the northern Atlantic Ocean populations. The birds that occur in Texas waters match the morphology and molt of the winter breeders of the western island groups in the eastern Atlantic Ocean, but formal description of this population has not yet occurred. Bolton (2007) refers to these birds as "Grant's" Storm-Petrel. Photographic evidence suggests that at least one other form of Band-rumped Storm-Petrel may occur in Texas waters.

ORDER PHAETHONTIFORMES

Family Phaethontidae: Tropicbirds

TBRC Review Species

RED-BILLED TROPICBIRD
Phaethon aethereus Linnaeus

Casual with most of the 13 documented records from off-shore. Records of Red-billed Tropicbirds have increased since 2000; whether this is a result of increased pelagic trips out of South Padre Island, Cameron County, or an actual increase in abundance is unknown. Three records refer to stranded, weak individuals that later died during rehabilitation and were preserved as specimens. One of these was a bird found in a Zapata County yard more than 140 miles from the Gulf. This species was not known in the state until 1985 when one was found stranded in Houston, Harris County, on 13 November (*TCWC 11,576). **Timing of occurrence:** The documented records all fall between 29 April and 13 November. **Taxonomy:** The records for Texas belong to *P. a. mesonauta* Peters.

ORDER CICONIIFORMES

Family Ciconiidae: Storks

TBRC Review Species

JABIRU *Jabiru mycteria* (Lichtenstein)

Casual. All 10 documented records of Jabirus in Texas have been of postbreeding wanderers, and all have been found on the Coastal Prairies or in the Lower Rio Grande Valley. The first individual discovered in Texas was present from 11 August to 8 September 1971 at Escondido Lake, Kleberg County (TPRF 22; Haucke and Keil 1973). The closest breeding populations of Jabirus are on the Yucatán Peninsula and in the state of Tabasco, Mexico. **Timing of occurrence**: All have occurred between 25 July and 29 October except one very early record from 10 June. **Taxonomy**: Monotypic.

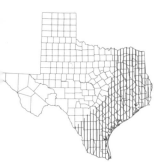

WOOD STORK *Mycteria americana* Linnaeus

Uncommon to locally common postbreeding visitor to the coast and inland to the eastern third of the state. Most Wood Storks are found east of a line from Dallas to San Antonio to Zapata. West of this line, due to drier conditions with fewer wetland options, they are very rare and generally absent. Wood Storks are very rarely found in Texas during the winter, though there are a few records from the coast and the South Texas Brush Country. During fall migration as many as 5,000 birds have been reported passing over hawk watches in a single day along the central and upper coasts. Half a century ago, Wood Storks nested in Chambers, Jefferson, and Harris Counties (Oberholser 1974), and an adult stork was observed carrying sticks at Caddo Lake in the 1990s, but nesting could not be confirmed. **Timing of occurrence**: Wood Storks arrive in Texas as early as May, but the majority of sightings begin in early June and continue through mid-October, with smaller numbers of birds lingering into early November. **Taxonomy**: Monotypic.

ORDER SULIFORMES

Family Fregatidae: Frigatebirds

MAGNIFICENT FRIGATEBIRD
Fregata magnificens Mathews

Uncommon summer and fall visitor throughout the coast. Magnificent Frigatebirds can be locally common in the Gulf and in larger bays, where high counts of more than 100 individuals have been noted. They are very rare inland in the eastern half of the state, with records north to Bowie and Denton Counties. These inland occurrences are typically associated with severe weather, such as tropical storms, along the coast. There is one breeding record from Aransas County from 6 June 1931 (Oberholser 1974). **Timing of occurrence:** Magnificent Frigatebirds arrive in the state as early as late March and are rare until mid-April, when the numbers start to increase through early May. Numbers peak between late June and September, although a few individuals linger through October and rarely as late as early December. There are a small number of reports of frigatebirds from late December to February, and birds found during this period should be carefully documented to exclude other species of frigatebird. **Taxonomy:** Monotypic.

Family Sulidae: Boobies and Gannets

MASKED BOOBY *Sula dactylatra* Lesson

Uncommon to rare migrant and summer visitor offshore and rare to very rare along the immediate coast. The Masked Booby is also a very rare winter visitor, with adults making up a very small percentage of the birds present. Individuals found on coastal beaches are frequently injured or sick. The plumage similarity between subadult or immature Masked Boobies and Northern Gannets often causes confusion and is an underappreciated identification challenge. **Timing of occurrence:** Although the Masked Booby has been found during every month of the year, the majority of individuals found in Texas are present between late March and early October. **Taxonomy:** The subspecies that occurs in Texas is *S. d. dactylatra* Lesson.

TBRC Review Species   **BLUE-FOOTED BOOBY** *Sula nebouxii* Milne-Edwards

Accidental. There is one documented record for this species in Texas. Remarkably, this bird occurred well inland in the center of the state. An immature bird appeared at Granite Shoals, Burnet County, in early June 1993 and remained for almost 16 months. It perched on a diving board throughout much its stay before it eventually disappeared. It reappeared at Lake Bastrop, Bastrop County, about 80 miles downstream along the Colorado River drainage, where it again remained for almost four months. An immature sulid photographed on 5 October 1976 at South Padre Island, Cameron County, has previously been reported as a Blue-footed Booby; however, a reexamination of the photographs showed it to be a Masked Booby. **Taxonomy:** There are two subspecies of Blue-footed Booby. Although it cannot be stated with certainty, the Texas record likely refers to *S. n. nebouxii* Milne-Edwards.

2 JUNE 1993–6 OCT. 1994, LAKE LYNDON B. JOHNSON, GRANITE
SHOALS, BURNET/LLANO CO. AND 10 DEC. 1994–12 APR. 1995,
LAKE BASTROP, BASTROP CO. (TBRC 1993-110; TPRF 1168)

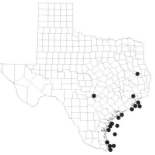

TBRC Review Species   **BROWN BOOBY** *Sula leucogaster* (Boddaert)

Very rare visitor to offshore waters and along the coast and accidental inland. There are 37 documented records, including three specimens. Two of these are of adults found well inland: one at Lake Sam Rayburn, San Augustine County, on 7 July 2012, and one at Canyon Lake, Comal County, from 25 August to 3 September 2012. There are 12 documented records of Brown Booby from 2004 to 2008, which is a very high frequency of occurrence and suggests an increase in abundance in Texas waters; however, since then the species seems to have returned to its former level of occurrence. The majority of Texas records are of immature birds. The first documented record for the state was one found on North Padre Island, Kleberg County, on 19 August 1967. There are 13 reports of Brown Booby from prior to the development of the Review List in 1988 for which there is no documentation on file. **Timing of occurrence:** There are records of Brown Booby from every month of the year, but the majority of occurrences have been between late April and late October. **Taxonomy:** The subspecies that occurs in Texas is *S.l. leucogaster* (Boddaert).

TBRC Review Species

RED-FOOTED BOOBY *Sula sula* (Linnaeus)

Accidental. There are three documented records for the state. The first of these was photographed on a petroleum platform in 1983, offshore Galveston. The most recent two records were each discovered on the beach and preserved as specimens (Arnold and Marks 2009). A bird reported to have been mounted by a Rockport taxidermist prior to 1910 has since been lost (Oberholser 1974). **Taxonomy:** The subspecies that occurs in Texas is *S. s. sula* (Linnaeus).

27 MAR. 1983, OFF GALVESTON, GALVESTON CO.
(TBRC 1988-258; TPRF 758)
2 OCT. 2002, ROCKPORT, ARANSAS CO.
(*TCWC 14,626; TBRC 2009–27; TPRF 2712)
12 JUNE 2008, GALVESTON, GALVESTON CO.
(*TCWC 14,601; TBRC 2009–05; TPRF 2899)

NORTHERN GANNET *Morus bassanus* (Linnaeus)

Uncommon to common migrant and winter resident offshore and along the immediate coast. This species is often found in large numbers, with reports of as many as 1,500 observed in a single day. Despite their relative abundance in the Gulf, and having been found in every coastal county, there are no records from inland locations in Texas. This species has been found increasingly during the summer, when it is a rare visitor. Immature Northern Gannets are sometimes confused with other members of the family, Masked Booby in particular. **Timing of occurrence:** Northern Gannets are primarily found between early November and early May. They appear to be annual in small numbers in the summer and earlier in the fall, primarily immature or subadult individuals. **Taxonomy:** Monotypic.

Family Phalacrocoracidae: Cormorants

NEOTROPIC CORMORANT
Phalacrocorax brasilianus (Gmelin)

Uncommon to common resident throughout the Coastal Prairies and south to the Lower Rio Grande Valley. Neotropic Cormorants are rare to locally uncommon summer visitors at scattered inland locations through the eastern half of the state. During the summer large concentrations have been found on reservoirs in the eastern half of the

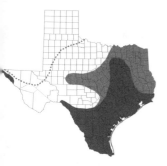

state, suggesting local or possibly inland movements from the coastal populations. These inland populations are increasing, and isolated breeding locations have been documented. This species is also a summer visitor westward through the eastern Edwards Plateau and the Colorado River drainage west to the Concho Valley, although breeding sites have not been discovered. Farther west, there are breeding colonies in Hudspeth and El Paso Counties. Neotropic Cormorants are uncommon and more local during the winter at these inland locations. They are casual visitors to the South Plains but are undocumented from the Panhandle. **Taxonomy**: The subspecies that occurs in Texas is *P. b. mexicanus* (Brandt).

DOUBLE-CRESTED CORMORANT
Phalacrocorax auritus (Lesson)

Uncommon to abundant migrant throughout the state and a common to abundant winter resident along the coast, becoming uncommon to locally abundant in winter inland throughout most of the state. In the Panhandle and northern Rolling Plains, this species is a rare and local winter resident. Double-crested Cormorants were reported to be a more regular nesting species in the early 1900s. Holm, Irby, and Inglis (1978) found a nesting colony at Toledo Bend Reservoir, Sabine County, in 1977, which they determined was the first nesting documented in the state in almost four decades. There have been additional inland breeding records since 1977 from scattered locations around the state, but the only known consistent nesting locations are in Delta, Hansford, Hopkins, Hudspeth, and Wood Counties. These cormorants are rare summer visitors along the upper and central coasts, with very little evidence of nesting (Ortego et al. 2011). **Timing of occurrence**: Double-crested Cormorants arrive in the northern portions of the state in early September, and the bulk of the winter population is in place by mid-November. Spring migrants begin to leave in late March, although some linger into early May. **Taxonomy**: Two subspecies have been reported in Texas; however, it appears that only one of these still occurs (Hatch 1995).

P. a. auritus (Lesson)
    Occurrence as is listed above.

P. a. floridanus (Audubon)
> Formerly considered a rare resident on the upper coast (AOU 1957; W. Davis 1940).

Family Anhingidae: Darters

ANHINGA *Anhinga anhinga* (Linnaeus)

Uncommon to locally common summer resident from most of the South Texas Brush Country and the Coastal Prairies northward through the eastern third of the state. Like populations of other colonial waterbirds, the inland populations are less common and more local. Anhingas are very rare to rare winter residents along the coast and in the Lower Rio Grande Valley. They are also rare inland during the winter and generally found only in the southern half of the Pineywoods. This species is a very rare postbreeding stray to areas west of the breeding range. Anhingas have been found on four occasions in both the Panhandle and the Trans-Pecos. **Timing of occurrence**: Anhingas begin migrating northward through Texas in late February. Northern breeding populations have moved south by late October, but a few individuals linger as late as December. **Taxonomy**: The subspecies that occurs in Texas is *A. a. leucogaster* (Vieillot).

ORDER PELECANIFORMES

Family Pelecanidae: Pelicans

AMERICAN WHITE PELICAN
Pelecanus erythrorhynchos Gmelin

Uncommon to common migrant over the eastern half of the state and decidedly less common in the western half. American White Pelicans are common winter residents in the southern half of the state, particularly along the coast, and locally on inland reservoirs in the northern half of the state as well as in the eastern Trans-Pecos. During migration, flocks of 4,000–11,000 birds have been reported.

There are only two nesting colonies of American White Pelicans in Texas: one on Pelican Island, Nueces County, and the other near South Bird Island, Kleberg County. Nonbreeding summer birds may be encountered, sometimes in fairly large numbers, at numerous locations statewide. **Timing of occurrence:** Fall migrants start to arrive in the state in late September, and spring migrants can be encountered until late May. **Taxonomy:** Monotypic.

BROWN PELICAN *Pelecanus occidentalis* Linnaeus

Common to uncommon resident along the entire coast. Every year some Brown Pelicans wander as far as 150 miles inland, primarily in the late summer and fall. They are casual to accidental farther inland, with records from virtually all areas of the state, and are almost annual in occurrence in the El Paso area. Most of these wandering individuals are immature or hatch-year birds, but adults are also involved and appear to be increasing. Large nesting colonies are present on the upper and central coasts, although none are known to breed along the lower coast. As the population continues to increase, colonies may become established farther south. Brown Pelicans were once nearly extirpated by the use of pesticides, and their recovery—which includes being delisted in 2009—represents one of the greatest success stories in the United States for an Endangered Species. By the late 1970s, the total population in Texas may have dropped to as low as 12–15 individuals. The conservation efforts and tighter controls on pesticide use (especially DDT) have given the species the opportunity to flourish, and there are once again thousands of these birds along the coast. **Taxonomy:** The subspecies of Brown Pelican found in Texas is *P. o. carolinensis* Gmelin. It is estimated that as many as 10 percent of the birds at some locations in Texas exhibit the "red pouch" that is reminiscent of *P. o. californicus* Ridgway. These birds are generally not thought to be part of that subspecies, which is supported by a red-pouched individual photographed as an adult in February 2013 that had been banded as a nestling in Cameron Parish, Louisiana. However, Brown Pelicans are known to move between the Pacific and Caribbean, which would allow for gene flow into the Texas population.

Family Ardeidae: Bitterns and Herons

AMERICAN BITTERN *Botaurus lentiginosus* (Rackett)

Rare to locally uncommon migrant in the eastern third of the state and generally rare west to the Trans-Pecos, where very rare to casual. In winter, this species is rare to uncommon on the Coastal Prairies and at scattered localities farther inland, as far north as the Panhandle. American Bitterns are very rare summer visitors locally throughout the state. There are breeding records from Chambers, Galveston, and Wilbarger Counties and indications of possible breeding from Delta, Grayson, and El Paso Counties. **Timing of occurrence**: Fall migrants start to arrive in the state in late September, and spring migrants can be encountered until early May. **Taxonomy**: Monotypic.

LEAST BITTERN *Ixobrychus exilis* (Gmelin)

Locally common summer resident on the Coastal Prairies and the eastern Lower Rio Grande Valley. Least Bitterns are rare to locally uncommon summer residents inland through the eastern half of the state, west to the Balcones Escarpment, and south through the eastern half of the South Texas Brush Country. Despite the widespread occurrence in the eastern portions of the state, few consistently used breeding areas are known. Least Bittern is a very rare and local summer resident in the western half of Texas, probably due to a lack of habitat. Isolated breeding populations have been found in El Paso, Hudspeth, and Tom Green Counties, and there are isolated breeding records from Hemphill, Lubbock, Midland, Presidio, Reeves, and Val Verde Counties. This bittern is a rare to uncommon migrant throughout the state and a very rare and secretive winter resident on the Coastal Prairies. **Timing of occurrence**: Spring migrants begin arriving in the state in late March, and the majority of the breeding population is present by late April. Fall migrants begin departing in early September and have left breeding areas by late October. **Taxonomy**: The subspecies that occurs in Texas is *I. e. exilis* (Gmelin). It is possible that *I. e. pullus* van Rossem of northwestern Mexico might occur as a vagrant to the Trans-Pecos.

TBRC Review Species

BARE-THROATED TIGER-HERON
Tigrisoma mexicanum Swainson

Accidental. There is only one record for the United States, a bird discovered in Texas; a near-adult plumaged bird was present at Bentsen–Rio Grande SP, Hidalgo County, during the winter of 2009–10 (Nirschl and Snyder 2010). The nearest known resident population of Bare-throated Tiger-Herons is in southern Tamaulipas, Mexico, approximately 150 miles south of the Rio Grande. The Texas bird was most frequently observed foraging in areas of dense tall grass with open light scrub and feeding on vertebrates, including hispid cotton rats (*Sigmodon hispidus*). **Taxonomy**: Monotypic.

21 DEC. 2009–20 JAN. 2010, BENTSEN–RIO GRANDE VALLEY SP,
   HIDALGO CO. (TBRC 2009–106; TPRF 2786)

GREAT BLUE HERON *Ardea herodias* Linnaeus

Common to uncommon migrant and summer resident throughout the state. Great Blue Herons are especially common along the Coastal Prairies and immediate coast. This species is less common, particularly in winter, away from the Coastal Prairies. Large rookeries containing hundreds of nests have been found at scattered locations across the state, especially on coastal islands. Most inland rookeries are much smaller, typically containing from 5 to 30 nesting pairs, and normally do not include other communal nesting birds. **Taxonomy**: Three subspecies are known to occur within the state, following Dickerman (2004b), although the first two listed are sometimes combined.

A. h. herodias Linnaeus

Uncommon to common in the western two-thirds of the state. This subspecies appears to occur east to the Oaks and Prairies region and south to the coast.

A. h. wardi Ridgway

Uncommon to common resident in the eastern third of the state. Exact delineation between ranges of *A. h. herodias* and this taxon is not clear.

A. h. occidentalis Audubon

The occurrence and status of this subspecies are uncertain. There are at least eight documented occurrences involving adults of this phenotype (all-white plumage but not albinistic) from the upper and central coasts. There are also at least four known occurrences of white nestlings with typical dark nestlings present in the same nest. In two of these instances both adults were dark Great Blue Herons, and the phenotype of the adults in the third is unknown. The nestlings did not exhibit signs of albinism but were not documented as adults. At least one white nonalbinistic adult nested with a dark adult (all nestlings were dark) annually between 2006 and 2010 in Aransas County (Huckabee, Moore, and Dorn 2010). This subspecies was formerly known as Great White Heron and was once considered a separate species (AOU 1957).

GREAT EGRET *Ardea alba* Linnaeus

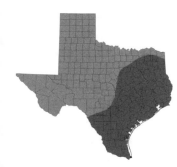

Common resident along the Coastal Prairies and locally common to uncommon resident in the eastern and central portions of the state to northeast Texas. The species is an uncommon year-round visitor to the Trans-Pecos, with sporadic nesting at McNary Reservoir, Hudspeth County (Peterson and Zimmer 1998). Great Egrets are rare to common postbreeding visitors to the rest of the state and can often be found in very large flocks in late summer and early fall. These flocks may include upward of 700 individuals at inland wetlands, and even larger groups have been noted along the Coastal Prairies. They are common to uncommon winter residents throughout the state. **Taxonomy:** The subspecies of Great Egret found in Texas is *A. a. egretta* Gmelin.

SNOWY EGRET *Egretta thula* (Molina)

Uncommon to common summer resident throughout much of the state to the eastern edge of the Rolling Plains and south along the Coastal Prairies to the Lower Rio Grande Valley. Snowy Egrets are rare to locally common summer residents in isolated areas of the Trans-Pecos and South Plains. They are rare summer visitors to the Panhandle, with two nesting records at Buffalo Lakes NWR, Randall County. In winter this species is an uncommon resident along the Coastal Prairies, generally becoming progressively rarer farther inland. Snowy Egrets are rare to uncommon migrants throughout the state. **Timing of occurrence:** Although Snowy Egrets are year-round residents in many areas of the state, there is a clear migration period in the west and along the Texas coast. Spring migrants are evident from late March through mid-May. Fall migrants and postbreeding wanderers are found between late July and mid-October. **Taxonomy**: Two subspecies occur within the state.

E. t. brewsteri Thayer & Bangs

Locally uncommon summer resident and rare winter resident in the El Paso area and possibly elsewhere in the Trans-Pecos.

E. t. thula (Molina)

The remainder of the range as described above. The subspecific identification of birds at Balmorhea Lake, Reeves County, is unknown.

LITTLE BLUE HERON *Egretta caerulea* (Linnaeus)

Common summer resident in the eastern third of the state, becoming locally abundant along the Coastal Prairies to the Lower Rio Grande Valley. The Little Blue Heron appears to be declining throughout much of its US range and is a rare to locally uncommon postbreeding wanderer west to the eastern edge of the Rolling Plains and Edwards Plateau. These birds are vagrants to the western two-thirds of the state, occurring most often in the late summer and early fall. In winter small numbers of Little Blue Herons are present along the Coastal Prairies and are very rare inland east and south of the Edwards Plateau to north-central and northeast Texas. This species has been noted in large numbers as migrants over the open waters of the Gulf. **Timing of occurrence**: Migrant Little Blue Herons begin to arrive at breeding locations in early March, and the majority of the population is present by early April. Fall migrants depart these areas between early September and early November. **Taxonomy**: Monotypic.

TRICOLORED HERON *Egretta tricolor* (Statius Müller)

Common summer resident along the immediate coast and locally common on the Coastal Prairies. This species is an uncommon to rare visitor to the remainder of the eastern third of the state in late summer and fall, then becomes increasingly rarer farther west. Spring occurrences from the eastern portion of the state away from the coast are unusual but have resulted in sporadic nesting records from as far north as north-central and northeast Texas. Tricolored Herons are casual visitors to the South Plains and accidental in the Panhandle, primarily as postbreeding wanderers. This heron has occurred almost annually in the Concho Valley and westward through the Trans-Pecos during all seasons. This species is uncommon to locally common along the Coastal Prairies during the winter and rare and local farther inland to Bexar County. Tricolored Herons are very rare farther inland during this season, although there are several winter records from northeast Texas. **Timing of occurrence:** Away from the Coastal Prairies where the species is present all year, Tricolored Herons occur most frequently between early July and early October. There is limited spring migration along the immediate coast; some birds are present in small numbers among northbound flocks of other small herons and egrets between late March and early May. **Taxonomy:** The subspecies that occurs in Texas is *E. t. ruficollis* Gosse.

REDDISH EGRET *Egretta rufescens* (Gmelin)

Uncommon to locally common resident along the coast that is most numerous from Matagorda Bay southward. Reddish Egret is a casual to very rare postbreeding visitor inland through the eastern third of the state and generally considered accidental farther west. An interesting phenomenon is the almost annual occurrence, with records from all seasons, at reservoirs in the Trans-Pecos, particularly Balmorhea Lake, Reeves County. In contrast, there are only two records each from the High Plains and the western Rolling Plains. The majority of the inland records involved first-year birds. Reddish Egret is a species of conservation concern and is listed as Threatened by the Texas Parks and Wildlife Department. Nesting locations in Texas are limited to a small number of coastal islands that are vulnerable to tropical storms and erosion from ship traffic. **Timing of occurrence:** Inland occurrences of Reddish Egret are primarily between early July and late October. However, there are records from as early as February, and fall-arriving birds have lingered into January. **Taxonomy:** The subspecies that occurs in Texas is *E. r. rufescens* (Gmelin).

CATTLE EGRET *Bubulcus ibis* (Linnaeus)

Common to abundant summer resident throughout most of the state but locally distributed on the High Plains, Rolling Plains, Edwards Plateau, and Trans-Pecos. During

migration, Cattle Egrets are locally common to abundant in the eastern half of the state and along the coast and uncommon in the west. In winter they are uncommon to rare along the Coastal Prairies south through the Lower Rio Grande Valley, becoming rare to very rare inland and generally absent from the High Plains and northern Rolling Plains. This species was unknown in Texas until late November 1955, when one was found in Nueces County (Oberholser 1974). The first breeding record for the state was discovered three and a half years later at Rockport, Aransas County, on 10 May 1959. Cattle Egrets are colonial nesters, and very large rookeries can be found at scattered locations through the eastern half of the state. Those in urban settings can sometimes create a nuisance. In the western half of the state, breeding is very localized and usually involves only a few pairs. Overall, the population of Cattle Egrets in Texas has been declining since a peak in the 1980s and early 1990s. **Timing of occurrence:** Spring migrants are found from mid-March through early May, and the majority of the population is present by mid-April. Inland populations are bolstered by postbreeding wanderers moving away from the coast beginning in early July. Fall migrants depart inland areas during October, but some linger through November. **Taxonomy:** The subspecies that occurs in Texas is *B. i. ibis* (Linnaeus).

GREEN HERON *Butorides virescens* (Linnaeus)

Common summer resident throughout the eastern two-thirds of the state, becoming uncommon westward across the Edwards Plateau and High Plains. Green Herons are uncommon and local in the Trans-Pecos during the summer. As migrants, they are uncommon to common statewide. Most Green Herons leave the state during the winter and become locally uncommon to rare along the coast and in the Lower Rio Grande Valley. This species is very rare to rare during the winter inland and is generally absent from the Panhandle, South Plains, and Trans-Pecos away from the Rio Grande. Unlike many other herons, this species does not breed communally. **Timing of occurrence:** Inland, spring migrants are recorded from late March through mid-May. Fall migrants are recorded from mid-August through late October. **Taxonomy:** Two subspecies occur in the state.

B. v. anthonyi (Mearns)

Uncommon and local summer resident in the Trans-Pecos. Casual in winter along the Rio Grande in the Trans-Pecos.

B. v. virescens (Linnaeus)

The remainder of the range as described above.

BLACK-CROWNED NIGHT-HERON

Nycticorax nycticorax (Linnaeus)

Common resident along the Coastal Prairies. Inland, this species is a locally common to uncommon summer resident west of the Pineywoods, the exception being the South Plains northward through the Panhandle, where the birds are fairly common to locally abundant. In the Pineywoods, they are rare to uncommon and very local summer residents. Black-crowned Night-Herons are rare to locally uncommon winter residents inland except in the Panhandle, where they are casual visitors, and in the Pineywoods, where they are very rare to locally uncommon. Given its nocturnal nature, this species often goes undetected and is probably more common in any given region than is readily apparent. Immature night-herons provide a difficult identification challenge for observers, which might also result in this species going undetected due to misidentifications. **Timing of occurrence**: Spring migrants arrive at inland breeding locations in late March and early April and generally depart during September. **Taxonomy**: The subspecies that occurs in Texas is *N. n. hoactli* (Gmelin).

YELLOW-CROWNED NIGHT-HERON

Nyctanassa violacea (Linnaeus)

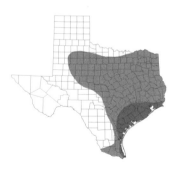

Uncommon to locally common summer resident along the Coastal Prairies and through the eastern third of the state and westward through the southern Rolling Plains and eastern Edwards Plateau. Yellow-crowned Night-Herons are locally common winter residents along the Coastal Prairies, most commonly from Matagorda Bay southward. They are rare to casual inland during winter. Yellow-crowned Night-Herons are casual to very rare, primarily during spring and summer, in the Panhandle south through the western Edwards Plateau and west through the Trans-Pecos. Like the Black-crowned Night-Heron, this species is often present in larger numbers than is realized. Yellow-crowned Night-Herons nest communally in loose colonies, most often in riparian areas, but usually do not nest alongside other species. **Timing of occurrence:** Spring migrants arrive in breeding areas from mid-March through early May and depart between mid-August and mid-September, with some lingering through October. This species is a casual to very rare visitor to the Panhandle and Trans-Pecos between late April and early September. **Taxonomy:** The subspecies that occurs in Texas is *N. v. violacea* (Linnaeus).

Family Threskiornithidae: Ibis and Spoonbills

WHITE IBIS *Eudocimus albus* (Linnaeus)

Common to abundant resident along the immediate coast and Coastal Prairies. White Ibis are particularly abundant along the upper coast during the summer. They are uncommon and local, but increasing in numbers, during the summer inland through the eastern third of the state north through northeast and north-central portions. The majority of White Ibis retreat from the northern half of the state in the winter, with only a few lingering through the season. They are casual visitors to the eastern Edwards Plateau and accidental on the South Plains and in the Trans-Pecos. **Timing of occurrence:** Spring migrants arrive in breeding areas away from the coast beginning in mid-March and depart in late October and early November. **Taxonomy:** Monotypic.

GLOSSY IBIS *Plegadis falcinellus* (Linnaeus)

The status of Glossy Ibis in Texas is poorly understood. This species was unknown in the state prior to November 1983, but by the early 1990s Glossy Ibis were being found with increasing regularity on the upper and central coasts. Currently, Glossy Ibis is a rare to locally uncommon resident on the central coast. The first nesting was discovered in 1997 on Sundown Island in Matagorda Bay, Matagorda County. Colonial waterbird surveys have noted nesting Glossy Ibis on Snake Island, Calhoun County, in Matagorda Bay since 2002, when a high count of 23 pairs was made. They are reported with regularity on the upper coast throughout the year, but nesting has not been confirmed. They are rare to very rare on the lower coast and through the eastern portions of the state west through the Oaks and Prairies region. There are documented records scattered across the state, including from as far west as El Paso, but there are only single records from the Panhandle and South Plains. **Taxonomy**: Monotypic.

WHITE-FACED IBIS *Plegadis chihi* (Vieillot)

Common to uncommon resident along the coast. The population along the Coastal Prairies increases as migrants and wintering birds arrive. White-faced Ibis is a rare and localized breeder away from the coast as far north as the Panhandle. This species is a casual to rare winterer inland through most of the state, but White-faced Ibis have not been recorded from the Panhandle and South Plains past November. They are uncommon to common migrants in most regions of the state but are rare to uncommon in the Pineywoods. White-faced Ibis populations decreased dramatically in the 1960s and 1970s due to habitat loss and widespread use of chemical pesticides. However, conservation efforts have been successful, and the species numbers are now increasing. **Timing of occurrence**: Spring migrants are recorded between early April and early June, and fall migrants are present from mid-July through late October. **Taxonomy**: Monotypic.

ROSEATE SPOONBILL *Platalea ajaja* (Linnaeus)

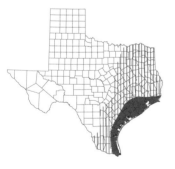

Locally common summer resident along the coast, becoming uncommon in the winter. Roseate Spoonbills are well known as postbreeding wanderers to inland locations. They are rare to uncommon visitors inland to the eastern third of the state during the summer and fall. Farther west, spoonbills are casual to accidental, with scattered records west to the Trans-Pecos and north to the Panhandle. Postbreeding wanderers are encountered with greater frequency in the southern half of the state. **Timing of occurrence**: Spring migrants start to arrive in mid-March, and most are on the breeding grounds by mid-April. Postbreeding wanderers and fall migrants are found from early July through mid-October. **Taxonomy**: Monotypic.

ORDER ACCIPITRIFORMES

Family Cathartidae: New World Vultures

BLACK VULTURE *Coragyps atratus* (Bechstein)

Common to locally abundant resident in the eastern two-thirds of the state. Breeding Bird Survey data from the last 40 years indicate that this species has increased in Texas by an annual rate of almost 5 percent. Black Vultures are casual summer and fall visitors to the South Plains and most of the Rolling Plains. This species is a locally uncommon resident along the Rio Grande from eastern Val Verde County to northern Presidio County in the Trans-Pecos. Some Black Vultures retreat from the northern portions of their range in Texas during the winter, although the species may continue to be locally common as migrants from more northern populations arrive to take their place. **Taxonomy:** Monotypic.

TURKEY VULTURE *Cathartes aura* (Linnaeus)

Common to locally abundant summer resident throughout the state. Turkey Vultures withdraw from most of the western half of the state during the winter, although small numbers remain on the eastern Edwards Plateau. A small number of Turkey Vultures now winter along the Rio Grande in El Paso County, which is the only reliable wintering area in the Trans-Pecos. They are common during winter in the eastern half of the state and in the South Texas Brush Country. In late fall, vast numbers of migrating Turkey Vultures pass through the eastern half of Texas on their way to wintering areas farther south. **Timing of occurrence:** Spring migrants arrive in the western portions of the state in mid- to late February and are common by late March. Most have departed these areas by mid-October, with a few lingering as late as mid-November. **Taxonomy:** Two subspecies occur in the state.

C. a. *aura* (Linnaeus)
   Distribution as given above.

C. a. *septentrionalis* Wied-Neuwied
   Status uncertain. There is one documented record: a specimen from Tarrant County from 21 May 1955 (Pulich 1979).

Family Pandionidae: Osprey

OSPREY *Pandion haliaetus* (Linnaeus)

Uncommon to rare migrant throughout the state. Ospreys are common to uncommon winter residents along the coast and northward through the eastern third of the state. They are locally uncommon to rare winter visitors west to the High Plains and eastern Trans-Pecos. This species is a very rare and local breeder in the Pineywoods and along the upper coast, typically near larger reservoirs. Nonbreeding individuals are rare summer visitors along the coast and through the eastern third of the state. Like many other wetland-associated birds, the population of this species has increased significantly since the early 1970s following the ban on DDT. **Timing of occurrence:** The migration periods for this species extend from mid-March to late May and from early September to mid-November. **Taxonomy:** The subspecies that occurs in Texas is *P. h. carolinensis* (Gmelin).

Family Accipitridae: Eagles, Kites, and Hawks

HOOK-BILLED KITE
Chondrohierax uncinatus (Temminck)

Rare and local resident in Hidalgo and Starr Counties of the Lower Rio Grande Valley. Hook-billed Kites are casual in Cameron and Zapata Counties, and there is a spectacular record from Smith Point, Chambers County, on 29 October 2011. Hook-billed Kites were first found in Texas and the United States in 1964 when a pair nested at Santa Ana NWR, Hidalgo County (Fleetwood and Hamilton 1967). This species was not reported in Texas again until 1976, when another nesting pair was discovered. Brush (1999b) reported that Hook-billed Kites might be somewhat nomadic, following changes in the populations of their favored food item, tree snails (*Rabdotes alternatus*). This may explain the variability in the number of individuals reported from year to year, especially during extended periods of drought. **Taxonomy:** The subspecies that occurs in Texas is *C. u. uncinatus* (Temminck).

SWALLOW-TAILED KITE
Elanoides forficatus (Linnaeus)

Rare to uncommon migrant through the Coastal Prairies and eastern third of the state. Swallow-tailed Kites are casual migrants west to the eastern Edwards Plateau, where they appear to be more regular during fall migration. They are accidental to casual migrants west to the Trans-Pecos, and there are records from Floyd, Midland, Lubbock, and Oldham Counties on the High Plains. This species is a rare to locally uncommon summer resident in the southern portion of East Texas west to Harris and Brazoria Counties. Swallow-tailed Kites were formerly common summer residents in the eastern half of the state and nested as far west as Bastrop and Medina Counties until about 1915 (Attwater 1892; Simmons 1925; Oberholser 1974). **Timing of occurrence**: The migration periods are between early March and early May and from mid-August to mid-October. **Taxonomy**: The subspecies that occurs in Texas is *E. f. forficatus* (Linnaeus).

WHITE-TAILED KITE *Elanus leucurus* (Vieillot)

Uncommon to common resident along the Coastal Prairies, the southern half of the Post Oak Savannah, and in the eastern South Texas Brush Country. This species occasionally forms large winter roosts in the South Texas Brush Country, sometimes including more than 100 individuals. White-tailed Kites are rare to locally uncommon summer residents in the northern half of the Post Oak Savannah and very rare and sporadic in the Pineywoods. This species is a rare and local summer resident along the Rio Grande to Val Verde County, with isolated breeding records from Real and Uvalde Counties. In the Trans-Pecos, White-tailed Kites are very rare nesting birds in El Paso County and rare visitors at all seasons elsewhere in the western portion of the region. They are also very rare to casual visitors west of the breeding range; nesting has been documented as far northwest as Kent County. **Timing of occurrence**: Spring migrants arrive on the northern and western breeding areas in early March and are present through late September. **Taxonomy**: The subspecies that occurs in Texas is *E. l. majusculus* Bangs & Penard.

SNAIL KITE *Rostrhamus sociabilis* (Vieillot)

Accidental. Texas has four documented records of this tropical species. The first was an immature bird found in Jim Wells County during the summer of 1977; this individual reportedly returned in 1978, but supporting documentation is lacking. There are two records from the Lower Rio Grande Valley, and the most recent bird was in Harris County on the upper Coastal Prairies. The northern extent of the Snail Kite's range in Mexico is in central Veracruz. **Taxonomy:** The individuals found in Texas are presumed to belong to the subspecies in northern Mexico, *R. s. major* Nelson & Goldman.

22–26 JULY 1977, LAKE ALICE, JIM WELLS CO. (TPRF 127)
17–29 MAY 1998, NEAR BENTSEN–RIO GRANDE VALLEY SP,
    HIDALGO CO. (TBRC 1998-119; TPRF 1724)
14 JULY 2007, NEAR PORT ISABEL, CAMERON CO.
    (TBRC 2007–58; TPRF 2497)
17 JUNE 2011, EL FRANCO LEE PARK, HARRIS CO.
    (TBRC 2011–062; TPRF 2964)

DOUBLE-TOOTHED KITE
Harpagus bidentatus (Latham)

Accidental. Texas has a single documented record of this tropical species. This unexpected first occurrence in the United States was an immature bird molting into full adult plumage. Double-toothed Kites are well known to soar up to high elevations and can wander locally within their range, including crossing areas of inhospitable habitat. **Taxonomy:** Considering that neither subspecies of Double-toothed Kite has a defined migratory pattern, it seems more likely that the Texas bird would belong to the subspecies in Mexico, *H. b. fasciatus* Lawrence.

4 MAY 2011, HIGH ISLAND, GALVESTON CO.
    (TBRC 2011–067; TPRF 2970)

MISSISSIPPI KITE *Ictinia mississippiensis* (Wilson)

Common to uncommon migrant throughout the state. Mississippi Kites are common summer residents on the High Plains and Rolling Plains. They are also uncommon, but increasing, summer residents southward through the Oaks and Prairies region to the upper coast, as well as in El Paso County in the Trans-Pecos. This species is generally rare during the summer in the Pineywoods but can be locally common. The population on the Coastal Prairies is primarily found in urban environments where large trees are present and appears to be expanding southward to the central coast. Mississippi Kites sometimes migrate in large flocks, as was the case on 18 April 1998 when approximately 16,000–18,000 were observed in northern Kenedy County. **Timing of occurrence**: Migration periods are between early April and mid-May and late August to mid-October. **Taxonomy**: Monotypic.

BALD EAGLE *Haliaeetus leucocephalus* (Linnaeus)

Rare and local summer resident primarily in the eastern third of the state. Current population estimates indicate there are more than 300 breeding pairs present (B. Ortego, pers. comm.). In the last decade nesting pairs have been found over a wider area of the state, including sites in the Panhandle and on the Edwards Plateau. Postbreeding dispersal of the nesting population is unclear, as Bald Eagles are even more rarely encountered between mid-June and early October, and a nestling banded in Matagorda County was later discovered nesting in Arizona. During migration and winter, Bald Eagles are more widely distributed in Texas, but they are locally common only on large reservoirs in the eastern third of the state. In the Trans-Pecos and southern South Texas Brush Country, Bald Eagles are casual midwinter visitors. **Timing of occurrence**: Fall migrants start to arrive in the state in early October, and wintering birds normally depart by mid-March. **Taxonomy**: Two subspecies have been reported for the state.

H.l. leucocephalus (Linnaeus)
  Distribution as given above.

H.l. washingtoniensis (Audubon)
  Status uncertain. Oberholser (1974) reported six specimens of this northern subspecies. Most authorities do not include the southern United States within the range of *H.l. washingtoniensis* (= *H.l. alascanus*).

NORTHERN HARRIER *Circus cyaneus* (Linnaeus)

Common to uncommon migrant and winter resident in all parts of the state. Northern Harriers are particularly common on the Coastal Prairies in winter. This species is a rare summer visitor to most regions, with breeding activity noted on very rare occasions. The majority of breeding

records come from native grasslands on the Coastal Prairies and in the Panhandle. Nesting has been suspected in the Trans-Pecos as well, but confirmation is lacking. Adult male Northern Harriers make up a larger percentage of the overall wintering population in the northern half of the state but are still outnumbered by female and immature birds in all regions. Northern Harriers are known to form communal roosts in the winter that can include as many as 50 individuals. **Timing of occurrence:** Fall migrants reach the state in late August, and the majority of the winter population is present by late October. Spring migrants are noted between late March and early May. **Taxonomy:** The subspecies that occurs in Texas is *C. c. hudsonius* (Linnaeus).

SHARP-SHINNED HAWK *Accipiter striatus* Vieillot

Uncommon to common migrant and uncommon winter resident throughout the state. Sharp-shinned Hawks are very rare and local summer residents in the southeastern Pineywoods (Shackelford, Saenz, and Schaefer 1996) and at higher elevations in the Guadalupe, Davis, and Chisos Mountains. There are isolated nesting records from Wise, Hays, and Hidalgo Counties and suspected nesting at other locations. **Timing of occurrence:** Fall migrants begin to arrive in mid-August, and most have arrived in the state by early October. Spring migrants are present from late March through early May. **Taxonomy:** The subspecies that occurs in Texas is *A. s. velox* (Wilson). The subspecies of birds breeding in the mountains of the Trans-Pecos needs to be investigated. The birds could belong to the subspecies found in the mountains of northern Mexico, *A. s. suttoni* van Rossem.

COOPER'S HAWK *Accipiter cooperii* (Bonaparte)

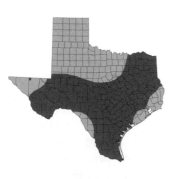

Uncommon to rare migrant and winter resident throughout the state. This species is a rare to locally uncommon summer resident in all areas of the state except the High Plains, western Trans-Pecos, and the Coastal Prairies south to Matagorda Bay, although there are isolated summer records in these areas that may signal an expansion into appropriate nesting habitat. Urban populations in the state seem to be expanding, possibly coinciding with the expansion of White-winged (*Zenaida asiatica*) and Eurasian Collared-Dove (*Streptopelia decaocto*) populations. **Timing of occurrence**: Fall migrants arrive in Texas from late August, with most arriving by early October. Spring migrants depart beginning in mid-March and continuing through late April. **Taxonomy**: Monotypic.

TBRC Review Species

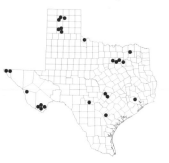

NORTHERN GOSHAWK *Accipiter gentilis* (Linnaeus)

Very rare winter visitor to the state. There are 24 documented records of Northern Goshawks in Texas. The majority of these records are from the Trans-Pecos and Panhandle, with others scattered across the northern two-thirds of the state. There are more than 70 undocumented reports of Northern Goshawks, including several involving specimens supposedly collected between 1885 and 1934. Interestingly, despite the frequency with which this species has been reported, only eight have been documented with photographs. The identification difficulties associated with this species are reflected by the 26 submissions that have been deemed unacceptable by the TBRC. There are 64 reports of Northern Goshawks from prior to the development of the Review List in 1988 for which there is no documentation on file. **Timing of occurrence**: All but two of the documented records occur between 1 November and 20 March. The exceptions include one at Big Bend NP, Brewster County, on 9 September 1999 and one in northern Presidio County on 8 May 2003. **Taxonomy**: The subspecies that occurs in Texas is *A. g. atricapillus* (Wilson).

TBRC Review Species

CRANE HAWK *Geranospiza caerulescens* (Vieillot)

Accidental. The lone record of Crane Hawk for Texas and the United States was found on 20 December 1987 at Santa Ana NWR, Hidalgo County. The Crane Hawk's range extends only to the central portions of Tamaulipas, Mexico (Howell and Webb 1995; J. Arvin, pers. comm.). **Taxonomy**: The Crane Hawks in northeastern Mexico belong to the subspecies *G. c. nigra* (Du Bus de Gisignies). This subspecies occurs through Mexico south to Panama and is the most likely to have occurred in Texas.

20 DEC. 1987–9 APR. 1988, SANTA ANA NWR, HIDALGO CO. (TBRC 1988-87; TPRF 595)

COMMON BLACK-HAWK

Buteogallus anthracinus (Deppe)

Rare and local summer resident, primarily in riparian areas, in the Davis Mountains in the Trans-Pecos. There are also small nesting populations along the Rio Grande in southern Brewster and adjacent Presidio Counties, along the Devils River in central Val Verde County, and in the Concho Valley. A pair of Common Black-Hawks nested unsuccessfully in Lubbock County in 1982 and 1983, and a single bird was found in Potter County on the northern High Plains in April 1999. Historically, this species was resident in the Lower Rio Grande Valley upriver to Webb County, but in recent years Common Black-Hawks have been very rare visitors to this area, with most records occurring in the winter. There are scattered records away from breeding areas within the Hill Country, including as far northeast as San Saba County. Vagrants have been reported west to El Paso, north to Palmer and Potter Counties, and east to Fayette County. **Timing of occurrence:** Spring migrants begin to arrive on nesting grounds in mid-March and are generally present by mid-April. They depart in September and early October. Winter sightings in South Texas are generally between early November and mid-March. **Taxonomy:** The subspecies found in Texas is *B. a. anthracinus* (Deppe).

HARRIS'S HAWK *Parabuteo unicinctus* (Temminck)

Common to uncommon resident in the South Texas Brush Country north to the southern edge of the Edwards Plateau and east to the Goliad–Victoria County line. This hawk is found with decreasing frequency up the coast from Refugio and western Calhoun Counties to the upper coast, where it is very rare. Harris's Hawk is a locally uncommon to rare resident along the Rio Grande from Big Bend NP northwest to El Paso County. There is also a resident population in the southern High Plains from Lubbock and Winkler Counties southeastward to Schleicher County on the northwestern Edwards Plateau. Harris's Hawks are vagrants elsewhere in the state, although the provenance of some of these birds has been questioned because of the popularity of this species among falconers. This hawk is known for its social behavior, with family groups of six or more often sighted. **Taxonomy:** The subspecies found in Texas is *P. u. harrisi* (Audubon).

TBRC Review Species

ROADSIDE HAWK *Buteo magnirostris* (Gmelin)

Casual visitor along the Rio Grande from Hidalgo County north to Zapata County. The first documented record for Texas and the United States is from Cameron County on 2 April 1901. Since that date, there have been eight additional documented records. This small *Buteo* has been found in areas of native thorn-scrub woodland. Roadside Hawks are very common from southern Tamaulipas in northeastern Mexico southward through northern South America. **Timing of occurrence:** All of the Texas records have been between 7 October and 2 April. **Taxonomy:** The subspecies found in Texas is *B. m. griseocauda* (Ridgway).

RED-SHOULDERED HAWK *Buteo lineatus* (Gmelin)

Common to uncommon resident throughout the eastern two-thirds of the state, becoming rare to casual farther west, where it is found primarily along wooded drainages. This resident population ranges west to the Concho Valley and western Edwards Plateau and south to the Nueces River. Red-shouldered Hawks are primarily winter residents south of the Nueces River, although there are records from all seasons. There is one recent nesting record from Starr County in 2001 (Patrikeev 2009), and this species may have formerly bred more widely in the Lower Rio Grande Valley (McKinney 1998). There are isolated nesting records out of the normal range, including as far north as Dickens County. One of the more interesting records for this hawk involved an unsuccessful nesting attempt with a Gray Hawk in Big Bend NP in 1989. Annually a few individuals wander west out of the breeding range and have been documented west to Jeff Davis and Presidio Counties and north to Roberts County. **Timing of occurrence:** In South Texas where this species is present only in winter, the birds arrive in late September and are present until late March. This is also the timing of occurrence of vagrants west of the breeding range, although some may linger into May. **Taxonomy:** Three subspecies are known to occur in the state.

B.l. lineatus (Gmelin)

Rare migrant and winter resident in the eastern third of the state south to the coast.

B.l. alleni Ridgway

Common resident from the western Edwards Plateau eastward through the Pineywoods. Uncommon to rare winter resident in the South Texas Brush Country south to the Lower Rio Grande Valley.

B.l. texanus Bishop

Common resident from the central coast (Calhoun and Victoria Counties) west to Bexar and Medina Counties south of the Balcones Escarpment. This subspecies is a rare winter resident south to the Lower Rio Grande Valley.

BROAD-WINGED HAWK *Buteo platypterus* (Vieillot)

Common to abundant migrant throughout the eastern half of the state. On ideal days during migration, it has been possible to see more than 100,000 of these small buteos as they migrate along the Coastal Prairies. A small percentage of migrating birds pass over the eastern portion of the Edwards Plateau, where this species is uncommon to rare, but farther west they are casual to rare at best. Broad-winged Hawks are uncommon to common summer residents in the Pineywoods and locally uncommon to rare west to north-central Texas and the eastern edge of the Edwards Plateau and Rolling Plains. There are a few scattered winter records from along the coast and the Lower Rio Grande Valley. The numbers were particularly high during the winter of 2007–8. **Timing of occurrence**: Typical migration dates are between mid-March and mid-May and from late August to mid-October, but a few lingering birds are often present into early November. **Taxonomy**: The subspecies found in Texas is *B. p. platypterus* (Vieillot).

GRAY HAWK *Buteo plagiatus* (Schlegel)

Rare to locally uncommon resident along the Rio Grande corridor from Hidalgo County to Webb County. Gray Hawks are rare and local summer residents in the Trans-Pecos in the Davis Mountains and at Big Bend NP and very rare west along the Rio Grande in Presidio County. This species was documented nesting in Val Verde County in 2012 and might be expected to occur at other locations along the Rio Grande where suitable habitat is available. Gray Hawks are rare summer residents north along the Coastal Prairies to southern Kleberg County. Out-of-range records exist for Bandera, El Paso, and Refugio Counties. **Timing of occurrence**: In the Trans-Pecos, Gray Hawks arrive in mid-March and generally have departed by mid-October. There is a single occurrence of two birds remaining through the winter in Big Bend NP. **Taxonomy**: Monotypic. Formerly included under *B. nitidus* but was elevated to species status in 2012 (Chesser et al. 2012).

TBRC Review Species

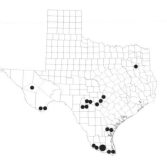

SHORT-TAILED HAWK *Buteo brachyurus* Vieillot

A very rare visitor to the Lower Rio Grande Valley and casual northward to the Edwards Plateau. There are 41 documented records for this tropical hawk in the state, and it has become an annual visitor to the Lower Rio Grande Valley over the past decade. The first for Texas was an adult found at the Santa Margarita Ranch, Starr County, in 1989. The increase in occurrence in the state mirrors a northward expansion of the species in Mexico, and it is becoming a rare summer resident in Arizona. Away from Lower Rio Grande Valley there are three records from the Trans-Pecos, nine from the Edwards Plateau, and one from Nueces County in the southern Coastal Prairies. By far the most unexpected occurrence in the state was a dark morph bird found injured on a roadside near Troup, Smith County, on 6 October 2008. It was immediately taken to a nearby raptor rehabilitator and eventually released in the Lower Rio Grande Valley. **Timing of occurrence**: The documented occurrences in the state fall between 15 February and 12 October, with a distinct peak between late March and late April. **Taxonomy**: The subspecies found in Texas is *B. b. fuliginosus* Sclater.

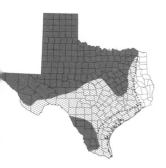

SWAINSON'S HAWK *Buteo swainsoni* Bonaparte

Common to locally abundant migrants throughout the state, except the Pineywoods region, where they are rare to very rare. Swainson's Hawks can occasionally be seen during the fall in very large migrating flocks from the Panhandle southward through the central portions of the state. They are uncommon to common summer residents from the Panhandle, south through the Rolling Plains to the northern Edwards Plateau, and west through most of the

Trans-Pecos. They are also rare summer residents across north-central Texas and in the western South Texas Brush Country south to Hidalgo and northern Starr Counties. This species is an irregular summer visitor east of the normal breeding range into the Oaks and Prairies region. There are isolated breeding records within this region from the Oklahoma border south to Bexar County and east to northeastern Harris County. Swainson's Hawks have also been reported during summer with increasing frequency along the Coastal Prairies, and breeding has been reported at a few scattered locations on the upper coast. This species is a rare to very rare winter resident along the southern Coastal Prairies from Calhoun County southward to the Lower Rio Grande Valley, with the vast majority being immature birds. The northernmost documented winter record was an immature bird photographed in Tom Green County on 2 February 1993 (TPRF 1958). **Timing of occurrence:** Spring migrants are seen between mid-March and mid-May, although small numbers are occasionally seen as late as mid-June. During the fall, migrants are recorded from early August to early November. **Taxonomy:** Monotypic.

WHITE-TAILED HAWK *Buteo albicaudatus* Vieillot

Uncommon to locally common resident in the Coastal Prairies and southeastern South Texas Brush Country. The breeding population of White-tailed Hawks is found primarily south of Matagorda Bay along the central and lower coasts. This species is a very rare visitor inland at all seasons to the central South Texas Brush Country and the Oaks and Prairies region north to McLennan County. Unexpected records have come from as far west as Jeff Davis County and as far north as Lubbock and Delta Counties. This species appears to have been more widespread in the state prior to 1920, with reported nesting records from Comal and Medina Counties before 1915 and an exceptional one from Taylor County in 1894. **Taxonomy:** The subspecies found in Texas is *B. a. hypospodius* Gurney.

ZONE-TAILED HAWK *Buteo albonotatus* Kaup

Uncommon and local summer resident in the mountains of the central Trans-Pecos, east through the southern Edwards Plateau to western Travis and Williamson Counties.

Zone-tailed Hawks are rare summer visitors to the Guadalupe Mountains and very rare visitors to the northwestern Trans-Pecos. This species is a rare migrant and winter resident in the Lower Rio Grande Valley and southern South Texas Brush Country and an irregular visitor there during the summer. Zone-tailed Hawks are rare and local winter visitors to the Edwards Plateau and Concho Valley eastward through Central Texas. Although there is no consistent wintering area east of the Balcones Escarpment, there are numerous records from Bastrop County south to Bexar, Colorado, and Victoria Counties. Vagrants have also been noted as far north as Rains and Terry Counties and east to Walker County. **Timing of occurrence:** The primary period of occurrence in summering areas of the state is from late March through mid-October. Winter records fall primarily between early November and mid-April. **Taxonomy:** Monotypic.

RED-TAILED HAWK *Buteo jamaicensis* (Gmelin)

Common resident virtually statewide. Red-tailed Hawks are common to uncommon summer residents throughout most of the state, except the southern South Texas Brush Country, where they are uncommon, and the Lower Rio Grande Valley, where they are absent as breeders. They are common migrants and winter residents in all areas. **Taxonomy:** Five subspecies are known to occur in the state.

B. j. calurus Cassin
Common resident from El Paso east to the Guadalupe Mountains. Common migrant and winter resident in the western third of Texas but rare farther east.

B. j. borealis (Gmelin)
Uncommon to common resident in the eastern half of the state west to the Balcones Escarpment and south to the coast. This subspecies is also present through the Rolling Plains and High Plains of northwest Texas. Common in winter in the South Texas Brush Country to the Rio Grande.

B. j. harlani (Audubon)
Rare to locally uncommon migrant and winter resident in North and Central Texas to the central Coastal Prairies. Very rare in the Trans-Pecos and Edwards Plateau south through the South Texas Brush Country. This subspecies is generally present in Texas between mid-October and mid-March.

B. j. kriderii Hoopes

Rare migrant and winter resident throughout the state except the Trans-Pecos. Generally present in Texas between mid-October and mid-March. This subspecies is increasingly considered as the pale extreme in *B. j. borealis* (Wheeler 2003; Liguori 2005).

B. j. fuertesi Sutton & van Tyne

Uncommon to rare resident from the southern Trans-Pecos eastward across the Edwards Plateau and southern Rolling Plains (north to Tarrant County) and south through the northern three-quarters of the South Texas Brush Country.

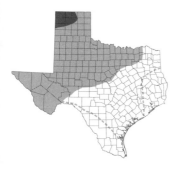

FERRUGINOUS HAWK *Buteo regalis* (Gray)

Common to locally uncommon winter resident in the Panhandle and South Plains and locally uncommon in the Trans-Pecos. Ferruginous Hawks are declining as migrant and winter visitors in the remainder of the state, where they are rare to casual, including the Coastal Prairies and northern Pineywoods. Ferruginous Hawks are rare summer residents in the northwestern portion of the Panhandle. Formerly this species was a much more common breeder in this region (Strecker 1912). Ferruginous Hawks have declined throughout their range, and the primary wintering range in Texas has shifted northward. **Timing of occurrence**: Fall migrants start to arrive in Texas in early October. The wintering population peaks from early December through mid-February and departs between mid-February and mid-March. **Taxonomy**: Monotypic.

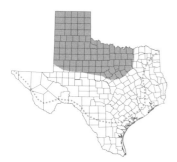

ROUGH-LEGGED HAWK *Buteo lagopus* (Pontoppidan)

Uncommon to rare winter resident in the Panhandle and South Plains, becoming increasingly rare farther east and south. This species is casual in the Trans-Pecos and south of the Rolling Plains. Rough-legged Hawks are casual wanderers to the Coastal Prairies and are accidental in the South Texas Brush Country and the Pineywoods. As the Ferruginous Hawks are, Rough-legged Hawks are declining in North America, and the primary wintering range in Texas has shifted northward. Many, though certainly not all, reports of dark morph individuals away from the core wintering range probably pertain to other similarly plumaged raptors, particularly "Harlan's" and, in the south, immature White-tailed Hawks. **Timing of occurrence:** Rough-legged Hawks arrive late in the fall, and the first birds reach the state in mid-October (Seyffert 2001b). Spring migrants have generally departed by early March. The latest record for the Panhandle was on 19 April 1970 (Seyffert 2001b). **Taxonomy:** The subspecies found in Texas is *B.l. sanctijohannis* (Gmelin).

GOLDEN EAGLE *Aquila chrysaetos* (Linnaeus)

Rare to locally uncommon resident in the Panhandle and western and central Trans-Pecos. There have been isolated breeding records as far east as central Val Verde County. Migrants have been found in all areas of the state. They are rare to uncommon winter residents from the Panhandle south through the South Plains and Trans-Pecos, as well as the Rolling Plains and western Edwards Plateau. This species is rare to very rare in winter eastward to Colorado and Fort Bend Counties and very rare to casual throughout the remainder of the state. The breeding range of the Golden Eagle has been much reduced in the last century. Oberholser (1974) cites breeding records eastward through the Edwards Plateau. **Timing of occurrence:** Away from breeding areas, Golden Eagles are encountered in Texas between late October and late March. **Taxonomy:** The subspecies found in Texas is *A. c. canadensis* (Linnaeus).

ORDER GRUIFORMES

Family Rallidae: Rails, Gallinules, and Coots

YELLOW RAIL *Coturnicops noveboracensis* (Gmelin)

Rare migrant through the eastern portions of the state. Although migrants are rarely detected, clouding the actual status, they have been found from the eastern Panhandle and South Plains south through the center of the state to the Coastal Prairies. They are rare to locally common winter residents on the upper and central coasts. Surveys in appropriate habitat, including rice fields, have uncovered large numbers of this species. Its status away from the Coastal Prairies is poorly known, although Yellow Rails have been found during winter as far west as the High Plains and from Big Bend NP, Brewster County, on 31 January 1976. On 16 October 1960, Pulich (1961b) picked up 13 dead Yellow Rails at the base of a television tower in Dallas County. These unexpected records underscore how little is known about this species in Texas. **Timing of occurrence:** Fall migrants have been detected as early as mid-September, although it is believed that the primary migration occurs in October. Yellow Rails depart their wintering ground during April, but there are records into early May. **Taxonomy:** The subspecies that occurs in Texas is *C. n. noveboracensis* (Gmelin).

BLACK RAIL *Laterallus jamaicensis* (Gmelin)

Rare migrant in the eastern third of the state, east of the Balcones Escarpment, but there is only one record from the Pineywoods. Migrants are rarely detected, clouding the actual status, but they have been reported as far west as the High Plains in Bailey, Crosby, Lubbock, and Randall Counties. Black Rails are rare to locally uncommon residents on the upper and central coasts. Vocalizing birds have been found at several locations along the lower coast during the spring and summer, which may suggest some localized breeding. An exceptional record involved a calling individual in the Panhandle, Hutchinson County, on 1 July 1979 (TBSL 31-1). Many reports of Black Rails seen out of proper habitat or during the breeding season probably pertain to dark chicks of King and Clapper Rails. **Timing of occurrence:** Fall migrants have been detected inland in Texas from early August through early October, and winter residents appear to arrive by the end of this period. Spring migrants have been found from early April through early May. **Taxonomy:** The subspecies that occurs in Texas is *L. j. jamaicensis* (Gmelin).

CLAPPER RAIL *Rallus longirostris* Boddaert

Common resident primarily in brackish and salt marshes along the coast, although Clapper Rails do occur in freshwater marshes that are very near the immediate coast. There is only one well-documented inland record of Clapper Rail: one was netted and photographed in Tom Green County on 20 August 1986 (TPRF 402; Burt et al. 1987).

Clapper and King Rails are very similar in appearance, and some individuals can be very difficult to identify with certainty. This identification challenge clouds inland reports of Clappers. **Taxonomy**: The subspecies that occurs in Texas is *R.l. saturatus* Ridgway.

KING RAIL *Rallus elegans* Audubon

Uncommon to locally common resident in freshwater marshes, irrigation ditches, and weedy lakes on the Coastal Prairies from the Louisiana border south to Kenedy County. King Rails are uncommon southward to the Lower Rio Grande Valley. This species is very rare and local elsewhere in the eastern half of the state during the summer. There are documented breeding records from the Panhandle (Hutchinson County in 1950) and the Edwards Plateau (Crockett County in 1977). This species formerly bred across north-central Texas; the only recent records are from Delta and Kaufman Counties. There are interesting summer records in the Trans-Pecos from El Paso County and Balmorhea Lake, Reeves County. The species is a rare or at least seldom observed migrant throughout the state. **Timing of occurrence**: Fall migrants have been detected in Texas during September and October. Apparent spring migrants have been found from early March through early May. **Taxonomy**: The subspecies that occurs in Texas is *R. e. elegans* Audubon.

VIRGINIA RAIL *Rallus limicola* Vieillot

Uncommon to common migrant in all parts of the state. Virginia Rails are uncommon to locally common residents in the northern half of the Panhandle (Seyffert 2001b). There are isolated nesting records from various locations across the state, mostly in the western third, suggesting a more widespread breeding range. Virginia Rails are uncommon to locally common winter residents along the coast and uncommon to rare and local inland north to the southern Panhandle. **Timing of occurrence**: Fall migrants are found from late September through October, and spring migrants can be found from mid-March through April. **Taxonomy**: The subspecies that occurs in Texas is *R.l. limicola* Vieillot.

SORA *Porzana carolina* (Linnaeus)

Uncommon to locally common migrant throughout the state. Soras are common to locally abundant winter residents along the Coastal Prairies. There was an extraordinary count of up to 622 individuals from a single rice field on the central coast on 14 October 1998. This species is rare to uncommon inland during the winter. Soras are also sporadic, although always very rare, summer residents throughout the state. Breeding records for this species are rare, and nesting activity appears to be irregular, even at known locations along the Coastal Prairies and elsewhere. **Timing of occurrence**: Fall migrants occur from late August through October, and spring migrants can be found from mid-April through May. **Taxonomy**: Monotypic.

TBRC Review Species

PAINT-BILLED CRAKE *Neocrex erythrops* (Sclater)

Accidental. There are only two documented records from the United States. The Texas record was first and involved a single bird captured in a mammal trap (Arnold 1978a). Paint-billed Crakes are found discontinuously from Costa Rica south to northern Argentina. **Taxonomy**: The subspecies of this specimen was not determined, but *N. e. olivascens* Chubb is thought to be partially migratory and the likely candidate for vagrancy.

17 FEB. 1972, NEAR COLLEGE STATION, BRAZOS CO. (*TCWC 8,930)

TBRC Review Species

SPOTTED RAIL *Pardirallus maculatus* (Boddaert)

Accidental. There is one documented record of this tropical rail for Texas. This individual, found in a weakened condition, was captured alive along a small creek, but it died the next day and represents the second record for the United States (Parkes, Kibbe, and Roth 1978). Spotted Rails are known for dispersing long distances from population centers (AOU 1998). They occur from central Mexico and the Greater Antilles southward to northern Argentina. **Taxonomy**: The specimen from Texas belongs to the subspecies *P. m. insolitus* (Bangs & Peck).

9 AUG. 1977, NEAR BROWNWOOD, BROWN CO. (*TCWC 10,400)

PURPLE GALLINULE *Porphyrio martinicus* (Linnaeus)

Rare to uncommon migrant in the eastern half of Texas and very rare to accidental in the western half. This species is a rare to locally common summer resident in the eastern third of the state, west to Red River, Smith, Gonzales, and Hidalgo Counties. Purple Gallinules are most common during the summer along the upper and central coasts. They are casual winter visitors along the coast and in the Lower Rio Grande Valley and casual summer visitors to Midland (TPRF 10) and Jeff Davis Counties. **Timing of occurrence**: Spring migrants begin to arrive in the state in early April, and the bulk of the breeding population is present by the end of the month. Fall migration begins in late August, with most departing the state by early October. **Taxonomy**: Monotypic.

COMMON GALLINULE *Gallinula galeata* (Lichtenstein)

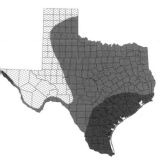

Uncommon to locally abundant resident from the Lower Rio Grande Valley and Coastal Prairies inland to Bexar and Brazos Counties. Common Gallinules are rare to locally uncommon summer residents throughout the eastern three-quarters of the state, including the Panhandle. In the Trans-Pecos, they are uncommon residents along the Rio Grande in El Paso and Hudspeth Counties. There are winter records for this species throughout the breeding range in Texas. This species was formerly considered conspecific with Common Moorhen (*G. chloropus*), which occurs in Europe, Africa, and Asia. **Timing of occurrence**: Common Gallinules arrive on the breeding grounds beginning in late March and are present through late October. **Taxonomy**: The subspecies that occurs in Texas is *G. g. cachinnans* Bangs.

AMERICAN COOT *Fulica americana* Gmelin

Uncommon to common resident in nearly all regions of the state. Despite the widespread occurrence of American Coots in Texas, breeding has been reported only from scattered locations statewide. They are common to abundant winter residents statewide, and some water impoundments support large populations, sometimes numbering in the thousands. **Taxonomy**: The subspecies that occurs in Texas is *F. a. americana* Gmelin.

Family Gruidae: Cranes

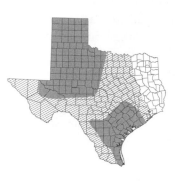

SANDHILL CRANE *Grus canadensis* (Linnaeus)

Common to uncommon migrant throughout much of the state. Migrants can be found throughout the state, although this species is very rare in the Pineywoods. There are two population centers of wintering Sandhill Cranes in Texas: the High Plains and Rolling Plains to the north and the Coastal Prairies to the south. This species winters locally elsewhere throughout the state, although Sandhill Cranes are rare in the Trans-Pecos during midwinter and casual in the Pineywoods. There are records of individual birds lingering into the summer and, on very rare occasions, remaining through the summer. These summer occurrences are mostly in the Panhandle and South Plains and may pertain to injured individuals. **Timing of occurrence:** Sandhill Cranes arrive in early October, and most depart by mid-March. This species was formerly a rare summer resident in the coastal marshes of southeast Texas (Strecker 1912). **Taxonomy:** Four subspecies have been known to occur in the state, but only three are still found in Texas.

G. c. canadensis (Linnaeus)
> Common to uncommon migrant over most of the state. This subspecies is common in winter on the Coastal Prairies and uncommon on the South Plains.

G. c. rowani Walkinshaw
> Common to uncommon migrant over the eastern half of the state. This subspecies is common in winter on the Coastal Prairies.

G. c. tabida (Peters)
> Common to abundant migrant and winter resident from the Panhandle south to the northwestern Edwards Plateau. There are specimens from the Coastal Prairies, but the status of this taxon from that region is poorly understood.

G. c. pratensis Meyer
> Formerly a rare resident in the marshes along the upper coast (Strecker 1912). Reported to have nested south to Matagorda County as late as 1895.

WHOOPING CRANE *Grus americana* (Linnaeus)

This critically Endangered Species is an uncommon winter resident in Aransas and Calhoun Counties. Individuals have wintered well away from the primary winter range, including as far north as the Panhandle and south to Willacy County. Exceptional was the presence of up to nine that wintered around Granger Lake in eastern Williamson County during 2011–12. This occurrence was attributed to poor habitat conditions on the central coast due to an extended drought. On rare occasions, this species has over-summered on the Texas coast, but most of these records pertained to sick or injured birds. Whooping Cranes are rarely encountered, as they migrate along a narrow corridor down the middle of the state. Migrants have been found, often with Sandhill Cranes, just west and east of the traditional migration route. These out-of-range birds have been reported west to Uvalde County and east to Jefferson and Smith Counties. As of this writing, the wintering population in Texas consists of approximately 245 birds. There were only 14 individuals in the wild in the late 1930s when concerted conservation efforts were initiated. **Timing of occurrence:** Fall migrants arrive in late October to early November, but stragglers come in as late as mid-December. Northward migration begins in mid-March; most birds depart by mid-April, but some occasionally linger into May. **Taxonomy:** Monotypic.

ORDER CHARADRIIFORMES

Family Burhinidae: Thick-knees

TBRC Review Species **DOUBLE-STRIPED THICK-KNEE**
Burhinus bistriatus (Wagler)

Accidental. Texas has one record for this species. This unusual nocturnal shorebird that resembles a giant plover was an unexpected find and represents the only record for the United States. Double-striped Thick-knees have been found irregularly as far north as southern Tamaulipas in northeastern Mexico and could potentially occur again in

the grasslands or farmlands of southern Texas. **Taxonomy**: The subspecies that has occurred in Texas is *B. b. bistriatus* (Wagler).

5 DEC. 1961, KING RANCH, KLEBERG CO.
  (*USNM 478,866; TPRF 2557)

Family Recurvirostridae: Stilts and Avocets

BLACK-NECKED STILT
Himantopus mexicanus (Statius Müller)

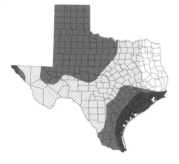

Common summer resident along the Coastal Prairies and locally common inland to the South Texas Brush Country and southern Blackland Prairies. Isolated breeding populations are also present in north and north-central Texas. Black-necked Stilts are uncommon to rare summer residents on the High Plains, western Rolling Plains, and northeastern Trans-Pecos. They are uncommon to common near the Rio Grande in El Paso and Hudspeth Counties in summer. As migrants, Black-necked Stilts are rare to uncommon in all areas except the Pineywoods, where they are very rare. During winter, they are rare to locally uncommon along the Coastal Prairies and in El Paso and Hudspeth Counties and very rare at other inland locations. **Timing of occurrence**: Spring migrants begin arriving in Texas in late March, and most of the breeding population is present by the end of April. Fall migrants can be found between late August and late September, but lingering individuals can be found as late as early December. **Taxonomy**: The subspecies that occurs in Texas is *H. m. mexicanus* (Statius Müller).

AMERICAN AVOCET *Recurvirostra americana* Gmelin

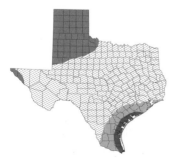

Common to locally abundant winter resident on the coast. American Avocets are typically absent from the northern half of the state in winter, and individuals present there after mid-November are very rare. This species is a common summer resident on the High Plains and near the Rio Grande in El Paso and Hudspeth Counties. American Avocets are rare and local summer residents on the Coastal Prairies. They are common to uncommon migrants through the western half of the state, becoming generally uncommon farther east. In the Pineywoods, they are rare in spring and uncommon in fall. In general, migrant American Avocets occur in greater numbers in the fall than in the spring. Large concentrations of these birds can sometimes be found during peak migration; up to 14,000 birds were noted at one location in Galveston County on 27 February 1998. **Timing of occurrence**: Spring migrants occur between mid-March through early May, and stragglers are present through early June. In fall, migrants are seen from late July through early November. The wintering population arrives on the coast beginning in mid-August. **Taxonomy**: Monotypic.

Family Haematopodidae: Oystercatchers

AMERICAN OYSTERCATCHER

Haematopus palliatus Temminck

Locally common resident along the upper and central coasts. This species is rare to locally uncommon along the lower coast. American Oystercatchers nest primarily on or near shell ridges and are seldom found far from areas that provide an ample supply of saltwater mollusks on which they feed. Breeding activities are under way in late February, and young can be encountered beginning in March. **Taxonomy**: The subspecies that occurs in Texas is *H. p. palliatus* Temminck.

Family Charadriidae: Plovers

BLACK-BELLIED PLOVER

Pluvialis squatarola (Linnaeus)

Uncommon to common migrant through the eastern half of the state, becoming increasingly less common westward. Black-bellied Plovers are rare but regular west to through the High Plains and eastern Trans-Pecos and are very rare farther west. They are common to abundant on the coast and rare to uncommon inland in the southeastern quarter of the state in winter. They are also rare to uncommon in summer along the coast, although this species does not breed in the state. These nonbreeding individuals usually retain basic plumage. **Timing of occurrence**: Fall migrants begin arriving in mid-August and are encountered at inland sites through October. Spring migrants are found between late March and late May. **Taxonomy:** The subspecies that occurs in Texas is *P. s. cynosurae* (Thayer and Bangs).

AMERICAN GOLDEN-PLOVER
Pluvialis dominica (Statius Müller)

Common spring migrant in the eastern half of the state, becoming very rare to casual west of the Balcones Escarpment. American Golden-Plovers are rare fall migrants in the eastern half of the state. In the western half of Texas, these plovers occur with greater regularity in the fall than the spring, although they are still very rare to casual. **Timing of occurrence:** Spring migrants are present between early March and mid-May, with occasional late February records and a few from early June. Most fall migrants are seen between mid-August and late October, with occasional July records. On rare occasions, individuals will linger well into December on the Coastal Prairies. **Taxonomy:** Monotypic.

TBRC Review Species   **PACIFIC GOLDEN-PLOVER** *Pluvialis fulva* (Gmelin)

Accidental. There is one documented record for the state. This individual was a spring migrant molting into alternate plumage, which allowed the bird to be differentiated in the large concentration of American Golden-Plovers present at the time. Although the primary wintering range in North America is on the Pacific Coast, this mostly Asian species has been documented in many other locations, including several records from the extreme eastern United States, Bermuda, and Barbados. **Taxonomy:** Monotypic.

12–13 APR. 2006, EDNA, JACKSON CO.
(TBRC 2006–41; TPRF 2409)

TBRC Review Species   **COLLARED PLOVER** *Charadrius collaris* Vieillot

Accidental. Collared Plover is yet another tropical species for which Texas holds the only record for the United States. An adult was discovered during the spring 1992 TOS meeting in Uvalde, Uvalde County. Given that this species occurs as far north as southern Tamaulipas in northeastern Mexico, it is surprising that more Collared Plovers have not been discovered in the state. **Taxonomy:** Monotypic.

9–12 MAY 1992, UVALDE NATIONAL FISH HATCHERY, UVALDE CO. (TBRC 1992-70; TPRF 1099)

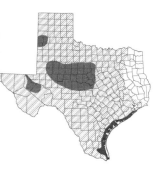

SNOWY PLOVER *Charadrius nivosus* (Cassin)

Uncommon summer resident along the immediate coast. Snowy Plovers are rare to locally uncommon summer residents, primarily on saline lakes and along major waterways, at scattered locations in the western half of the state, east to Burnet and Stephens Counties. They are rare to uncommon migrants through most of the state, becoming very rare to casual in the eastern quarter. Snowy Plovers are rare to locally uncommon winter residents along the coast and are very rare to casual inland to north-central Texas. **Timing of occurrence**: Spring migrants begin returning to the state in early March and are present through early May. Fall migrants pass through the state between late July and early October. **Taxonomy**: The subspecies that occurs in Texas is *C. n. nivosus* (Cassin). This species was separated from the Kentish Plover (*C. alexandrinus* Linnaeus) of Eurasia in 2011 (Chesser et al. 2011).

WILSON'S PLOVER *Charadrius wilsonia* Ord

Common migrant and summer resident along the immediate coast. Elsewhere, Wilson's Plovers are casual visitors inland, primarily to the southeastern quarter of the state. In winter, they are very rare and local along the upper coast and locally uncommon farther south to the Rio Grande. This mostly coastal plover is a rare to locally uncommon breeder up to 60 miles inland, primarily at salt lakes, in Willacy, Hidalgo, and western Kenedy Counties. **Timing of occurrence**: Summer residents begin returning to the coast in mid-February but are not common until early March. The majority of the population departs by late September, although some linger as late as early November. **Taxonomy**: The subspecies that occurs in Texas is *C. w. wilsonia* Ord.

SEMIPALMATED PLOVER

Charadrius semipalmatus Bonaparte

Common winter resident along the coast and rare to accidental inland. Semipalmated Plovers are uncommon to common migrants throughout the state, although they are generally least common in the western third. Semipalmated Plovers may at times congregate in large flocks along the coast during spring migration, sometimes including up to 1,000 individuals. This species is a rare summer visitor on the coast but does not breed in Texas. **Timing of occur-**

rence: The migration periods for this species are between early March and late May and from early July to late October. **Taxonomy:** Monotypic.

PIPING PLOVER *Charadrius melodus* Ord

Uncommon to locally common winter residents along the coast, where they linger through the summer on very rare occasions. Piping Plovers are not often observed during migration at inland locations, but most appear to pass east of the Balcones Escarpment. They are very rare to casual migrants in the western two-thirds of the state. Conservation efforts to protect Piping Plovers over the past 20 years have resulted in an increase in the population. The world's population in 2009 was estimated to be around 12,000 individuals; the Great Plains population equaled roughly 60 percent of the total. Piping Plover is listed as Endangered, and many seen on Texas beaches in the winter are color-banded. **Timing of occurrence:** Fall migrants begin arriving in Texas as early as the end of June, but the primary migration occurs between late July and early September. Spring migrants can be found between late March and early May. **Taxonomy:** The subspecies that occurs in Texas is *C. m. circumcinctus* (Ridgway).

KILLDEER *Charadrius vociferus* Linnaeus

One of the most ubiquitous birds in Texas. They are common to abundant residents in all parts of the state, increasing in numbers in the central and southern parts of the state during the winter. Killdeer populations fluctuate greatly during the winter in the Panhandle, where they can be common some years and virtually absent in others. **Taxonomy:** The subspecies that occurs in Texas is *C. v. vociferus* Linnaeus.

MOUNTAIN PLOVER *Charadrius montanus* Townsend

Very rare summer resident in open grasslands of the northwestern Panhandle. Mountain Plovers were very rare summer residents in the mid-elevation grasslands of the central Trans-Pecos into the early 1990s, but there are no recent reports of summer occurrence. They are rare and local winter residents from the southern Blackland Prairies south to the central Coastal Prairies. In recent years, they appear

to occur with more consistency on the Blackland Prairies, particularly in Bell, Williamson, and Comal Counties. This species also winters very locally in the Trans-Pecos, South Texas Brush Country, Concho Valley, and northwestern Edwards Plateau, where some flocks exceed 300 birds. Migrants are very rarely encountered away from the Panhandle and South Plains but have been documented east to the upper coast. Occasionally, Mountain Plovers are reported from unlikely locations when American Golden-Plovers are migrating through. These reports probably pertain to basic-plumaged American Golden-Plovers. A review of the population of Mountain Plovers in 2011 determined that the species was not threatened with extinction, and the total population was determined to be near 20,000 individuals in 2009 (Tipton, Doherty, and Dreitz 2009). **Timing of occurrence:** Fall migrants have been detected in the Panhandle from early August through October. It is possible that the earlier dates refer to birds present through the summer. The early date for returning wintering birds in Central Texas is 3 September. Spring migrants have been found from early March through early May. **Taxonomy:** Monotypic.

Family Jacanidae: Jacanas

TBRC Review Species **NORTHERN JACANA** *Jacana spinosa* (Linnaeus)

Very rare visitor, primarily during the winter, to the Lower Rio Grande Valley. The occurrence of this species is sporadic with occasionally long periods between records, such as between spring 1994 and fall 2006. Northern Jacanas have been found along the Coastal Prairies and inland north to Travis County. There are single records from the Edwards Plateau (Kerr County) and the Trans-Pecos (Brewster County). Thirty-six documented records exist for Texas, with only six of those since 2006. Northern Jacana appears to have been a rare resident in the Lower Rio Grande Valley prior to 1910. Interestingly, a resident population of more than 40 jacanas became established at Maner Lake, Brazoria County, between the winter of 1967 and April 1978. There are 44 reports of Northern Jacanas from prior to the development of the Review List in 1988 for which there is no documentation. **Timing of occur-**

rence: There are records from every month of the year, but the majority have been between November and April. **Taxonomy:** The subspecies that occurs in Texas is *J. s. gymnostoma* (Wagler).

Family Scolopacidae: Sandpipers, Phalaropes, and Allies

SPOTTED SANDPIPER *Actitis macularius* (Linnaeus)

Common migrant throughout the state. Spotted Sandpipers are also common winter residents in the southern half of the state but are absent from the Panhandle and very rare on the remainder of the High Plains. They are uncommon elsewhere in the northern half of the state, as well as the Pineywoods and Trans-Pecos. The distribution during winter fluctuates greatly depending on the harshness of the season. Spotted Sandpipers can be found during summer in the northern half of the state, although they are irregular in occurrence and in number. Even with a regular summer occurrence, they are only casual breeders in the Panhandle (Seyffert 2001b) and are believed to have nested on the Edwards Plateau (Lockwood 2001). Oberholser (1974) also cites a breeding record reported by John J. Audubon from Harris County in 1837. **Timing of occurrence:** Fall migrants are seen from early July through October. Spring migrants are recorded from late March to late May. **Taxonomy:** Monotypic.

SOLITARY SANDPIPER *Tringa solitaria* Wilson

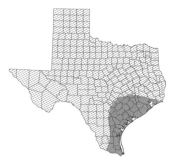

Uncommon to common migrant throughout the state. This species lives up to its name and is very seldom found in flocks of any size, though many may be found loosely scattered at any one location during peak migration. Solitary Sandpipers are rare winter residents along the coast and casual to very rare inland east of the Balcones Escarpment and north to at least Tarrant County. There are winter records from most of the remaining areas of the state, including the Panhandle and Trans-Pecos. **Timing of occurrence:** Spring migration extends from early March to early May, but some linger into early June. Fall migration is mainly from mid-July to late October, peaking from late August to mid-September. Fall migrants have been seen as early as late June. **Taxonomy**: Both subspecies have been documented in the state.

T. s. cinnamomea (Brewster)

Status uncertain but may be a rare migrant through the western half of the state. Specimens noted from as far east as Brazos (W. Davis 1940) and Tarrant Counties (Pulich 1979).

T. s. solitaria Wilson

Uncommon to common migrant throughout the state.

TBRC Review Species

WANDERING TATTLER *Tringa incana* (Gmelin)

Accidental. The lone Texas record was of a bird frequenting the large blocks of the seawall in the city limits of Galveston, Galveston County. This species normally frequents rocky shorelines, and the Galveston seawall approximates that habitat type. Wandering Tattlers are typically found along the Pacific Coast, and this is one of only a handful of documented records east of the Rocky Mountains. **Taxonomy**: Monotypic.

23 APR.–9 MAY 1992, GALVESTON, GALVESTON CO.
(TBRC 1992-64; TPRF 1090)

TBRC Review Species

SPOTTED REDSHANK *Tringa erythropus* (Pallas)

Accidental. There is one record of this Eurasian shorebird in Texas. Spotted Redshank is a rare, but regular, migrant through western Alaska and a casual vagrant farther south along the Pacific Coast to California. There are a few scattered records from the interior of North America, including one from Kansas and from along the Atlantic Coast as far south as North Carolina. **Taxonomy**: Monotypic.

28, 29, OR 30 SEPT. 2000, AUSTIN, TRAVIS CO.
(TBRC 2001-129; TPRF 1983)

GREATER YELLOWLEGS *Tringa melanoleuca* (Gmelin)

Uncommon to common migrant across the state. Greater Yellowlegs is an uncommon to locally common winter resident along the Coastal Prairies and inland in the eastern half of the state, becoming uncommon to rare farther west. Greater Yellowlegs are more common and widespread during the winter than the Lesser Yellowlegs. Migrant Greater Yellowlegs can be found in Texas during June and July, suggesting a summer occurrence, but they do not breed in the state. **Timing of occurrence**: Fall migrants begin to arrive in early July and are present through mid-November, although the majority of the winter population arrives by early October. Spring migration occurs mainly between mid-March and early May, although some linger into late June. **Taxonomy:** Monotypic.

WILLET *Tringa semipalmata* (Gmelin)

Common to abundant year-round along the coast. There are two populations of Willet, currently classified as subspecies, that occur in Texas. The breeding population belongs to the eastern subspecies, which winters primarily in Central and South America. The western population nests in the interior of North America and winters in Texas along the coast. These western birds are rare to uncommon migrants throughout the state. Willets, possibly early migrants or from the breeding population, are rare to casual wanderers inland during the late summer and early fall. **Timing of occurrence:** Western birds begin to arrive in the state in mid-July and can be found at inland locations through mid-September. They begin arriving on the wintering grounds in early August before the bulk of the breeding eastern birds depart for their wintering areas. In spring, migrating western birds begin to leave the coast in late March, but some remain there into early May. Migrating western birds are found in the interior of the state through mid-May. Eastern birds depart the breeding grounds during July and August. There are periods in the spring and late summer when both subspecies are present on the coast. **Taxonomy:** As previously mentioned, both subspecies are present in the state.

T. s. inornata (Brewster)
    Refers to the western population described above.

T. s. semipalmata (Gmelin)
    Refers to the eastern population described above.

LESSER YELLOWLEGS *Tringa flavipes* (Gmelin)

Common migrant throughout the state. This species is an uncommon to locally common winter resident on the coast, becoming rare to uncommon inland in southern Texas. Lesser Yellowlegs are casual to very rare in most of the northern half of the state and Trans-Pecos in win-

ter. During peak migration in the spring, concentrations of more than 1,000 birds have been noted with some regularity along the coast. **Timing of occurrence:** Fall migrants pass through the state between late June and late October. Early fall migrants have been detected in the Panhandle as early as mid-June. Spring migrants are seen from mid-March to mid-May, and some linger into early June. **Taxonomy:** Monotypic.

UPLAND SANDPIPER *Bartramia longicauda* (Bechstein)

Common spring migrant and uncommon to common fall migrant east of the Pecos River. This species is a rare spring and uncommon fall migrant in the Trans-Pecos. Once greatly reduced in numbers due to market hunting, this species has since rebounded somewhat but remains a species of conservation concern. This species was reported to have bred in the northern Panhandle as late as 1945 and eastward across north-central Texas to Cooke County during the 1890s. Occasionally, summering birds are found in those areas today—including an adult with two recently hatched chicks in July 1983—suggesting breeding may still occur on an irregular basis. **Timing of occurrence:** Spring migrants are found between mid-March and mid-May, but a few birds are seen into early June. Fall migrants arrive in mid-July and are seen through September, with a few recorded into early October. **Taxonomy:** Monotypic.

TBRC Review Species

ESKIMO CURLEW *Numenius borealis* (Forster)

Probably extinct. The last fully documented record of this small curlew in Texas was in March and April 1962, when the only photographs ever taken in the wild were obtained. The last documented record at any location in the world was in Barbados in November 1963. Eskimo Curlews were formerly regular spring migrants in the state, with records falling between 8 March and 23 April. They migrated along the east coast of North America during the fall, and as a result there are no records for this season in Texas. Available records suggest that the majority of Eskimo Curlews migrated east of the Edwards Plateau but west of the Pineywoods. There are 41 published reports of this small curlew from Texas between 1850 and 1905, including specimens collected (but not currently located) as far west as Pecos County in the Trans-Pecos (Oberholser 1974). This information suggests that large numbers migrated through Texas until the very early 1900s. Strecker (1912) listed them as a common migrant in the eastern half of the state. From these 41 reports the TBRC has located documentation of 19, and all but two pertain to specimens collected prior to 1900. Eskimo Curlews have been reported from Texas several times since the spring of 1962, but incontrovertible evidence of their presence has not been obtained. **Taxonomy**: Monotypic.

WHIMBREL *Numenius phaeopus* (Linnaeus)

Uncommon to rare migrant along the coast, becoming rare inland in the eastern half of the state except the Pineywoods, where Whimbrels are very rare. In the western half of the state, this species is a very rare to casual migrant. As a result, most Whimbrels detected during migration are within a narrow corridor through the center of the state, and they are most commonly encountered on the Coastal Prairies. Inland, Whimbrels are much more regular in spring than in fall. This species is a rare winter resident and summer lingerer on the coast. **Timing of occurrence**: Migration periods are between mid-March and late May and from mid-July to late October. **Taxonomy**: Two subspecies occur in Texas.

N. p. phaeopus (Linnaeus)
    Accidental. One record from Crystal Beach, Galveston County, from 29 April to 19 May 2013 (TBRC 2013–37; TPRF 3034).

N. p. hudsonicus Latham
    As described in main species account. Engelmoer and Roselaar (1998) suggested that the two breeding populations in North America be recognized as separate subspecies. If recognized, the western population, *N. p. rufiventris* Vigors, may also occur in the state.

LONG-BILLED CURLEW
Numenius americanus Bechstein

Common to locally abundant winter resident on the coast, where flocks of several hundred are not unusual. Wintering Long-billed Curlews are common to uncommon inland in the Coastal Prairies and uncommon to rare farther

inland to the southern Blackland Prairies, as far north as Bell County and west to El Paso and Hudspeth Counties. They are casual winter visitors in most other areas of the state, although very locally uncommon in the Panhandle and extremely rare in the Pineywoods. They are locally common summer residents in the northwestern Panhandle, and there are reports of occasional nesting along the upper coast. Long-billed Curlews are rare to locally uncommon summer visitors on the coast and elsewhere within the species' migratory pathway. Small flocks, sometimes containing up to 50 birds, have been encountered along the Coastal Prairies during the summer. Historical breeding records exist from Jeff Davis and Cameron Counties. This curlew can be found throughout the year at various locations, and in some areas the beginning or ending of migration can be difficult to discern. This species is an uncommon migrant throughout the state with the exception of the Pineywoods, where it is essentially absent. On the High Plains, flocks of several hundred or a thousand or more may be encountered. **Timing of occurrence**: Migrants are seen in the state mostly from mid-March to mid-May and from early July to mid-November. **Taxonomy**: Both subspecies occur in Texas.

N. a. parvus Bishop

> Uncommon migrant through the eastern half of the state and rare winter resident on the central and lower coasts.

N. a. americanus Bechstein

> Locally common summer resident in the northwestern Panhandle. Common migrant through the state west of the Pineywoods. Common to locally abundant winter resident along the coast and inland as described above.

TBRC Review Species

BLACK-TAILED GODWIT *Limosa limosa* (Linnaeus)

Accidental. The lone state record refers to a single individual that was discovered in spring 2012 on the upper coast and lingered well into the summer. This species is a very rare to casual spring migrant in extreme western Alaska and casual to accidental along the Atlantic Coast. Louisiana also has a record from spring 1994. **Taxonomy**: The subspecies of the Texas individual is currently undetermined. Structural and plumage characteristics exhibited by this bird suggest *L.l. limosa* (Linnaeus). The records from along the Atlantic Coast are thought to primarily pertain to *L.l. islandica* Brehm, while those from Alaska pertain to *L.l. melanuroides* Gould.

6 JUNE–10 AUG. 2012, BRAZORIA NWR, BRAZORIA CO.
(TBRC 2012–45; TPRF 2972)

HUDSONIAN GODWIT *Limosa haemastica* (Linnaeus)

Uncommon to rare spring migrant through the eastern half of the state. Hudsonian Godwits are encountered much more frequently along the Coastal Prairies of the upper and central coasts than at inland locations. They are casual spring migrants in the western half of the state, with fewer than 10 reports from the Trans-Pecos. On rare occasions, lingering spring migrants have remained along the immediate coast well into the summer. Exceptional was a group of seven that remained in Calhoun County until 7 July 2006. Hudsonian Godwits migrate southward along the east coast of North America and, as a result, are a casual fall migrant in Texas and accidental away from the coast. Interestingly, of the very few Texas Panhandle records, there are as many in the fall as in spring. **Timing of occurrence**: The migration window for this species is between mid-April and mid-May, and late or lingering birds are often present into early June. The very few fall records are between mid-August and late September. Some of these records may pertain to birds that lingered through the summer rather than true fall migrants. **Taxonomy**: Monotypic.

MARBLED GODWIT *Limosa fedoa* (Linnaeus)

Common to uncommon winter resident along the Coastal Prairies. This species is a rare to locally uncommon migrant throughout the state. Due to a general lack of habitat, Marbled Godwits are very rarely encountered in the Pineywoods or on the Edwards Plateau. They are rare summer visitors along the coast. Inland, there are scattered records from June and July that may represent late or early migrants rather than birds actually spending the summer. **Timing of occurrence**: Fall migrants begin passing through the state in late June, with the primary migration period between early August and late September. Lingering migrants have been noted at inland locations through November. Spring migrants can be found between late February and mid-May. **Taxonomy:** The subspecies that occurs in Texas is *L. f. fedoa* (Linnaeus).

RUDDY TURNSTONE *Arenaria interpres* (Linnaeus)

Common migrant and winter resident along the coast. This species is a rare summer visitor on the coast, and most of these individuals are in basic or incomplete alternate plumage. Migrant Ruddy Turnstones are rare at inland locations in the eastern half of the state and casual in the west. Like that of so many other shorebird species, the path of migration is most obvious along a corridor through the center of the state. **Timing of occurrence:** Fall migrants are seen from late July to mid-October. Spring migrants are noted from late March to late May, but a few linger into early June. **Taxonomy:** The subspecies that occurs in Texas is *A. i. morinella* (Linnaeus).

RED KNOT *Calidris canutus* (Linnaeus)

Uncommon migrant along the coast, especially the upper coast. Red Knots are very rarely detected inland and are a very rare migrant through the eastern half of the state, although they are at best casual in the Pineywoods. In the western half of the state this species is a casual migrant, with the possible exception of the eastern Panhandle, where Red Knots have been reported with slightly higher frequency. Inland migrants are much more frequently encountered in the fall than in the spring. Red Knots are rare and local winter residents on the coast, where they are also very rare summer visitors. Large numbers used to be encountered in migration on the Bolivar Peninsula in Galveston County, but such concentrations are a thing of the past. **Timing of occurrence:** Spring migrants are found between late March and late May. Fall migrants are found from early August to early November. **Taxonomy:** The subspecies that occurs in Texas is *C. c. rufa* (Wilson).

TBRC Review Species   ### SURFBIRD *Calidris virgata* (Gmelin)

Casual. There are 11 documented records for Texas of this west coast species, and all are from coastal locations. Surfbirds are typically associated with rocky coastlines. Almost

all Texas records are from jetties, but one present from 15 to 21 April 1995 frequented the sand beaches of North Padre Island, Kleberg County (TBRC 1994-67; TPRF 1291). **Timing of occurrence:** All records are from the spring, with the dates of occurrence ranging from 12 March to 9 May and seven of the records from April. **Taxonomy:** Monotypic.

TBRC Review Species   **RUFF** *Calidris pugnax* (Linnaeus)

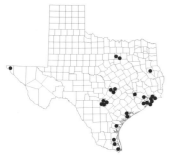

Very rare migrant and accidental winter visitor. There are 34 documented records in Texas. The majority of these sightings are from coastal counties; however, there are a number of inland records. Ruffs have been recorded multiple times in Bexar, Tarrant, and Travis Counties, and there is a single record from El Paso County. **Timing of occurrence:** There are 13 spring records for the state, occurring between 29 March and 14 May. The 19 fall records are between 25 July and 5 October. Three of these records involved individuals that remained well into the winter. There are also two records of overwintering birds that were not discovered until December. **Taxonomy:** Monotypic.

TBRC Review Species   **SHARP-TAILED SANDPIPER**

Calidris acuminata (Horsfield)

Accidental. There are three documented records in the state. The breeding ranges of Sharp-tailed Sandpipers and Pectoral Sandpipers overlap in Siberia, and it has been speculated that Sharp-tailed Sandpipers found in North America migrate with Pectorals to that species' wintering grounds in South America. **Taxonomy:** Monotypic.

17–18 MAY 1991, ARLINGTON, TARRANT CO.
  (TBRC 1991-56; TPRF 993)
21–22 SEPT. 1996, SAN ANTONIO, BEXAR CO. (TBRC 1996-128)
14–18 NOV. 2004, SPRINGLAKE, LAMB CO.
  (TBRC 2004–91; TPRF 2241)

STILT SANDPIPER *Calidris himantopus* (Bonaparte)

Uncommon to locally common migrant statewide. Stilt Sandpipers are rare to uncommon winter residents on the coast and are most common along the lower coast and in

the Lower Rio Grande Valley. They are rare in winter inland to Central Texas and casual farther north. As many as 4,000 Stilt Sandpipers are found each winter at Laguna Atascosa NWR, Cameron County. **Timing of occurrence**: Fall migrants are passing through from early July to late October, although some birds are found north of the winter range as late as early December. Spring migrants can be found from late March to late June. **Taxonomy**: Monotypic.

TBRC Review Species **CURLEW SANDPIPER** *Calidris ferruginea* (Pontoppidan)

Casual. There are 12 documented records in Texas. Records exist for both spring and fall migrants and one odd occurrence of a basic-plumaged bird present at Bolivar Flats, Galveston County, from 24 June to 7 July 1994. The majority of these birds have been found at shorebird stopover habitat along the coast, with just two inland records, from Travis and Bexar Counties. **Timing of occurrence**: The seven spring records fall between 28 April and 3 June. Three of the four fall records occur between 17 August and 9 September, with the remaining record from 24 November. **Taxonomy**: Monotypic.

TBRC Review Species **RED-NECKED STINT** *Calidris ruficollis* (Pallas)

Accidental. There are two records for Texas of this Eurasian shorebird. Both records are of adults in alternate plumage found during the summer. These finds contribute to a rather short list of sightings from the interior of North America. Red-necked Stints are rare breeders in western and northern Alaska, and there are numerous records of migrants from both coasts of North America, where almost all are of adults in the fall. **Taxonomy**: Monotypic.

17–22 JULY 1996, FORT BLISS SEWAGE PONDS, EL PASO CO.
  (TBRC 1996-94; TPRF 1447)
26 JUNE–8 JULY 2011, BOLIVAR FLATS, GALVESTON CO.
  (TBRC 2011–070; TPRF 2966)

SANDERLING *Calidris alba* (Pallas)

Common to abundant migrant and winter resident along the immediate coast. Sanderlings are also common non-

breeding summer visitors along the coast. They are rare migrants inland through the eastern two-thirds of the state and casual spring and very rare fall migrants farther west. **Timing of occurrence:** Spring migrants are found from late March through late May, with a few still passing through inland sites in early June. Fall migrants occur from mid-July through mid-October, but stragglers are found as late as early December. **Taxonomy:** Monotypic.

DUNLIN *Calidris alpina* (Linnaeus)

Common to abundant winter resident on the coast and rare inland throughout most of the state. Dunlins are casual in winter on the High Plains and in the Trans-Pecos. This species is a rare to uncommon migrant through the eastern half of the state, becoming very rare in the western third. Dunlins are more frequently encountered at inland sites during the fall migration than in the spring. There are records of individuals lingering through the summer, particularly along the coast, although these occurrences are very rare. **Timing of occurrence:** Fall migrants are passing through from late August to mid-November. Spring migrants are found from mid-March to late May. **Taxonomy:** The subspecies that occurs in Texas is *C. a. hudsonia* (Todd). It is possible that *C. a. pacifica* (Coues) and *C. a. arcticola* (Todd) may occur as a rare migrant through the Trans-Pecos.

TBRC Review Species

PURPLE SANDPIPER *Calidris maritima* (Brünnich)

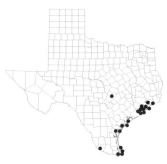

Very rare winter visitor on the immediate coast and accidental inland and in the early spring. There are 26 documented records from the state; all but two occur on the coast. Purple Sandpipers have been nearly annual in occurrence in the state between 2002 and 2012, and the records almost always involve single individuals. Separating winter-plumaged Purple and Rock Sandpipers (*C. ptilocnemis*) may not be possible in the field. The TBRC considers all records from Texas to be of Purple Sandpipers, an east coast species, until such time as either a Rock Sandpiper, from the west coast, is documented in or near Texas or field identification criteria are developed that clearly separate the two in basic plumage. **Timing of occurrence:** All records fall between 12 November and 4 May. All but four

of the records were first detected between November and late February. **Taxonomy:** The subspecies of Purple Sandpiper that has occurred in Texas is unknown due to a lack of specimens. Two subspecies may account for these records: *C. m. belcheri* Engelmoer & Roselaar, which nests around Hudson Bay, and *C. m. maritima* (Brünnich), which breeds slightly farther north on Baffin Island and in southern Greenland.

BAIRD'S SANDPIPER *Calidris bairdii* (Coues)

Uncommon to common migrant through most of the state during spring and fall, although Baird's Sandpipers are considered rare to locally uncommon in the Pineywoods. The primary migration route for this species is through the Great Plains and Texas. Baird's Sandpipers winter in South America and are not expected in Texas during winter, although a presumed very early migrant was at San Antonio, Bexar County, on 25 February 1988 (TPRF 609), and an injured individual was at Brazos Bend SP, Fort Bend County, on 26 December 1993. **Timing of occurrence:** Fall migrants pass through the state from mid-July to mid-October, but stragglers are occasionally found as late as early December. Spring migrants are present from mid-March and mid-May, with a few straggling into June. **Taxonomy:** Monotypic.

LEAST SANDPIPER *Calidris minutilla* (Vieillot)

Probably the most widespread migrant shorebird in Texas. Least Sandpipers are common to abundant as migrants statewide and as winter residents on the coast. They are uncommon to locally common as winter residents north of the coast to north-central Texas and west to the Trans-Pecos, becoming rare in the Panhandle. They are also rare summer visitors, most commonly along the coast, but can occur in other areas of the state. Least Sandpipers do not nest in Texas. **Timing of occurrence:** Fall migrants occur from early July to mid-November. Spring migrants are recorded from mid-March to mid-May, but a few linger as late as early June. **Taxonomy:** Monotypic.

WHITE-RUMPED SANDPIPER
Calidris fuscicollis (Vieillot)

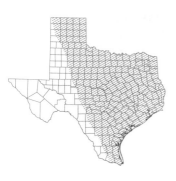

Uncommon to common spring migrant in the eastern half of the state, becoming increasingly uncommon farther west and rare to very rare in the Trans-Pecos. White-rumped Sandpipers migrate southward along the east coast of North America and, as a result, are casual to accidental fall migrants, with few documented occurrences anywhere in the state. They are also casual in summer on the coast, where individuals have been known to linger to the end of July. These summering birds may be injured or first-year birds and are not in breeding condition. **Timing of occurrence:** This species is a late-spring migrant, with most passing through the state from late April through early June and some stragglers recorded into very late June. The timing of true fall migrants is obscured by summering birds, but the few fall records are from late July to early October. **Taxonomy:** Monotypic.

BUFF-BREASTED SANDPIPER
Calidris subruficollis (Vieillot)

Rare to uncommon migrant in the eastern half of the state and casual west to the Panhandle south to the Concho Valley. Buff-breasted Sandpipers are a grassland species and often found well away from water, much like the Upland Sandpiper. Market hunting in the late 1800s severely reduced the population of Buff-breasted Sandpipers, which has still not fully recovered. Occasionally, huge concentrations are found at sod farms in the Coastal Plain, such as 1,150 at Calallen, Nueces County, on 3 May 2009. **Timing of occurrence:** Spring migrants pass through the state from early April to mid-May, rarely lingering into early June. Fall migrants can be found from late July to early October, and a few linger into late October. **Taxonomy:** Monotypic.

PECTORAL SANDPIPER *Calidris melanotos* (Vieillot)

Locally common migrant through the eastern two-thirds of the state. Pectoral Sandpipers are rare to uncommon in the western third of the state in the spring. They are more readily found in fall migration, when they are uncommon

to common throughout the state. These sandpipers are rare to very rare during winter on the Coastal Prairies, with most individuals found along the immediate coast. There is one documented record of a wintering bird in north-central Texas: a photographed bird present from 29 January to 4 February 1983 in Tarrant County. **Timing of occurrence:** Fall migrants pass through the state from mid-July to mid-October, with late birds often found into November. Spring migrants pass through from early March to early June. **Taxonomy:** Monotypic.

SEMIPALMATED SANDPIPER
Calidris pusilla (Linnaeus)

Uncommon to locally common migrant east of the Trans-Pecos. Semipalmated Sandpipers are very uncommon to rare migrants in the Trans-Pecos. The only documented winter record from Texas was an obviously sick bird (large growth on underparts) that lingered until 2 December in Bexar County, otherwise unrecorded between early December and late February. **Timing of occurrence:** Spring migrants are found from early March to early June. Migration peaks between early April and mid-May. Fall migrants are found from early July through mid-October, but some linger to early November. **Taxonomy:** Monotypic.

WESTERN SANDPIPER *Calidris mauri* (Cabanis)

Common to abundant migrant along the coast, where it is also an uncommon to common winter resident. Migrating Western Sandpipers are uncommon to locally common inland throughout the state. They are casual to rare as a winter resident in the northern half of the state, as well as the Pineywoods and Trans-Pecos. They are uncommon inland in the remainder of the southern half of the state during the winter. This species is an uncommon to rare summer visitor along the coast, becoming very rare inland. **Timing of occurrence:** Western Sandpipers have been recorded in the state in every month. Fall migrants arrive as soon as early July and can be found through early November. Spring migrants pass through the state from mid-March to late May and occasionally straggle into June. **Taxonomy:** Monotypic.

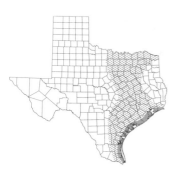

SHORT-BILLED DOWITCHER
Limnodromus griseus (Gmelin)

Uncommon to rare migrant throughout the eastern half of the state. This species is a casual to rare migrant through the western half of the state and very rarely seen in the Pineywoods. Short-billed Dowitchers are apparently encountered more often in the fall than in the spring, possibly due to the more easily distinguished juveniles. This species shows a strong preference for salt water while in Texas; thus, it is a locally common winter resident along the immediate coast while being accidental father inland. In general, the status of this species is poorly understood both in terms of seasonal movements and abundance because of the paucity of inland sightings and identification issues involving the very similar Long-billed Dowitcher (*L. scolopaceus*). **Timing of occurrence**: Fall migrants are found from early July to mid-October, and spring migrants are found from late March to mid-May. **Taxonomy**: The occurrence of the subspecies of Short-billed Dowitcher in Texas is very poorly understood. There are conflicting accounts concerning the relative abundance of those that have been reported, and identification of these taxa is fraught with difficulties. Two subspecies have been reported in the state, but a third from southern Alaska and western Canada, *L. g. caurinus* Pitelka, has the potential to occur as a vagrant, especially in the far west.

L. g. hendersoni Rowan
> As is given above. This subspecies nests across central Canada and should be the predominant taxon to occur in Texas.

L. g. griseus (Gmelin)
> Status uncertain. This subspecies has been reported as being a regular migrant and winter resident in the state, but its breeding range in eastern Canada would suggest that if it does occur in the state, it should be much less common than *L. g. hendersoni*.

LONG-BILLED DOWITCHER
Limnodromus scolopaceus (Say)

Uncommon to common migrant throughout the state. Long-billed Dowitchers are common winter residents along the coast, as well as locally common inland in the southeastern portion of the state. This species becomes

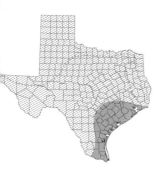

increasingly less common northward during the winter and is casual in the Panhandle during that season. **Timing of occurrence**: Long-billed Dowitchers migrate through Texas between early March and late May and from early July to mid-November. During the spring, a few individuals routinely linger into early June; as a result, they can be found during every month of the year. **Taxonomy**: Monotypic.

WILSON'S SNIPE *Gallinago delicata* (Ord)

Common migrant throughout the state. This species is a common winter resident in the eastern two-thirds of the state, particularly in the southern half. In the western third of the state, Wilson's Snipe is uncommon to rare during the winter. There are a few records from the Panhandle in mid-summer, but the species is not known to nest in the state. **Timing of occurrence**: Fall migrants have been detected in the state as early as late July and early August, but the primary migration period is from early September through early November. Spring migrants are found from mid-March through mid-May, and individuals occasionally linger as late as early June. **Taxonomy**: Monotypic. This species was formerly considered conspecific with *Gallinago gallinago* (Linnaeus) of Eurasia (Banks et al. 2002), which retained the common name of Common Snipe.

AMERICAN WOODCOCK *Scolopax minor* Gmelin

Rare to locally common winter resident in the eastern half of the state, becoming rare to very rare farther west and south as far as Hidalgo County. This species is a rare, but regular, breeder in the eastern third of the state north of the Coastal Prairies; nesting activities generally occur between January and April. There are nesting records as far west as the Edwards Plateau (Kostecke, Sperry, and Cimprich 2006). It is probable that breeding and wintering in Texas are more widespread and common than is currently known. American Woodcocks are secretive and nocturnal and are usually found in dense woodlands during the day but frequently feed in more open areas at night. Their secretive habits make determining the actual status at any season difficult, and larger numbers may be present than records indicate. Dry winters in the Pineywoods appear to push birds farther south to the Coastal Prairies. **Timing of**

occurrence: Migration periods are hard to define because of the habits of this species, but American Woodcocks have been recorded during every month of the year. Fall migrants have been detected in early October, but the bulk of the population arrives in mid-November. Spring migrants have been found from mid-February to mid-April. There are several reports of birds found during the summer months, which are difficult to categorize. One present at Estero Llano Grande sp, Hidalgo County, from 18 July to 28 August 2012 is particularly noteworthy. **Taxonomy:** Monotypic.

WILSON'S PHALAROPE *Phalaropus tricolor* (Vieillot)

Common to abundant migrant across the state. Wilson's Phalaropes are rare to casual winter visitors to the state. Winter records are primarily from the lower coast and Lower Rio Grande Valley, but there are records from as far north as the Panhandle and as far west as El Paso County. The only consistent wintering area has been at La Sal del Rey nwr, Hidalgo County, where an amazing concentration of more than 200 was present during the winter of 2002–3. Wilson's Phalarope is a casual summer resident in the Panhandle and is more common during wet years. The first nesting was discovered in Carson County in 1980 (Seyffert 1985b), and there have been five additional breeding records from the Panhandle since that time (Seyffert 2001b). **Timing of occurrence:** Fall migrants are found from late June to late October, with individuals rarely lingering to mid-November. Spring migrants are found from late March to late May. **Taxonomy:** Monotypic.

RED-NECKED PHALAROPE
Phalaropus lobatus (Linnaeus)

Uncommon to rare fall migrant throughout the western third of the state. Red-necked Phalaropes are rare to very rare throughout most of the remainder of the state except the Pineywoods, where they are casual. This species is more numerous in fall and is considered very rare to casual in spring. Red-necked Phalaropes have occasionally been reported from coastal and offshore waters during the late summer and fall. Four Red-necked Phalaropes were at La Sal del Rey nwr, Hidalgo County, during the winter of 2002–3, providing the first record for that season. **Timing** of occurrence: Fall migrants are found from mid-August through mid-October. Individuals have been found as early as mid-July and as late as mid-November. Spring records generally fall between mid-April and late May. **Taxonomy:** Monotypic.

TBRC Review Species   **RED PHALAROPE** *Phalaropus fulicarius* (Linnaeus)

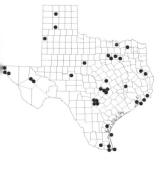

Very rare fall and accidental spring migrant. There are 42 documented records for Texas. Of these, 33 are from the fall, and there is a single winter record of a bird found in Austin, Travis County, in January 1975 (TPRF 167). There are 37 reports of Red Phalaropes from prior to the development of the Review List in 1988 for which there is no documentation on file. **Timing of occurrence:** The fall dates of occurrence are between 15 July and 17 November. Spring records are between 20 April and 4 June. **Taxonomy:** Monotypic.

Family Stercorariidae: Jaegers and Skuas

TBRC Review Species   **SOUTH POLAR SKUA**
Stercorarius maccormicki Saunders

Accidental. There is one record of South Polar Skua for Texas and the Gulf of Mexico. One was studied at length on a deepwater pelagic trip out of South Padre Island in the fall of 2004. Although this species breeds in the Antarctic, these skuas regularly occur off both coasts of North America. They are present off the North Carolina coast from May through September. **Taxonomy:** Monotypic.

1 OCT. 2004, OFF SOUTH PADRE ISLAND, CAMERON CO. (TBRC 2004–81; TPRF 2224)

POMARINE JAEGER
Stercorarius pomarinus (Temminck)

Uncommon migrant and winter resident at sea in the Gulf and a rare migrant and winter resident along the immediate coast. Pomarine Jaegers are very rare to locally rare summer visitors along the coast and in the Gulf. This species is a very rare to casual migrant inland; there are more reports during the fall than spring. Pomarine Jaegers are occasionally reported from inland locations after hur-

ricanes or strong tropical storms. During fall migration, large numbers have been reported from petroleum platforms in the Gulf of Mexico. These jaegers are occasionally seen on beaches and, more frequently, following fishing and shrimp boats. **Timing of occurrence**: This species has been found in Texas during every month of the year. Pomarine Jaegers are found in Texas waters primarily between late September and late April, but a few linger or straggle into early May. They are more frequently encountered inland during the fall, when most records are between mid-October and late December. **Taxonomy**: Monotypic.

PARASITIC JAEGER *Stercorarius parasiticus* (Linnaeus)

Uncommon migrant and rare winter resident in offshore waters and a rare migrant and very rare winter visitor on the immediate coast. Parasitic Jaegers are casual to accidental inland during migration periods, where they are much less common than Pomarine Jaegers. The southern United States, including the Gulf of Mexico, is the northernmost portion of this species' wintering range. A January record from Lake Tawakoni, Rains County, was unexpected. Parasitic Jaegers are very rare to casual summer visitors and are much less likely to be encountered in summer than are Pomarines. This species is also occasionally reported inland after hurricanes or strong tropical storms. **Timing of occurrence**: Fall migrants have been noted from early October through late November. Spring migrants have been found between early March and mid-May. **Taxonomy**: Monotypic.

TBRC Review Species

LONG-TAILED JAEGER *Stercorarius longicaudus* Vieillot

Casual migrant and accidental in winter, with 23 documented records. Long-tailed Jaegers have been most frequently documented as fall migrants. The majority of fall migrants found in Texas are first-year birds, although intermediate-aged birds have been recorded and adults have been found as spring migrants. Note that while this species in Texas is much rarer than Parasitic Jaeger offshore and on the coast, it is the more likely species inland. **Timing of occurrence**: Records of fall migrants are from 16 August through 6 November, which coincides with the primary fall migration period of this species along the east coast of North America. Long-tailed Jaegers are typically late-spring migrants, and the four such records for the state reflect that pattern, falling between 6 and 17 June. There is one winter record from the Texas City Dike, Galveston County, on 1 December 2012. **Taxonomy:** The subspecies that occurs in Texas is *S.l. pallescens* Løppenthin.

Family Laridae: Gulls, Terns, and Allies

TBRC Review Species

BLACK-LEGGED KITTIWAKE
Rissa tridactyla (Linnaeus)

Very rare winter visitor and casual migrant along the coast and at scattered inland localities over the eastern third of the state, becoming casual to accidental farther west. The vast majority of records refer to first-winter birds. This species was removed from the Review List in 1999 but placed back on the list in 2005. There are 89 accepted records, of which 17 were documented since the species was placed back on the Review List (see map). Overall, there are 12 documented records from the western third of the state, ranging from El Paso to Randall and Reagan Counties. **Timing of occurrence:** Fall migrants reach the state in late October and early November. There is one sight record with details from Galveston County on 10 September 1995. The majority of records for the state are from December to February. Apparent spring migrants are found during March and April, with a late date of 28 April. **Taxonomy:** The taxonomy of Black-legged Kittiwake appears uncertain. It is most often considered monotypic, but two subspecies have been described. Those found in Texas are presumed to belong to the Atlantic population, *R. t. tridactyla* (Linnaeus), but it is possible that some of the westernmost records pertain to the Pacific population, *R. t. pollicaris* Ridgway.

SABINE'S GULL *Xema sabini* (Sabine)

Rare to casual fall migrant, with records from all regions of the state. Sabine's Gull was removed from the Review List in 1999. At that time, there were almost 60 documented records for the state, only five involving adult birds. This species has continued to be found annually in the state,

with the number of individuals fluctuating greatly from year to year. The fall of 2007 was a particularly good year, with 15 separate Sabine's Gulls reported. **Timing of occurrence**: Fall migrants passing through Texas have been found from late August through early November. There are four records of late migrants from late November and December. Unexpected records include two documented from the spring (6 May 1995 and 22 May 2010), one from the summer (13–15 July 1996), and one from midwinter (7–15 January 2011). **Taxonomy**: Monotypic.

BONAPARTE'S GULL
Chroicocephalus philadelphia (Ord)

Uncommon to common migrant and winter resident along the coast and inland on large reservoirs. This species is most common in the eastern two-thirds of the state and is a rare to uncommon migrant and winter visitor in the western third. Bonaparte's Gulls are casual to accidental in summer along the immediate coast. **Timing of occurrence**: Fall migrants typically start to arrive in late October to early November, and most depart by late March, although some linger as late as May. **Taxonomy**: Monotypic.

TBRC Review Species   ## BLACK-HEADED GULL
Chroicocephalus ridibundus (Linnaeus)

Very rare winter visitor. There are 27 documented records for Texas, almost all from the northeastern quarter of the state. Two individuals have returned to the same location for several years. One at Cooper Lake, Delta County, was present for five consecutive winters. The most amazing record was of a bird banded as a juvenile in Finland in 1996 and discovered at the Village Creek Drying Beds, Tarrant County, in January 1998, where until March it shared the site with a second, adult bird. This banded individual returned to Village Creek for three consecutive winters. **Timing of occurrence**: Documented records of this species are between 26 October and 25 March. **Taxonomy**: Monotypic.

LITTLE GULL *Hydrocoloeus minutus* (Pallas)

Rare winter resident to north-central Texas and very rare in the remainder of the eastern half of the state. There are

only two records from the western half of the state: one near Lubbock, Lubbock County, on 9 April 1983 (TBRC 1984-9); and one at Imperial Reservoir, Pecos County, on 10 November 2000 (TBRC 2000-130; TPRF 1904). Little Gull was removed from the Review List in 2012, by which time there were 73 documented records. Interestingly, 30 of the documented sightings are from Dallas and Tarrant Counties alone. Unlike some other rare gulls, adults have frequently been found in Texas. The presence of this species in the state is strongly associated with large flocks of migrating or wintering Bonaparte's Gulls. **Timing of occurrence**: Based on documented records, the winter occurrence in the state has been from 11 November to 31 March, and there are six additional records from the spring, the latest occurring on 29 April. **Taxonomy**: Monotypic.

LAUGHING GULL *Leucophaeus atricilla* (Linnaeus)

Abundant resident along the coast. A small breeding colony can be found inland at Falcon Reservoir, Zapata County. This species has nested at Lake Amistad, Val Verde County, but the occurrence of birds during the summer at this location appears to be sporadic. Otherwise, this species is an uncommon to casual visitor inland throughout the state as far north as the southern Panhandle. Most of these inland records are from the late summer and fall, reflecting postbreeding dispersal, but there are records from all seasons. Laughing Gulls have become uncommon late-summer and fall visitors to many of the large reservoirs in the eastern half of the state. They are less common inland during winter and spring but are becoming more frequent. Banding of this species in the Galveston area colonies has shown that many disperse south to the Pacific coast of Central America as far south as Panama, with one remarkable record of a Texas-banded Laughing Gull found in the Marquesas, French Polynesia (Eubanks, Behrstock, and Weeks 2006). It is likely that Laughing Gulls from north and east of Texas migrate in to replace some of the population that moved south into Mexico. **Taxonomy**: The subspecies that occurs in Texas is *L. a. megalopterus* (Bruch).

FRANKLIN'S GULL *Leucophaeus pipixcan* (Wagler)

Uncommon to common migrant in all areas of Texas. Franklin's Gulls are rare winter visitors along the coast and to inland reservoirs and landfills, particularly in the eastern two-thirds of the state. Migrating Franklin's Gulls are well known for moving in very large flocks, occasionally including 1,000 or more individuals. They are casual visitors to the Panhandle during the summer and accidental in other parts of the state at that time of year. **Timing of occurrence**: Fall migrants are found from mid-September to early December. Spring migrants pass through the state from late February to mid-May, but a few linger into early June. **Taxonomy**: Monotypic.

TBRC Review Species   **BLACK-TAILED GULL** *Larus crassirostris* Vieillot

Accidental. There are two documented records of this Asian gull for Texas. A near-adult-plumaged bird was discovered at the Brownsville landfill, Cameron County, in the spring of 1999 (D'Anna et al. 1999), and a second winter bird was found in Corpus Christi, Nueces County, in early spring 2004. In the last 15–20 years there have been numerous records of Black-tailed Gulls from both coasts of North America, including one as far south as Belize (Lethaby and Bangma 1998). **Taxonomy**: Monotypic.

11–13 FEB. AND 5–16 MAR. 1999, BROWNSVILLE, CAMERON CO.
  (TBRC 1999-10; TPRF 1745)
6 MAR. 2004, CORPUS CHRISTI, NUECES CO.
  (TBRC 2004–23; TPRF 2189)

TBRC Review Species   **HEERMANN'S GULL** *Larus heermanni* Cassin

Accidental. There are three records of the species in Texas, all of them first-winter birds. The first was at Big Lake, Reagan County, in early December 1975 (Maxwell 1977). Heermann's Gulls breed along the west coast of Mexico and locally in southern California and are known for their postbreeding wandering. They have been found as far east as Michigan, Ohio, and Florida and as far south as Guatemala. **Taxonomy**: Monotypic.

2–4 DEC. 1975, BIG LAKE, REAGAN CO. (TPRF 97)
8 FEB. 1983, MUSTANG ISLAND, NUECES CO.
  (TBRC 1988-137; TPRF 648)
24 SEPT. 2012, TORNILLO RESERVOIR, EL PASO CO.
  (TBRC 2012–61, TPRF 2973)

MEW GULL *Larus canus* Linnaeus

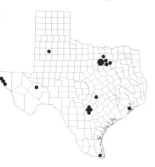

Very rare winter visitor to the state. There are 37 documented records in the state, with those sightings heavily weighted toward the El Paso area and north-central Texas. Since 1998 there has been at least one documented occurrence of Mew Gull in Texas every year except 2010. This recent trend of regular, although rare, occurrence in the state does not point to a larger number of birds in the state as much as it does the greater scrutiny of gulls by observers. Interestingly, there is only one coastal record for this species in Texas, despite an increase in gull watching along the immediate coast and at nearby landfills. **Timing of occurrence**: Records of Mew Gulls fall between 1 November and 31 March. **Taxonomy**: The Mew Gulls documented in the state appear to pertain to the North American subspecies, *L. c. brachyrhynchus* Richardson.

RING-BILLED GULL *Larus delawarensis* Ord

Common migrant throughout the state and a common to abundant winter resident along the coast and inland at larger cities, reservoirs, and landfills. Ring-billed Gulls are also uncommon to rare summer visitors throughout their winter range but are not known to nest in the state. The majority of summering individuals are immature. Ring-billed Gulls are the most widespread and common of the white-headed gulls found in Texas in winter. **Timing of occurrence**: Fall migrants reach Texas as early as late July, with the peak of migration from early October to early November. Spring migrants are seen from late March to early May, with a few straggling through late May. **Taxonomy**: Monotypic.

WESTERN GULL *Larus occidentalis* Audubon

Accidental. There are three records of the species in Texas: two adults and one first-winter bird. Western Gulls are found along the west coast of North America from southwestern British Columbia to Baja California, and at the northern end of their breeding range there is a dense hybrid zone with Glaucous-winged Gull. Dispersal inland is extremely rare, although there are records for Arizona and New Mexico. **Taxonomy**: There are two subspecies of this gull, and the Texas records cannot be definitively assigned to either.

14 MAY 1986, FORT BLISS, EL PASO CO. (TBRC 1987-3; TPRF 514)
6 APR. 1995, BOCA CHICA, CAMERON CO.
  (TBRC 1995-51; TPRF 1495)
14–16 NOV. 2004, LA SAL DEL REY NWR, HIDALGO CO.
  (TBRC 2004–93; TPRF 2243)

CALIFORNIA GULL *Larus californicus* Lawrence

Rare winter resident to El Paso and Hudspeth Counties. California Gulls are very rare to rare migrant and winter visitors to the upper and central coasts. This species is a casual visitor to the lower coast and at scattered inland locations east through the Oaks and Prairies region. California Gulls have become annual visitors to the western Trans-Pecos, and the number of sightings in the northwestern and north-central portions of the state has increased steadily over the past decade. During this period, this species has been found with increasing frequency at other seasons in the El Paso area, where these gulls are now casual to very rare except during winter. There were more than 60 accepted records for the state when this species was removed from the Review List in November 1999. **Timing of occurrence:** Fall migrants arrive in the state from late October through November, and spring migrants are found from late February through late April. There are scattered records from later in the spring and into the summer. **Taxonomy:** There are two subspecies of this gull, but to date, the Texas records have not been definitively assigned to either form. The dark mantle color of many of the adults found in Texas suggests the subspecies found in the Great Basin, *L. c. californicus* Lawrence.

HERRING GULL *Larus argentatus* Pontoppidan

Common migrant and winter resident along the coast. This species is generally a rare migrant and winter resident inland but is uncommon and local on the many reservoirs found in the northeastern quarter of the state. Herring Gulls are uncommon to rare summer visitors along the coast and very rare on inland reservoirs during this season. The vast majority of summering individuals are in first- or second-year plumage, which is often very worn and bleached so that aging and identification are difficult to determine. There are two breeding records for the state: a nest with eggs was discovered on a spoil island in the Laguna Madre, Cameron County, on 9 June 1989 (Farmer 1990) and again at the same location in 1990. **Timing of occurrence:** Fall migrants arrive as early as late September, but the bulk of the population does not arrive until late November. Birds wintering in Texas begin departing in early February, and many are gone by mid-March. However, they appear to be replaced into early April by birds that wintered in Mexico and/or birds pushed south by late-winter weather to the north of the state. Lingering or late-migrating subadult birds have been noted as late as early May. **Taxonomy:** Two subspecies have been documented to occur in the state.

L. a. smithsonianus Coues
> As described above.

L. a. vegae Palmén
> Accidental. There are four documented occurrences, all adults, of this Asian subspecies from the upper and central coasts. Some authorities regard these forms as separate species: American Herring Gull (*L. smithsonianus*) and Vega Gull (*L. vegae*). The TBRC has requested documentation for all reports of this taxon.

TBRC Review Species

YELLOW-LEGGED GULL *Larus michahellis* Naumann

Accidental. There are two documented records of this European gull for Texas. Both of these birds were found at the same location during the winter and spring of 2004, and both were in first-winter plumage. Yellow-legged Gulls are very rare to accidental winter visitors to the east coast of North America from Newfoundland (where annual in very low numbers) south to the Mid-Atlantic states. The differentiation of Yellow-legged Gull from Lesser Black-backed Gull (*L. fuscus* Linnaeus) is fraught with difficulties, compounded by the increasing number of suspected hybrids of that taxon with Herring Gull; extreme caution should be employed when attempting to identify this species. **Taxonomy:** The Yellow-legged Gulls documented in Canada and the United States are generally believed to belong to the subspecies *L. m. atlantis* Dwight of the Azores. One of the individuals found in Texas exhibited characteristics that suggested it belonged to *L. m. michahellis* Naumann of Europe, northwest Africa, and the Canary and Madeiran Islands.

24 JAN.–4 APR. 2004, CORPUS CHRISTI, NUECES CO.
(TBRC 2004–15; TPRF 2673)
4 MAR. 2004, CORPUS CHRISTI, NUECES CO.
(TBRC 2004–25; TPRF 2508)

THAYER'S GULL *Larus thayeri* Brooks

Rare to very rare migrant and winter resident, mostly in the eastern half of the state. There were 62 state records of Thayer's Gull when it was removed from the Review List in 2004. The majority of those records were from the coast, but there were also many inland records. Thayer's Gull was considered accidental in Texas prior to 1990, but since then this species has been found annually in the state. The taxonomic status of this species continues to be debated, and its relationship with Iceland Gull remains to be determined. **Timing of occurrence:** Based on the 62 records up until 2004, Thayer's Gulls have been found in Texas primarily between late October and late April. **Taxonomy:** Monotypic.

TBRC Review Species

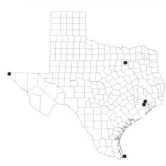

ICELAND GULL *Larus glaucoides* Meyer

Accidental. The five documented records for Texas have all been individuals in first-winter plumage. Plumage variation in Iceland and Thayer's Gulls plus potential intergrades makes documenting this species exceedingly difficult. Some studies have suggested that Thayer's and Iceland Gulls should be considered as one species (Snell 2002), while others have suggested that the North American population of Iceland Gull may represent a species separate from the nominate Greenland population (AOU 1998). **Taxonomy:** The subspecies that has been found in Texas is *L. g. kumlieni* Brewster.

15 JAN.–12 FEB. 1977, SOUTH PADRE ISLAND, CAMERON CO.
(TBRC 1989-245; TPRF 935)
4 FEB. 2001, DALLAS, DALLAS CO. (TBRC 2001–47)
20 DEC. 2006–4 APR. 2007, HOUSTON, HARRIS CO.
(TBRC 2006–06; TPRF 2466)
25 DEC. 2006–10 FEB. 2007, EL PASO, EL PASO CO.
(TBRC 2006–117; TPRF 2460)
6–15 MAR. 2008, HOUSTON, HARRIS CO.
(TBRC 2008–22; TPRF 2544)

LESSER BLACK-BACKED GULL

Larus fuscus Linnaeus

Uncommon to locally common winter resident on the upper coast. This species is rare to uncommon on the central coast south to Cameron County. Inland, Lesser Black-backed Gulls are generally very rare migrants and winter residents but have been found with increasing frequency in the northeastern quarter of the state as well as in the El Paso area. There are inland records from all regions of the state except the Edwards Plateau. They began to occur with greater frequency from about 1990, and that trend is continuing. A significant count of 75 birds was made in Brazoria County on 3 February 2012. Two interesting records involve individuals that wintered at the same locations for 12 years. This species was first documented summering in the state on the upper coast in 2004 and has slowly increased in occurrence during that season. **Timing of occurrence**: Lesser Black-backed Gulls have been found in Texas as early as mid-August, but the majority of the wintering population arrives between mid-September and mid-October. They depart during March and early April, with a very few lingering as late as mid-May. **Taxonomy**: The majority of Lesser Black-backed Gulls found in North America are believed to be the subspecies *L. f. graellsii* Brehm. However, there are many examples of birds found in Texas that do not fit that taxon well. They may represent other subspecies or be from the large intergrade population from the Netherlands between *L. f. graellsii* and *L. f. intermedius* Schiřler.

TBRC Review Species **SLATY-BACKED GULL** *Larus schistisagus* Stejneger

Accidental. The six documented records of this Asian gull for Texas cover almost all age classes from first-winter to adult. The first was a subadult at Brownsville, Cameron County, and this unexpected record for Texas was at the beginning of a trend of Slaty-backed Gull occurrences across North America. Virtually all accepted North American records are of birds old enough to show adult coloration on the upperparts. The identification of first-winter birds remains challenging and controversial, such that individuals of this age may be overlooked or left unidentified, leaving the true status of first-winter birds a mystery. **Taxonomy**: Monotypic.

GLAUCOUS-WINGED GULL
Larus glaucescens Naumann

Accidental. There is one record of an adult for Texas. Glaucous-winged Gulls nest along the perimeter of the northern Pacific Ocean and in winter are present along the west coast of North America south to Baja California. They are rarely found inland and are casual, at best, in the western two-thirds of the United States. Glaucous-winged Gulls hybridize extensively with Western Gulls in Washington and Herring Gulls in southern and central Alaska, producing a high percentage of hybrids in the overall population (Hoffman, Wiens, and Scott 1978). An individual presumed to be a hybrid with Western Gull was documented at Corpus Christi, Nueces County, from 14 February to 8 March 2004. **Taxonomy:** Monotypic.

6–8 JAN. 2004, FORT WORTH, TARRANT CO.
  (TBRC 2004–5; TPRF 2180)

GLAUCOUS GULL *Larus hyperboreus* Gunnerus

Rare winter resident along the coast and very rare to casual migrant and winter resident inland, with records from all regions of the state. The majority of the Glaucous Gulls reported in Texas have been either first- or second-year birds. This species was removed from the Review List in 1997, when there were more than 70 documented records. **Timing of occurrence:** Glaucous Gulls arrive in late November, although there is an exceptionally early record from San Luis Pass, Galveston County, on 28 September 1995 (TBRC 1995-154; TPRF 1410). Glaucous Gulls sometimes remain in the state through early April, and a few individuals linger as late as mid-May. There is also one record of a second-cycle individual that was present at Galveston, Galveston County, from 5 June to 19 August 2009. **Taxonomy:** Specimens of Glaucous Gulls from Texas have been identified as the subspecies *L. h. barrovianus* Ridgway. It is possible, if not likely, that some of the birds found in the state belong to *L. h. leuceretes* Schleep.

GREAT BLACK-BACKED GULL
Larus marinus Linnaeus

Very rare winter visitor along the coast. Just over half of the 49 documented records for the state come from the

upper coast. The majority of Texas records pertain to first-winter birds, but there are numerous records of older birds, including adults. There are only two documented inland records: a third-winter bird near Longview, Harrison County, on 30 December 1992 and presumably the same individual rediscovered at Lake O' the Pines, Marion County, in early January 1993, where it remained until March; and one at Lake Meredith, Hutchinson and Moore Counties, from 26 December 2005 through 22 January 2006. **Timing of occurrence**: Records of Great Black-backed Gull in Texas extend from late September to early May, with the majority falling between early November and early March. **Taxonomy**: Monotypic.

TBRC Review Species

KELP GULL *Larus dominicanus* Lichtenstein

Accidental. There are five records of this primarily Southern Hemisphere gull. The first involved an adult found at Galveston, Galveston County, from January to April 1996, which was completing a Southern Hemisphere molt schedule. Presumably, the same individual returned the following winter. Kelp Gulls were present in the Mississippi Delta of Louisiana from 1989 to 2000 and hybridized with Herring Gulls (Dittmann and Cardiff 2005). This species has also been documented in Colorado, Florida, Indiana, Maryland, and Ontario. There are additional records from the Yucatán Peninsula (Howell, Carrea S., and Garcia B. 1993) and northern Tamaulipas, Mexico (Gee and Edwards 2000). All but one of the Texas (and North American) records pertain to adults or near adults; the identification of younger birds is difficult and controversial, such that individuals of this age may be overlooked or left unidentified. **Taxonomy**: The subspecies that has been found in Texas is *L. d. dominicanus* Lichtenstein.

15 JAN.–5 APR. 1996, GALVESTON, GALVESTON CO.
 (TBRC 1996-17; TPRF 1393)
4 MAY 1996, NORTH PADRE ISLAND, KLEBERG CO.
 (TBRC 1996-88; TPRF 1627)
30 NOV. 1996–21 APR. 1997, GALVESTON, GALVESTON CO.
 (TBRC 1996-180; TPRF 1523)
8 NOV.–24 DEC. 2008, QUINTANA, BRAZORIA CO.
 (TBRC 2008–94; TPRF 2681)
19 DEC. 2008, QUINTANA, BRAZORIA CO.
 (TBRC 2009–01; TPRF 2697)

TBRC Review Species

BROWN NODDY *Anous stolidus* (Linnaeus)

Casual to very rare summer visitor to offshore waters and the coast. Brown Noddy appears to be occurring with greater frequency in Texas waters over the past decade. There are 20 documented records, 13 of which are since June 2005. The closest breeding colonies of Brown Noddy are on islands off the Yucatán Peninsula, Mexico, and the Dry Tortugas, Florida. **Timing of occurrence:** The Texas records occur between 27 April and 18 September, with the majority occurring in June and July. **Taxonomy:** The subspecies that occurs in Texas is *A. s. stolidus* (Linnaeus).

TBRC Review Species

BLACK NODDY *Anous minutus* Boie

Accidental, with three documented records. All are from the spring and summer, as would be expected since this species winters at more tropical latitudes. A single individual was seen at opposite ends of Bolivar Peninsula, Galveston County, during the spring of 1998, but photos confirmed that only one individual was involved. Separating Black from Brown Noddy requires detailed study of both plumage and bill structure. **Taxonomy:** The subspecies that occurs in Texas is *A. m. americanus* (Mathews).

22 JUNE 1975, NORTH PADRE ISLAND, NUECES CO. (TPRF 77)
15 APR. AND 1 MAY 1998, GALVESTON CO.
  (TBRC 1998-63; TPRF 1692)
27 JULY 1998, SAN JOSE ISLAND, ARANSAS CO.
  (TBRC 1998-100; TPRF 1693)

SOOTY TERN *Onychoprion fuscatus* (Linnaeus)

Rare and local summer resident along the central and lower coasts and uncommon to rare visitor to offshore waters in late summer and fall. Sooty Terns nest in small numbers on islands in the Laguna Madre and previously nested in Galveston County (Meitzen 1963). This species has occasionally been found along the coast loafing in tern colonies or on the beach. Larger numbers are routinely found along the immediate coast following hurricanes and tropical storms. These disturbances out of the Gulf of Mexico have pushed Sooty Terns far inland, with records from as far west as Brewster and Jeff Davis Counties and as far north as Cooper Lake, Delta County. **Timing of occurrence:** The breeding population arrives in Texas in mid-April and is present until early September. The offshore occurrence appears to be from mid-May through September. A very early date for the state is 5 March 2013, and an equally unexpected late occurrence was on 14 November 2005. Both records are from South Padre Island, Cameron County. **Taxonomy**: The subspecies that occurs in Texas is *O. f. fuscatus* (Linnaeus).

BRIDLED TERN *Onychoprion anaethetus* (Scopoli)

Uncommon to rare summer and fall visitor in offshore waters. Prior to 1988 there were no records of this pelagic tern for the state. The first Bridled Tern records were associated with the passage of Tropical Storm Frances. Since that time, deepwater pelagic trips have revealed that this species is more common and regular than previously thought. The highest number of individuals observed on a single trip was more than 200 off Port O'Connor, Calhoun County, on 28 June 1997. Despite their regular occurrence along the Continental Shelf in the Gulf of Mexico, Bridled Terns are still very rare to casual visitors along the coast. There are two reports of Bridled Terns being pushed inland with a hurricane: two at Cooper Lake, Delta County, on 3 September 2008; and one at Calaveras Lake, Bexar County, on 7–8 September 2010. **Timing of occurrence**: Bridled Terns are present offshore from mid-May through early November, although most are recorded from June to September. **Taxonomy**: The subspecies that occurs in Texas is *O. a. melanopterus* (Swainson).

LEAST TERN *Sternula antillarum* Lesson

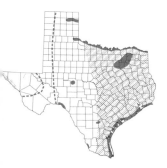

Common summer resident along the coast and rare to locally uncommon summer resident at scattered inland locations (see map). Least Terns are uncommon to rare migrants in the eastern two-thirds of the state and become increasingly rare westward except along the Rio Grande near El Paso, where they are rare. They are also very rare to casual winter residents along the coast. **Timing of occurrence**: Spring migrants are found in the state from late March through mid-May, with a few straggling into early June. Fall migrants are seen from mid-July to late September, and some linger or straggle into October. **Taxonomy**: Uncertain. The two breeding populations, one coastal and the other inland, have been considered separate subspecies. There is mounting evidence, though, that only one taxon is involved (Thompson et al. 1992; Draheim et al. 2010). Currently, the inland population is listed as an Endangered Species, and the US Fish and Wildlife Service recognizes any nesting birds at least 50 miles or greater from the coastline as being Interior Least Terns.

S. a. athalassos (Burleigh & Lowery)

Known as the Interior Least Tern and known to breed along the Red River to Hall County, along the Canadian River to Roberts County, locally in north-central Texas, and at reservoirs around San Angelo, Tom Green County; Lake Amistad, Val Verde County; and Falcon Reservoir, Zapata County. Least Tern may also nest at Choke Canyon Reservoir, Live Oak/McMullen Counties, but the subspecies involved has not been determined. There are isolated breeding records from reservoirs at various other locations across the state, perhaps most frequently in northeast Texas.

S. a. antillarum Lesson

Common summer resident and very rare to casual winter visitor along the immediate coast.

GULL-BILLED TERN *Gelochelidon nilotica* (Gmelin)

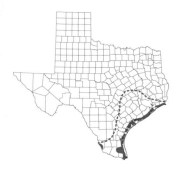

Uncommon to common resident along the coast, although generally much less numerous in the winter. Gull-billed Terns are casual visitors away from the coast, most often associated with tropical storms. There are numerous reports from the South Texas Brush Country as well as the Oaks and Prairies region, with records as far inland as Granger Lake, Williamson County, and Lake Waco, McLennan County. Oberholser (1974) lists sight records from Dallas and Denton Counties, but Pulich (1988) notes that these lack details. Interestingly, there are small resident populations at Falcon Reservoir, Zapata County, and Lake Casa Blanca, Webb County. **Taxonomy:** The subspecies that occurs in Texas is *G. n. aranea* (Wilson).

CASPIAN TERN *Hydroprogne caspia* (Pallas)

Common resident along the coast. Caspian Terns are also rare to uncommon spring and rare fall migrants and winter visitors to the eastern half of the state away from the coast and casual migrants through the western half. These migrant birds nest in central Canada and winter along the Gulf Coast, including Mexico. The few midsummer inland records may either be postbreeding wanderers from the coast or very early migrants. **Timing of occurrence:** Migrants are found from late April to mid-June and from as early as mid-July to late October. **Taxonomy:** Monotypic.

BLACK TERN *Chlidonias niger* (Linnaeus)

Uncommon to common migrant in all parts of the state, including far offshore, and at times abundant along the coast. This species is locally uncommon during the summer on the coast and rare to very rare at inland locations at this time. However, these summering birds do not breed even though small numbers are in alternate plumage. Black Terns are very rare to locally rare winter visitors along the coast and casual inland along the Coastal Prairies at this time. **Timing of occurrence:** Spring migrants begin to appear in mid-March and are present through late May. Fall migrants are found from early July through late October, with a strong peak in migrants during late August. **Taxonomy:** The subspecies that occurs in Texas is *C. n. surinamensis* (Gmelin).

TBRC Review Species

ROSEATE TERN *Sterna dougallii* Montagu

Accidental. Texas has two documented records, both adults. Oberholser (1974) reported three specimens from Texas, but only the one taken by Frank Armstrong in the spring of 1901, which is housed at the Royal Ontario Museum, has been located. The closest breeding colonies to Texas are found in the Florida Keys. Separating Roseate Terns from Forster's Tern (*S. forsteri*) and Common Tern (*S. hirundo*) is an identification challenge that has made documenting this species difficult. **Taxonomy:** The subspecies that has been found in Texas is *S. d. dougallii* Montagu.

10 APR. 1901, CORPUS CHRISTI, NUECES CO.
  (TBRC 1996-111; TPRF 1535)
25 JUNE 1995, COOPER LAKE, DELTA CO. (TBRC 1995-92)

COMMON TERN *Sterna hirundo* Linnaeus

Common fall and uncommon spring migrant along the coast and offshore. This species is a rare migrant through the eastern half of the state and a casual migrant through the western half. Fall migrants generally outnumber those found in spring, both on the coast and inland. Common

Terns are casual winter visitors on the coast. Small numbers of basic-plumaged birds are found in summer, primarily along the upper and central coasts. There are numerous reports of breeding colonies along the coast from the 1880s to the 1930s. Dresser (1865) reported the species as an abundant nester in Galveston Bay in 1864, while Pemberton (1922) reported thousands of individuals in Cameron County in 1921. The veracity of these reports has often been questioned, in part because of the large numbers of nesting pairs reported and the dearth of nesting records since that time. It may be that Forster's Terns were mistaken for Common Terns in these reports. **Timing of occurrence:** Spring migrants pass through the state from early March to early June, with the bulk of migration during April. Fall migrants are found from mid-August to early November, with the peak of migration in late September and early October. There are a small number of records of Common Terns straggling to very early December, of which an unknown percentage attempt to winter locally. **Taxonomy:** The subspecies that occurs in Texas is *S. h. hirundo* Linnaeus.

TBRC Review Species

ARCTIC TERN *Sterna paradisaea* Pontoppidan

Casual. Texas has nine documented records of this species: six spring and three fall migrants. Arctic Tern is likely to occur as a very rare migrant, particularly during the late spring, in the Gulf of Mexico (Lee and Cardiff 1993). Surprisingly, five of the Texas records are from inland lakes. Separating Arctic from Common Terns can be difficult; most sight records and one photograph taken prior to 1996 have been deemed insufficient to document the species. **Timing of occurrence:** Spring records are from 10 May to 11 June, with three from June. The fall records range from 26 July to 18 October. **Taxonomy:** Monotypic.

FORSTER'S TERN *Sterna forsteri* Nuttall

Common resident along the coast and common to uncommon migrant in nearly all parts of the state. Forster's Terns are locally common winter residents on inland lakes and reservoirs east of the Pecos River except on the High Plains. They are casual winter visitors in the Trans-Pecos.

They have been found nesting in small numbers at Choke Canyon Reservoir, Live Oak/McMullen Counties, and at Lake Amistad, Val Verde County. Forster's Tern is the common species of tern found wintering on inland reservoirs statewide. Along the coast, they are found in much higher numbers during winter than in summer. **Timing of occurrence:** Migrants are found away from the coast from early March through mid-May and from late July through October. **Taxonomy:** Monotypic.

ROYAL TERN *Thalasseus maximus* (Boddaert)

Common resident usually confined to coastal habitats. This species is casual to accidental inland in the eastern third of the state as far north as Lake Tawakoni, Rains/Van Zandt Counties, mostly during the summer and fall. Many of the inland reports are associated with the passage of hurricanes or tropical storms. This was particularly true with the passage of Hurricane Ike in fall 2008, when Royal Terns were found in at least eight counties with as many as 12 on Lake Livingston, Polk County. One at Fort Hancock Reservoir, Hudspeth County, on 13 September 2009 defies explanation and is the only record for the western half of the state. **Taxonomy:** The subspecies that occurs in Texas is *T. m. maximus* (Boddaert).

SANDWICH TERN *Thalasseus sandvicensis* (Latham)

Common summer resident and rare to uncommon winter resident along the coast. Inland records of Sandwich Terns in Texas are exceptional, and all are of birds found after hurricanes or tropical storms. The earliest record was of two found near Austin, Travis County, on 12 September 1961. Additional inland records include two at Falcon Reservoir, Zapata County, on 22 September 1967; one in Burleson County on 11 October 1982 (*TCWC 11,274); and one at Lake Alcoa, Milam County, on 12 September 1998. Hurricane Ike deposited one at Cooper Lake, Delta County, on 13 September 2008 and one at Lake Livingston, Polk and San Jacinto Counties, on 15 September. There are no records from the western half of the state. **Taxonomy:** The subspecies that occurs in Texas is *T. s. acuflavidus* (Cabot); however, there have been occasional individuals with more than typical amounts of yellow in the bill that have led to speculation that those birds belong to the subspecies *T. s. eurygnathus* (Saunders) of the southernmost Caribbean and eastern South America. Some authorities consider these two American subspecies together to be a separate species, Cabot's Tern (*T. acuflavidus*).

TBRC Review Species

ELEGANT TERN *Thalasseus elegans* (Gambel)

Accidental. There are three documented records of this species for Texas, involving two adult-types and one probable first-winter. The first was collected by Frank Armstrong in 1889 at Corpus Christi, Nueces County. In the Northern Hemisphere, Elegant Terns are found along the west coast of North America, with most of the breeding population restricted to the Gulf of California, Mexico. **Taxonomy**: Monotypic.

25 JULY 1889, CORPUS CHRISTI, NUECES CO.
  (*BMNH 91-10-20-92; TPRF 1460)
23 DEC. 1985, BALMORHEA LAKE, REEVES CO. (TPRF 397)
4–18 NOV. 2001, GALVESTON, GALVESTON CO.
  (TBRC 2001-134; TPRF 1988)

BLACK SKIMMER *Rynchops niger* Linnaeus

Locally common resident along the coast. Black Skimmers are very rare to accidental inland, occurring with greater regularity within 100 miles of the coast. This species has been found as far inland as the Dallas–Fort Worth area and Cooper Lake, Delta County, to the north and as far west as Midland County and Balmorhea Lake, Reeves County. Dispersal far away from the coast is largely limited to summer and early fall and is often associated with the passage of a hurricane or strong tropical storm, although spring records do exist. **Taxonomy**: The subspecies that occurs in Texas is *R. n. niger* Linnaeus.

ORDER COLUMBIFORMES

Family Columbidae: Pigeons and Doves

ROCK PIGEON *Columba livia* Gmelin

Common to abundant resident throughout the state, primarily in urban areas. Rock Pigeons were introduced from

Europe to Nova Scotia in the early 1600s and quickly spread to all corners of the continent. In Texas, this feral species has become an established resident around human environs and is uncommon to rare in more remote rural areas. Natural populations of Rock Pigeons nest on cliffs, which explains their presence in areas with tall buildings or highway overpasses that mimic their natural habitat. Formerly known as Rock Dove. **Taxonomy:** The population in Europe belongs to *C.l. livia* Gmelin, which was the source of the original introduction into North America.

RED-BILLED PIGEON *Patagioenas flavirostris* (Wagler)

Locally uncommon to rare summer resident in the western Lower Rio Grande Valley. Red-billed Pigeons are rare and very local winter residents within the breeding range. This tropical pigeon is found primarily in close association with the Rio Grande. The center of abundance for this species in Texas is Starr and Zapata Counties, as these pigeons have declined significantly in Hidalgo County. Red-billed Pigeons have been found upriver to northern Webb and southern Maverick Counties, where they are rare to locally uncommon. There are several reports away from the known range, particularly along the lower coast north to Victoria County. One of these is documented, a single bird at Corpus Christi, Nueces County, from 3 to 5 March 1988 (TPRF 615). There is also a documented record in the Hill Country from near Lost Maples State Natural Area, Real County, during 10–12 April 2012. **Timing of occurrence:** The breeding population is present from early March through late September. **Taxonomy:** The subspecies that occurs in Texas is *P. f. flavirostris* (Wagler).

BAND-TAILED PIGEON *Patagioenas fasciata* (Say)

Uncommon to rare summer and rare winter resident at the higher elevations of the Davis and Chisos Mountains. This species is rare during summer in the Guadalupe Mountains and has not been reported in the winter in recent decades. Band-tailed Pigeons appear to be declining in the Davis and Guadalupe Mountains, and the status in the latter range requires closer examination. These pigeons are very rare visitors, primarily in the fall, to the lowlands of the Trans-Pecos and casual to accidental visitors east of the Pecos River. There are 10 records from the Panhandle and South

Plains and four from the western Edwards Plateau. Most of these are from the fall and early winter, although there is a single spring and two summer records. This species is also an accidental vagrant to the Coastal Prairies, with records from Aransas, Chambers, and Galveston Counties. **Timing of occurrence**: The breeding population in the Guadalupe Mountains is present from early April through mid-October. More observations in the winter are needed to determine the status during that season. **Taxonomy**: The subspecies that occurs in Texas is *P. f. fasciata* (Say).

EURASIAN COLLARED-DOVE
Streptopelia decaocto (Frivaldszky)

Common to locally abundant in urban areas throughout the state. The source of the North American population is reported to be from a release of birds from Europe in the Bahamas in the 1970s, which then spread to Florida (P. Smith 1987). This species was first reported in Texas in March 1995 and by 1996 had reached the northwestern portion of the state in Randall County. Eurasian Collared-Doves became established breeders on the upper coast by 1999. Since then populations have become established in all areas of the state. Eurasian Collared-Doves are present in every county in the state and seem to be particularly common in smaller towns. They have the potential to be found at any location, including farmsteads and other rural locations. (See African Collared-Dove account in appendix C for the status of this similar species that was formerly known as Ringed Turtle-Dove.) **Taxonomy**: The population in Europe belongs to *S. d. decaocto* (Frivaldszky), which was the source of the original introduction.

WHITE-WINGED DOVE *Zenaida asiatica* (Linnaeus)

Common to locally abundant summer resident throughout the southern half of the state. White-winged Doves are uncommon to locally common in the northern half of the state. The population of White-winged Doves in Texas continues to expand and is now found throughout the western two-thirds of the state. They are slowly colonizing urban areas in the Pineywoods and are established in some areas of northeast Texas. The highest densities continue to be found in large urban centers, particularly in San Antonio and Austin. White-wingeds generally depart from rural areas in the northern portions of their Texas range in winter. However, they are beginning to winter in increasing numbers in urban areas as far north as the northern Panhandle. Studies have shown that migrants from the majority of the state are thought to retreat to wintering areas from southern Mexico to northwestern Costa Rica, while birds in the western Trans-Pecos migrate to western Mexico. However, the rapid growth of the breeding population and the increasing numbers of wintering birds suggest that there has been a northward expansion of the wintering areas. **Timing of occurrence:** Spring migrants are seen from late March to early May. The fall migration period is from mid-September through early November. **Taxonomy:** Two subspecies occur in Texas. With the rapid expansion of White-winged Dove populations in the state, there is a large area of intergrades between these taxa in the western half of the state (Pruett et al. 2000).

Z. a. mearnsi (Ridgway)

Common resident in the western Trans-Pecos.

Z. a. asiatica (Linnaeus)

Common to locally abundant resident in the eastern Trans-Pecos and eastward through the remainder of the state.

MOURNING DOVE *Zenaida macroura* (Linnaeus)

Common to abundant summer and winter resident throughout the state. Large numbers of Mourning Doves from outside the state replace most of the summer resident populations from the northern half of the state during the winter. This species is the most abundant game bird in North America (Tomlinson et al. 1994); however, there is anecdotal evidence from southern Texas that the population in some areas has shrunk because of growing competition with colonizing White-winged Doves. This appears to be the case in urban habitats in many areas of the state. **Timing of occurrence:** Mourning Doves have a protracted spring migration that begins in February and extends through late May. The fall migration period is from late August through early November. **Taxonomy:** Two subspecies occur in Texas.

Z. m. marginella (Woodhouse)

Common to abundant resident in the western two-thirds of the state, east through the Oaks and Prairies region.

Z. m. carolinensis (Linnaeus)

   Common resident in the eastern third of the state.

PASSENGER PIGEON *Ectopistes migratorius* (Linnaeus)

Extinct. The Passenger Pigeon was once a regular migrant and winter resident in the northeastern quarter of Texas. Occasionally, large numbers would invade the state, and birds could be found as far west as the Rolling Plains and western Edwards Plateau. Although the Passenger Pigeon was primarily a winter resident, there are several reports of nesting in the state. The only known specimen from Texas is a single egg collected near Colmesneil, Tyler County, on 3 May 1877. Large numbers of Passenger Pigeons were reported from Texas until the early 1880s, when the population declined abruptly (Casto 2001). Oberholser (1974) considered a report from upper Galveston Bay from March 1900 as the last reliable report of its presence in the state. Estimated at 3–5 billion individuals, the Passenger Pigeon is thought to have once been the most common bird in North America (Schorger 1955). **Timing of occurrence:** Migrants are reported to have arrived in September and generally departed by early March. There are reports of birds still present well into May. **Taxonomy:** Monotypic.

INCA DOVE *Columbina inca* (Lesson)

Common resident in the southern two-thirds of the state, becoming uncommon to rare and local farther north, including up to the Red River and in the Pineywoods. Inca Doves underwent a range expansion to the north and east through the 1990s, but the populations in northwestern Texas are now rapidly declining. This equals a similar range reduction in Oklahoma that has occurred over the past eight years. Small breeding populations persist as far north as Dalhart, Dallam County, and in other smaller towns in the Panhandle. Inca Doves are much more frequently encountered in cities and towns than in rural areas. **Taxonomy:** Monotypic.

COMMON GROUND-DOVE

Columbina passerina (Linnaeus)

Uncommon to locally common resident in the South Texas Brush Country, Coastal Prairies, and southern Trans-Pecos

along the Rio Grande but uncommon to rare away from the river. Common Ground-Doves are uncommon and local summer residents along the eastern and southern edges of the Edwards Plateau west to Presidio County. This species is locally uncommon to rare along the upper coast east to Galveston Bay. These doves have declined significantly in the southeastern United States, including in the eastern part of the Texas range. Common Ground-Doves occur casually north of the breeding range at all seasons, although fall records are more numerous. These records are from virtually all areas of the state, including the northernmost reaches. **Taxonomy:** Two subspecies occur in Texas.

C. p. *passerina* (Linnaeus)
Uncommon to rare resident from Galveston Bay west to Travis County and south to Nueces County.

C. p. *pallescens* (Baird)
Uncommon to locally common resident in the South Texas Brush Country and lower Coastal Prairies, and from the eastern edge of the Hill Country west through the southern Trans-Pecos.

TBRC Review Species ## RUDDY GROUND-DOVE
Columbina talpacoti (Temminck)

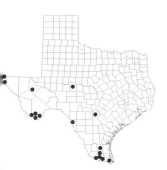

Casual visitor to the Lower Rio Grande Valley, the Big Bend region, and El Paso County. Texas has 22 documented records of Ruddy Ground-Dove. Single records from Bexar, Brooks, and Maverick Counties represent the only ones for the South Texas Brush Country north of the Lower Rio Grande Valley. A male that was near San Angelo, Tom Green County, from 11 December 2008 to 29 March 2009 is an exceptional record. **Timing of occurrence:** All records fall between 5 October and 5 May, although the bulk of records from South Texas are in the fall and winter and those from the Trans-Pecos are scattered from winter into spring. **Taxonomy:** Two subspecies have occurred in Texas.

C. t. *eluta* (Bangs)
The individuals documented in the Trans-Pecos appear to belong to this pallid western subspecies; however, there are no specimens from West Texas to confirm this identification.

C. t. *rufipennis* (Bonaparte)
Records from South Texas pertain to this subspecies.

WHITE-TIPPED DOVE *Leptotila verreauxi* Bonaparte

Common resident in the Lower Rio Grande Valley. White-tipped Doves are uncommon and local north through the South Texas Brush Country to Maverick, Zavala, and Atascosa Counties. This species is slowly expanding its range northward. In the west, White-tipped Doves were first encountered in La Salle and Dimmit Counties in the late 1980s and are now casual visitors as far north as Calhoun, Guadalupe, Uvalde, and Val Verde Counties. The occurrence of White-tipped Doves has increased on the Edwards Plateau as well, with multiple records from along the Devils River in Val Verde County and along the Frio and West Nueces Rivers in Uvalde County. Farther out of range, there have been two summer reports from Big Bend NP, Brewster County. **Taxonomy:** The subspecies that occurs in Texas is *L. v. angelica* Bangs & Penard.

TBRC Review Species   **RUDDY QUAIL-DOVE** *Geotrygon montana* (Linnaeus)

Accidental. Texas has one record of this tropical dove, a juvenile. There are six documented records for the United States, and the Texas record is the only one outside Florida. Ruddy Quail-Doves are widespread in the Neotropics and Caribbean and an uncommon to rare resident in northeastern Mexico as far north as southern Tamaulipas. **Taxonomy:** The subspecies that has occurred in Texas is *G. m. montana* (Linnaeus).

2–6 MAR. 1996, BENTSEN–RIO GRANDE VALLEY SP, HIDALGO CO. (TBRC 1996-28; TPRF 1468)

ORDER CUCULIFORMES

Family Cuculidae: Cuckoos and Allies

TBRC Review Species   **DARK-BILLED CUCKOO**
Coccyzus melacoryphus Vieillot

Accidental. There is a single documented record of this South American species for Texas. An adult cuckoo was delivered to a wildlife rehabilitator in Hidalgo County on 10 February 1986. It is not known exactly where the bird

was found, but it is thought to have come from Hidalgo County. The bird later died, and the specimen was salvaged with a tentative identification of Black-billed Cuckoo (*C. erythropthalmus*). The identification as a Dark-billed Cuckoo occurred when the specimen was prepared and placed in the Museum of Natural Science at Louisiana State University. Dark-billed Cuckoos are long-distance austral migrants that have also been documented as vagrants north to Clipperton Island (off the western coast of Mexico), Grenada, and Panama. The ABA Checklist Committee accepted the record but placed it in the Origin Uncertain category, not on the main list (Robbins et al. 2003). **Taxonomy:** Monotypic.

10 FEB. 1986, HIDALGO CO. (TBRC 1995-115; TPRF 1600)

YELLOW-BILLED CUCKOO

Coccyzus americanus (Linnaeus)

Uncommon migrant and summer resident throughout the eastern two-thirds of the state. Yellow-billed Cuckoos are locally uncommon to rare migrants and summer residents in the Trans-Pecos and are accidental winter visitors along the coast. This species has declined significantly and is of conservation concern, particularly the populations in the western portion of the range, which are now protected by the Endangered Species Act. **Timing of occurrence:** Spring migrants in the eastern half of the state are seen from early April to late May. The western population arrives later, from mid-May to mid-June. Fall migrants are found from mid-August to mid-October, although a few individuals straggle into late November. **Taxonomy:** The Yellow-billed Cuckoo is most often considered monotypic, following Banks (1988). There have been two subspecies described, however, and both occur in Texas: *C. a. americanus* (Linnaeus) in the eastern two-thirds and *C. a. occidentalis* Ridgway in the far west.

TBRC Review Species   **MANGROVE CUCKOO** *Coccyzus minor* (Gmelin)

Casual visitor to the Coastal Prairies and accidental farther west in the Lower Rio Grande Valley. There are 13 documented records in the state. The records from the Lower Rio Grande Valley are from Cameron and Hidalgo Counties. Coastal records come from Brazoria, Cameron,

Galveston, Harris, Matagorda, and Nueces Counties. The Mangrove Cuckoo was first documented in the state in Galveston, Galveston County, on 30 December 1964. **Timing of occurrence**: Most records are scattered from April to August, but two are from December and one from January. **Taxonomy**: Mangrove Cuckoo is now considered monotypic (Banks and Hole 1991), but all of the Texas records pertain to buff-throated individuals typical of the population in eastern Mexico that were formerly known under the name *C. m. continentalis* Van Rossem.

BLACK-BILLED CUCKOO
Coccyzus erythropthalmus (Wilson)

Generally a rare spring and very rare to rare fall migrant in the eastern half of the state. Black-billed Cuckoos are most frequently encountered in woodlots and other stopover habitats along the Coastal Prairies, where they can be locally uncommon in spring. Black-billed Cuckoos are very rare to accidental migrants through the western half of the state. There are four reported nesting records for the state: three from Wise County in 1888 and one from Live Oak County in 1977 (Fischer 1979). The validity of these reports has been questioned (Pulich 1988), and there is no unequivocal documentation to prove nesting in the state. **Timing of occurrence**: Spring migrants are seen from mid-April to late May, with a few found as late as very early June. Fall migrants are seen from mid-August to late October. There is a very late record from the Sabal Palm Sanctuary, Cameron County, from 20 November 2004. **Taxonomy**: Monotypic.

GREATER ROADRUNNER
Geococcyx californianus (Lesson)

Common to locally uncommon resident throughout the state, except the Pineywoods and northeastern corner of the state, where this species is rare to locally uncommon. Greater Roadrunners breed from early March to early October. They are also highly territorial, and males defend the same area for years. In the last century, this species has expanded eastward as forests have been fragmented, resulting in more open habitat. **Taxonomy**: Monotypic.

GROOVE-BILLED ANI *Crotophaga sulcirostris* Swainson

Common summer resident in the Lower Rio Grande Valley west to Webb County. Groove-billed Anis are locally uncommon to rare north through the South Texas Brush Country to the Balcones Escarpment and east to Gonzales County. They are rare visitors along the Rio Grande west to Brewster County during the spring and summer. This species has occurred in virtually all areas of the state as a late summer and fall vagrant. There are isolated nesting records from as far north as Tom Green and Lubbock Counties (Maxwell 1980). They are uncommon fall and rare winter visitors in the eastern Lower Rio Grande Valley and along the lower and central coasts, especially on barrier islands, with records as far north as Jefferson County. Though the species can be found throughout the year, there is a significant withdrawal from the state during the winter, at which time the species becomes rare and local. **Timing of occurrence**: Spring migrants begin arriving in very late April, and the bulk of the breeding population arrives in early to mid-May. The breeding population departs in October and is rare by the end of the month. **Taxonomy**: The subspecies that occurs in Texas is *C. s. sulcirostris* Swainson.

ORDER STRIGIFORMES

Family Tytonidae: Barn Owls

BARN OWL *Tyto alba* (Scopoli)

Rare to locally uncommon resident throughout most of the state. Barn Owls are very rare and local in the forested parts of the Pineywoods and above 5,000 feet in the mountains of the Trans-Pecos. They are cavity nesters and, as their name implies, have adapted to take advantage of human-made structures. They can be locally common on the Coastal Prairies, as they occupy open habitats and tend to avoid closed-canopy forests. **Taxonomy**: The subspecies that occurs in Texas is *T. a. pratincola* (Bonaparte).

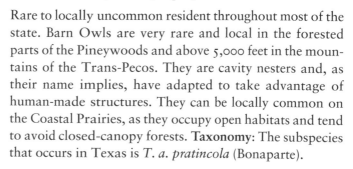

Family Strigidae: Typical Owls

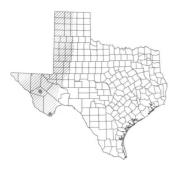

FLAMMULATED OWL *Psiloscops flammeolus* (Kaup)

Uncommon to rare summer resident in the Chisos, Davis, and Guadalupe Mountains above 6,000 feet. Flammulated Owls are migratory, although their global winter range is poorly known. Migrants have been found at various locations throughout the Trans-Pecos, and the migration route encompasses the western third of the state west to Crosby, Lubbock, and Midland Counties. Migrating Flammulated Owls have been found as far east as Hunt County in northeast Texas and the upper and central coasts. There are records from Galveston, Jefferson, and Nueces Counties, but most of the eastern records are in the fall. An unexpected record involved one photographed more than 50 miles from shore on a petroleum platform off Calhoun County on 10 October 1999 (TPRF 1812). Subsequently, two others have been found on platforms in Texas waters in the Gulf. **Timing of occurrence:** Spring migrants pass through the state from late March to early May. A migrant at Balmorhea SP, Reeves County, on 31 May 2011 provides a late date for the spring. Fall migrants have been detected from late September through mid-November, although the breeding population departs by mid-October. The latest fall record is a specimen from Lubbock, Lubbock County, on 5 December 1950, and there is one documented winter record from South Padre Island, Cameron County, from 8 January to 4 March 2013. **Taxonomy:** Monotypic.

WESTERN SCREECH-OWL
Megascops kennicottii (Elliot)

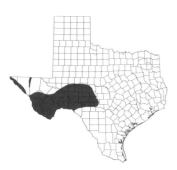

Common to uncommon resident in the southern Trans-Pecos east through the western Edwards Plateau, where Eastern Screech-Owls are uncommon. This species is rare and local in El Paso and southern Hudspeth Counties and uncommon in the Guadalupe Mountains. Western Screech-Owls are found east to western Kerr County on the Edwards Plateau and Tom Green County in the Concho Valley. There is only one confirmed record for the Texas Panhandle (Seyffert 2001b), even though this owl is resident in the western Oklahoma Panhandle and has not been reported from the South Plains (Hewetson, Kostecke, and

Best 2006). This species is found in a variety of woodland habitats in the Trans-Pecos and primarily occupies mesquite and live oak savannahs on the Edwards Plateau. **Taxonomy**: Two subspecies are known to occur in Texas, but the distribution of those taxa is poorly understood. The ranges described here are tentative.

M. k. aikeni Brewster
> Resident from El Paso County east to the Guadalupe Mountains. The lone specimen from Parmer Co. (UKNHM 678) in the Panhandle is also of this race.

M. k. suttoni (Moore)
> Resident from the southern Trans-Pecos east through the western Edwards Plateau and Concho Valley.

EASTERN SCREECH-OWL *Megascops asio* (Linnaeus)

Common resident throughout the eastern three-quarters of the state. Eastern Screech-Owls have a less continuous distribution on the High Plains, occupying riparian corridors and urban areas with numerous trees. This owl is resident, at least locally, as far west as the Pecos River drainage. Eastern Screech-Owls were uncommon residents as far west as southern Brewster County until at least the early 1970s. Since that time, scattered records have come from Brewster, El Paso, Hudspeth, and Jeff Davis Counties, but with no indication of resident populations. There are two color morphs, gray and red, although a third color morph, brown, has been suggested but appears to be a dull variant of the red morph (Gehlbach 1995). Gray morphs are more common the farther west in Texas, while red morphs are more common in the eastern portion of the state. **Taxonomy**: Four subspecies have been reported to occur in Texas, but the status of two of these taxa is uncertain.

M. a. maxwelliae (Ridgway)
> Status uncertain. Marshall (1967) reported this subspecies as being rare in the northern Panhandle. Current range descriptions, however, suggest that it occurs south only to central Kansas.

M. a. hasbroucki (Ridgway)
> Resident in the northern two-thirds of the state, south to the Balcones Escarpment.

M. a. mccallii (Cassin)

Resident through the South Texas Brush Country. The contact zone, if one exists, with *M. a. hasbroucki* is poorly understood.

M. a. floridanus (Ridgway)

Status uncertain. Reported to be the subspecies present in the pine forests of East Texas. Current range descriptions, however, suggest that it occurs only west to central Louisiana.

GREAT HORNED OWL *Bubo virginianus* (Gmelin)

Common resident throughout the state except in the Pineywoods, where this species is uncommon. Great Horned Owls are perhaps most common in the arid habitats of the Trans-Pecos, where they are one of the dominant avian predators. They are one of the more adaptable owls in the United States, which allows them to use a wide variety of habitats. In Texas, they are found in open woodlands, desert scrublands, and riparian corridors. The only habitat not routinely used is dense closed-canopy forest. **Taxonomy**: Two subspecies occur in Texas, and the distribution provided here follows Dickerman (2004a).

B. v. pallescens Stone

Common resident in the western third of the state, including the Panhandle south to Val Verde County. This race is also present along the Rio Grande south to the Lower Rio Grande Valley.

B. v. virginianus (Gmelin)

Common resident in the eastern two-thirds of the state west to Wichita, Sutton, and Edwards Counties and south to northern Kenedy and Brooks Counties. This distribution follows the extent of continuous woodland habitats in the state.

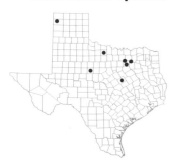

TBRC Review Species **SNOWY OWL** *Bubo scandiacus* (Linnaeus)

Casual. Texas has seven records of this Arctic species, four of which are specimens. The most recent records were part of a major movement of Snowy Owls into the United States during the winter of 2011–12, which included two birds documented in Dallas and Rockwall Counties in north-central Texas. As was the case that winter, Snowy Owls are known to move southward in large numbers when food resources become scarce in their normal wintering areas. A Snowy Owl was found dead near Waco, McLennan

38. Willets (*Tringa semipalmata*) found in Texas belong to two discrete populations that are now considered subspecies. Some authorities believe that these two taxa should be considered separate species. The eastern subspecies is the breeding population on the Texas coast, and the western birds are winter residents. This Willet is of the western subspecies and was on Mustang Island, Nueces County, on 15 September 2007. *Photograph by Mark W. Lockwood.*

39. This Eskimo Curlew (*Numenius borealis*) was one of two present on Galveston Island in March 1962. The photographs taken that day are the only ones of this species in the wild and were the last documented in Texas. The only unquestionable record since then is a bird collected in Barbados in 1963. *Photograph by Don Bleitz, © Western Foundation of Vertebrate Zoology.*

40. This Black-tailed Godwit (*Limosa limosa*) provided a first state record when it was discovered at Brazoria NWR, Brazoria County, on 6 June 2012. Although it was at times very difficult to find, it remained in the same general area through the summer and was last reported on 10 August. *Photograph by Michael L. Gray / Gulf Coast Bird Observatory.*

41. Surfbirds (*Calidris virgata*) have been documented in the state on 10 occasions, all during spring. All but one have been associated with jetties or other structures that mimic the rocky shoreline habitats this species frequents. Some of these birds were in full breeding plumage, such as this individual at the Packery Channel Jetties, Nueces County, from 22 March to 1 April 2012. *Photograph by Lee Pasquali.*

42. Ruff (*Calidris pugnax*) is a very rare migrant and winter visitor to the state. Despite multiple spring records, there has never been one discovered with the ornate ruff that is characteristic of the breeding-plumaged male. This bird was on western Galveston Island, Galveston County, on 1 September 2008. *Photograph by David McDonald.*

43. Curlew Sandpiper (*Calidris ferruginea*) is another shorebird that is a casual visitor to Texas. Most of the records have involved migrating birds that did not remain to be observed by large numbers of birders. This individual was one of the longer-staying birds. It was present at Corpus Christi, Nueces County, from 17 August to 4 September 2009. *Photograph by Christopher Taylor / Kiwifoto.com.*

44. Both of the Red-necked Stint (*Calidris ruficollis*) Texas records are of breeding-plumage birds. The second record was on Bolivar Flats, Galveston County, from 25 June to 8 July 2011, which allowed a lucky few to study it. *Photograph by Kerry Taylor.*

45. Although wintering Purple Sandpipers (*Calidris maritima*) are primarily found on rocky shorelines along the Atlantic Coast, they are casual visitors to the Gulf Coast. This individual was photographed at Port Mansfield, Willacy County, on 29 January 2012. *Photograph by Robert Epstein.*

46. The primary migration paths of the Red Phalarope (*Phalaropus fulicarius*) take them just offshore along the east and west coasts of North America to wintering grounds off South America. A few migrants are found in the interior, most often in fall. This adult was at Stillhouse Hollow Reservoir, Bell County, on 28 October 2006. *Photograph by Tony Frank.*

47. Although jaegers are seen regularly along the immediate coast as well as off-shore, there is only one record of South Polar Skua (*Stercorarius maccormicki*) for the state. This adult-type-plumaged bird (on the right) was found off South Padre Island, Cameron County, on 1 October 2004. This date coincides with the end of the normal departure time for birds found off the southern Atlantic Coast of the United States. *Photograph by Andy Garcia.*

48. Black-legged Kittiwake (*Rissa tridactyla*) was considered a Review Species from 1987 to 1999 before being removed from that list. But it was added back to the list in 2005. Since then there have been 16 additional documented records. This individual was one of two present at McNary Reservoir, Hudspeth County, from 16 November 2008 through 14 February 2009. *Photograph by Mark W. Lockwood.*

49. Little Gull (*Hydrocoloeus minutus*) is now a regular winter resident on reservoirs in north-central Texas. There are also a number of records from along the upper and central coasts, but this species is extremely rare elsewhere in the state. This first-winter bird was at San Jacinto SP, Harris County, on 20 November 2011. *Photograph by Michael Lindsey.*

50. The occurrence of Slaty-backed Gulls (*Larus schistisagus*) in North America away from Alaska has increased greatly in the past decade. They are now regularly found as far south as California and in the northeastern portion of the continent. This increase has included Texas, where there are now five records, four from the past decade. This second-year bird was at a private landfill near Houston, Harris County, on 22 February 2006. *Photograph by Martin Reid.*

51. This adult Glaucous-winged Gull (*Larus glaucescens*) represents the only record for Texas. This species is rarely documented in North America away from the west coast. It was at Fort Worth, Tarrant County, on 6–8 January 2004. *Photograph by Martin Reid.*

52. Great Black-backed Gulls (*Larus marinus*) have increased during winter in the state and particularly so on the upper and central coasts. This first-winter bird was at a private landfill near Houston, Harris County, on 22 February 2006. *Photograph by Martin Reid.*

53. Kelp Gulls (*Larus dominicanus*) are found primarily in the Southern Hemisphere. Since the early 1990s this species has occurred several times along the Gulf of Mexico, as well as in the Caribbean and elsewhere in North America. This individual was found in Quintana, Brazoria County, on 8 November 2008. *Photograph by Michael Lindsey.*

54. The Brown Noddy (*Anous stolidus*) is a bird of tropical waters and is rare in Texas. For unknown reasons, there was an influx of records from 2005 through 2009, including this individual on North Padre Island, Kenedy County, on 10 June 2009. *Photograph by Billy Sandifer.*

55. Although Arctic Tern (*Sterna paradisaea*) is only a casual migrant through the state, there are enough records to begin to define migration periods. Spring migrants have been found mid-May to mid-June. This adult was at Balmorhea Lake, Reeves County, from 1 to 11 June 2006. *Photograph by Mark W. Lockwood.*

56. Band-tailed Pigeons (*Patagioenas fasciata*) appear to be declining in the Guadalupe and Davis Mountains, and more research is needed to determine their true status. The population in the Chisos Mountains seems to be more stable. This adult was along the Window Trail in Big Bend NP, Brewster County, on 1 May 2010. *Photograph by Mark W. Lockwood.*

57. The range of Inca Doves (*Columbina inca*) has slowly expanded to include almost all of the state over the past century, but since 2005 there has been a noticeable contraction in the southern Panhandle and South Plains. The reason for this latest change is unclear but may be related to the harshness of the 2005–6 winter season. These were in Alpine, Brewster County, on 25 March 2009. *Photograph by Mark W. Lockwood.*

58. Flammulated Owl (*Psiloscops flammeolus*) is an uncommon to locally common summer resident in the upper elevations of the three highest mountain ranges in the state. This adult was in the Davis Mountains, Jeff Davis County, on 13 June 2005. *Photograph by Mark W. Lockwood.*

59. Snowy Owl (*Bubo scandiacus*) has been documented on seven occasions in Texas. This includes two found during the winter of 2011–12. One of those was photographed on an apartment balcony in downtown Dallas. This is the other individual, at Lake Ray Hubbard, Rockwall County, from 11 to 19 February 2012. *Photograph by Bruce Strange.*

60. Northern Pygmy-Owl (*Glaucidium gnoma*) has proven to be a difficult species to unequivocally document in the state. There have been many reports of calling birds in the Chisos and Guadalupe Mountains that could not be conclusively documented. As a result there are only four records for the state. This is one of two birds present in Pine Canyon in Big Bend NP from 17 August to 7 October 2007. *Photograph by Mark W. Lockwood.*

61. Spotted Owls (*Strix occidentalis*) have a very limited range in Texas, and little is known about them even within this area. The largest population, which is small, is in Guadalupe Mountains NP. This adult was found there on 2 July 2006. *Photograph by Mark W. Lockwood.*

62. The occurrence of Long-eared Owls (*Asio otus*) in Texas is somewhat enigmatic. The number of birds present and the extent of the winter range in a given year seem to always be in flux. There are a few nesting records from the western part of the state. This adult was in the Davis Mountains, Jeff Davis County, on 26 April 2009. *Photograph by Mark W. Lockwood.*

63. The status of Northern Saw-whet Owl (*Aegolius acadicus*) in Texas is poorly understood. There is some evidence that it may be a rare resident in the Guadalupe Mountains, and maybe the Davis Mountains, but there are very few records. They are most often discovered in the winter, as was the case with this adult found in Lubbock, Lubbock County, from 9 February to 14 March 2007. *Photograph by Brandon Best.*

64. Common Nighthawk (*Chordeiles minor*) is a widespread migrant and summer resident in the state. This species has also been reported rather frequently in midwinter when these birds should be in South America. There is a strong need to document these winter sightings to gain a clear understanding of the status during that season. This rufous-morph adult was near Lorenzo, Crosby County, on 9 June 2012. *Photograph by Mark W. Lockwood.*

65. Mexican Whip-poor-wills (*Antrostomus arizonae*) are known to be common summer residents in the upper elevations of the three major mountain ranges in the Trans-Pecos, but virtually nothing is known about the potential occurrence of this newly recognized species away from the nesting areas. This one was in the Chisos Mountains, Brewster County, on 18 May 2012. *Photograph by Mark W. Lockwood.*

66. Lucifer Hummingbirds (*Calothorax lucifer*) have long been considered a specialty of Big Bend NP, but we know that they also nest in the Christmas Mountains and possibly in the Chinati Mountains. This immature male was in the Christmas Mountains, Brewster County, on 30 June 2010. *Photograph by Mark W. Lockwood.*

67. The number of records for Costa's Hummingbirds (*Calypte costae*) has greatly increased since 2000. The majority of the birds found have not remained in one place long, but there have been exceptions. This male returned to the same yard in Alpine, Brewster County, for three consecutive fall seasons. This image was taken on 1 November 2008. *Photograph by Mark W. Lockwood.*

68. Intensive banding studies in the Trans-Pecos have shown that Allen's Hummingbird (*Selasphorus sasin*) is a regular, if not uncommon, fall migrant through the region. This adult male was in the Davis Mountains, Jeff Davis County, on 23 July 2005. *Photograph by Mark W. Lockwood.*

69. Broad-billed Hummingbird (*Cynanthus latirostris*) was once considered a very rare to casual visitor to Texas. Since 1992 there have been a remarkable number of occurrences, particularly in the Davis Mountains during the spring and summer. This adult male was in the Davis Mountains, Jeff Davis County, on 9 November 2008. *Photograph by Mark W. Lockwood.*

71. There has been a definite spike in the records of Violet-crowned Hummingbird (*Amazilia violiceps*) in Texas since 2006, when 12 of the 19 state records were documented. This one was in the Christmas Mountains, Brewster County, on 1 December 2011. *Photograph by Carolyn Ohl-Johnson.*

70. Berylline Hummingbird (*Amazilia beryllina*) is an accidental visitor to the state, with all but one of the records coming from the Davis Mountains. The most recent record was this adult in the Davis Mountains, Jeff Davis County, on 26 August 2007. *Photograph by Mark W. Lockwood.*

72. White-eared Hummingbird (*Hylocharis leucotis*) was considered an accidental visitor to Texas prior to the summer of 2005. Between that summer and the fall of 2010 the species appeared to be establishing itself as a summer resident. However, the severe drought of 2011 seems to have negatively impacted the population. This male was photographed in the Davis Mountains, Jeff Davis County, on 16 June 2008. *Photograph by Mark W. Lockwood.*

73. Although Elegant Trogon (*Trogon elegans*) is a common resident in the mountains of northeastern Mexico, this spectacular bird has rarely wandered into Texas. This adult male was at the Frontera Audubon Sanctuary in Weslaco, Hidalgo County, from 14 January to 12 May 2005, which enabled large numbers of birders to see it. *Photograph by Steve Bentsen.*

74. Amazon Kingfisher (*Chloroceryle amazona*) is a bird
that had been expected to eventually be found in Texas. That
prediction was fulfilled when this female was discovered in
Laredo, Webb County, on 24 January 2010. It is the only record
for the United States. *Photograph by Robert Epstein.*

75. The adult Red-headed Woodpecker (*Melanerpes
erythrocephalus*) is one of the most striking birds found in
Texas. Although these woodpeckers are often thought to be
a bird of the Pineywoods, they range westward to the eastern
Panhandle as well. This adult was in Hansford County on 15
July 2009. *Photograph by Greg W. Lasley.*

County, on 31 December 1974, but the location of the specimen is unknown, and it is not considered a proven record. **Timing of occurrence:** Although the number of records is small, most are between 28 January and 26 February. The exception is a bird present near Tye, Taylor County, from 22 March to 2 April 2002. **Taxonomy:** Monotypic.

TBRC Review Species

NORTHERN PYGMY-OWL *Glaucidium gnoma* Wagler

Accidental. The four documented records for the state are all from the Chisos Mountains. There are more than a dozen undocumented reports from the mountains of the Trans-Pecos, and most involve heard-only birds. Some of these reports may be valid, although confusion with Northern Saw-whet Owl may account for others. Some studies have suggested that the Northern Pygmy-Owl should be split into two species, *G. californicum* of the western United States and *G. gnoma* of the interior of Mexico southward (Heidrich, König, and Wink 1995). There are 11 reports of Northern Pygmy-Owls from prior to the development of the Review List in 1988 for which there is no documentation on file. **Taxonomy:** The two most recent records include audio recordings as part of the documentation, and in both cases the subspecies involved is *G. g. gnoma* Wagler.

12 AUG. 1982, BOOT SPRING, BIG BEND NP, BREWSTER CO. (TPRF 278)
25 APR. 1993, CHISOS MOUNTAINS, BIG BEND NP, BREWSTER CO. (TBRC 1993-73)
17 AUG.–7 OCT. 2007, PINE CANYON, BIG BEND NP, BREWSTER CO. (TBRC 2007–66; TPRF 2510)
28 MAR. 2008, PINNACLES TRAIL, BIG BEND NP, BREWSTER CO. (TBRC 2008–24; TBSL 242)

FERRUGINOUS PYGMY-OWL
Glaucidium brasilianum (Gmelin)

Uncommon and local resident on the Coastal Sand Plain. This species was a rare resident in the Lower Rio Grande Valley in riparian woodlands prior to the 2010 floods that severely impacted those habitats. It is uncertain whether any of that population still exists. There may be a small population in Zapata County that would not have been impacted by the 2010 flooding. There is a large population in the open live oak woodlands of Kenedy County and eastern Brooks County and a small number in mesquite and oak savannahs elsewhere in Brooks County. A sight record from Rio Grande Village, Brewster County, in Big Bend NP on 9 April 1999 (TBRC 1999-87) is the only one for the Trans-Pecos. **Taxonomy:** The subspecies that occurs in Texas is *G. b. ridgwayi* Sharpe. This taxon is sometimes considered part of *G. b. cactorum* van Rossem found in northwestern Mexico. The lone Trans-Pecos report could be of this western subspecies.

ELF OWL *Micrathene whitneyi* (Cooper)

Uncommon to locally common summer resident from the southern Trans-Pecos eastward onto the western Edwards Plateau and in the Lower Rio Grande Valley. This species was found on the southwestern Edwards Plateau in the late 1980s. In 2002 nesting was documented in Irion County, expanding the known range farther north. Elf Owls are also rare summer residents in the Guadalupe Mountains and possibly the Sierra Diablo, Culberson/Hudspeth Counties. This owl is a rare winter resident in the Lower Rio Grande Valley, and there is a single early-winter record from the Edwards Plateau. There is an apparent break in its distribution along the Rio Grande from southern Kinney County to northern Webb County. **Timing of occurrence:** Spring migrants have been recorded from early March through early April, when the breeding population appears to arrive in large numbers. Fall migrants occur from mid-September through mid-October. **Taxonomy:** Two subspecies occur in Texas.

M. w. whitneyi (Cooper)
   Uncommon to locally common summer resident in the Trans-Pecos and eastward through the western Edwards Plateau.

M. w. idonea (Ridgway)
   Uncommon summer and rare winter resident in the Lower Rio Grande Valley. Dickerman and Johnson (2013) suggest that this taxon occurs northward to include the eastern Trans-Pecos.

BURROWING OWL *Athene cunicularia* (Molina)

Uncommon to common summer resident and uncommon to rare winter resident in the western half of the state, east to Hardeman County. There are exceptional successful

nesting records from well away from known nesting areas, including from Kenedy County in 2006. Burrowing Owl is a rare to locally uncommon migrant and winter visitor in the Oaks and Prairies region south to Williamson County and rare to very rare migrant and winter resident farther south to the Lower Rio Grande Valley and Coastal Prairies. This species was formerly much more common and widespread in Texas, including areas away from current breeding sites. Conversion of prairie habitats to agriculture and the decline in populations of black-tailed prairie dogs (*Cynomys ludovicianus*) have caused a substantial decline in the number of resident Burrowing Owls. Simmons (1925) reported that Burrowing Owl was an uncommon and local breeding species as far southeast as Travis County. **Timing of occurrence:** Fall migrants have been recorded from mid-September through mid-October, and spring migrants from late March through late April, with a few individuals lingering into early May. **Taxonomy:** The subspecies that occurs in Texas is *A. c. hypugaea* (Bonaparte).

TBRC Review Species

MOTTLED OWL *Ciccaba virgata* (Cassin)

Accidental. There are two records for the United States, both from Texas. The first involved a road-killed individual that was photographed, but the specimen was not salvaged (Lasley, Sexton, and Hillsman 1988). The 2006 bird at Frontera Audubon Sanctuary was seen by more than one observer, but no photographs were obtained of this reportedly wary individual. Mottled Owls are found throughout the Neotropics and occur in northern Nuevo León, Mexico, within 75 miles of the Rio Grande. **Taxonomy:** The subspecies found in northeastern Mexico, *C. v. tamaulipensis* Phillips, is the most likely to occur in Texas.

23 SEPT. 1983, BENTSEN–RIO GRANDE VALLEY SP, HIDALGO CO. (TBRC 1988–18; TPRF 377)
5–11 JULY 2006, WESLACO, HIDALGO CO. (TBRC 2006–94)

SPOTTED OWL *Strix occidentalis* (Xántus de Vesey)

Rare and local resident in wooded canyons in the Guadalupe and Davis Mountains of the Trans-Pecos. This species has been known from the Guadalupe Mountains since

the early 1900s, but only a small number of pairs have been found. They have been reported from only three locations in the Davis Mountains (Bryan and Karges 2001) but may be more widespread than these few records indicate. Spotted Owls presumably wandering from populations in southern New Mexico have also been documented during the fall and winter in El Paso County. **Taxonomy:** The subspecies that occurs in Texas is *S. o. lucida* (Nelson).

BARRED OWL *Strix varia* Barton

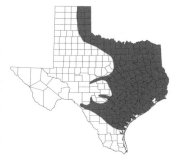

Uncommon to common resident in the eastern two-thirds of the state. Barred Owls are found as far west as the eastern Panhandle and eastern Val Verde County and as far south as Nueces County. Barred Owls occupy forests throughout the eastern third of the state but are restricted to riparian corridors on the Edwards Plateau, the eastern Panhandle, and the Coastal Prairies. Two Barred Owls are reported to have been collected in Cameron County in May 1900. These specimens represent the only documented records from south of the Nueces River, although there have been reports from Kleberg County. Thornton (1951) heard a Barred Owl in the Pecos River drainage in July 1949, which is assumed to be a nonbreeding bird that had wandered into that drainage. **Taxonomy:** Three subspecies occur in Texas.

S. v. varia Barton
> Reported as a rare winter visitor to northeast Texas, with specimens as far west as Cooke County.

S. v. georgica Latham
> Resident in eastern and Central Texas west to the eastern Panhandle and the Balcones Escarpment. This is probably the subspecies found in the Colorado River drainage, including the San Saba River.

S. v. helveola (Bangs)
> Resident in the southern Edwards Plateau and northern South Texas Brush Country.

LONG-EARED OWL *Asio otus* (Linnaeus)

Uncommon migrant and winter resident in the Panhandle and rare to locally uncommon migrant and winter resident on the South Plains and westward through the Trans-

Pecos. This species is rare and unpredictable during winter in the remainder of the state, with the exception of the Pineywoods, where it is typically absent or at least undetected. A few have been found as far south as the Lower Rio Grande Valley. These birds are well known for forming communal winter roosts, which can sometimes contain as many as 40 individuals. They formerly nested sporadically in Texas, but such records have been scarce in recent years. Two successful nests discovered in the Davis Mountains, Jeff Davis County, in 2013 were the first in the state since a pair in El Paso County in 2007. **Timing of occurrence:** Fall migrants have been detected from late September to early December, and spring migrants have been found from early March through mid-April, with a few remaining into early May. Nesting birds have remained in the state into early July. **Taxonomy:** Two subspecies are known to occur in Texas.

A. o. tuftsi Godfrey
Rare migrant and winter resident in the western Trans-Pecos, east to at least Jeff Davis and Brewster Counties. There are specimens of this race from as far east as Bell County.

A. o. wilsonianus (Lesson)
Uncommon to rare winter resident east of the Pecos River.

TBRC Review Species   **STYGIAN OWL** *Asio stygius* (Wagler)

Accidental. Two records of this tropical owl exist for the United States, both from Texas. The first was photographed at Bentsen–Rio Grande Valley SP, Hidalgo County, in 1994 (Cooksey 1998). Another Stygian Owl was discovered at the same location in 1996 (Wright and Wright 1997). The first record did not come to light until after the discovery of the second bird. Stygian Owls are rare and local residents through much of the Neotropics and parts of the Caribbean. In 2005 they were discovered to be rare residents as far north as near Alta Cima in southern Tamaulipas, Mexico. **Taxonomy:** The subspecies found in Texas and Tamaulipas is not known. The described subspecies closest to Texas, *A. s. lambi* Moore, occurs in northwestern Mexico.

9 DEC. 1994, BENTSEN–RIO GRANDE VALLEY SP, HIDALGO CO.
  (TBRC 1998-46; TPRF 1705)
26 DEC. 1996, BENTSEN–RIO GRANDE VALLEY SP, HIDALGO CO.
  (TBRC 1997-4; TPRF 1527)

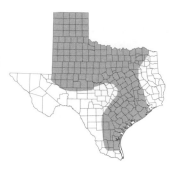

SHORT-EARED OWL *Asio flammeus* (Pontoppidan)

Rare to locally uncommon migrant and winter resident across the northern third of the state, the Oaks and Prairies region, and the Coastal Prairies. Short-eared Owls are casual to rare in the Edwards Plateau and west through the Trans-Pecos. This species is probably most commonly encountered along the prairies of the upper and central coasts and the prairies of the northeastern part of the state. In the Pineywoods, Short-eared Owls are very rare in open pastures. They are most common in areas where sufficient grassland habitat provides roosting sites and hunting areas. Their abundance at any given location varies greatly from year to year because it is largely dependent on grassland habitat conditions and the availability of prey. **Timing of occurrence:** Fall migrants have been detected from mid-October to early December, with a few records from early September. Spring migrants are found from early March through mid-April, and some birds linger into late May. There are scattered summer occurrences, mostly from the Coastal Prairies, which allows for the possibility that early-fall and late-spring records could pertain to oversummering individuals. **Taxonomy:** The subspecies found in Texas is *A. f. flammeus* (Pontoppidan).

TBRC Review Species

NORTHERN SAW-WHET OWL
Aegolius acadicus (Gmelin)

Possibly a very rare resident in the Guadalupe (Newman 1974) and Davis Mountains (Bryan and Karges 2001). A single summer record exists from the Chisos Mountains from 1997 (TBRC 1997-86). There are 30 documented records for the state. Northern Saw-whet Owls are considered casual migrants and winter visitors elsewhere in the state. Except for single birds in Denton, Liberty, and Somervell Counties, all other documented records are from the Trans-Pecos and High Plains. **Timing of occurrence:** Excluding the summer records from the Trans-Pecos, Northern Saw-whet Owls have been primarily found between 4 September and 6 May. **Taxonomy:** The subspecies found in Texas is *A. a. acadicus* (Gmelin).

ORDER CAPRIMULGIFORMES

Family Caprimulgidae: Nighthawks and Nightjars

LESSER NIGHTHAWK
Chordeiles acutipennis (Hermann)

Common to uncommon migrant and summer resident from El Paso County across much of the Trans-Pecos and east through the South Texas Brush Country. Lesser Nighthawks are generally absent in habitats found above 4,300 feet in the Trans-Pecos. They are rare migrants and summer visitors to the southern Edwards Plateau and range north on the Coastal Prairies to Calhoun County. This species is a very rare to rare migrant to the upper coast, and vagrants have been found as far north as Bailey and Lubbock Counties. Lesser Nighthawks are rare to locally uncommon winter residents in the western half of the South Texas Brush Country. **Timing of occurrence**: Spring migrants have been found as early as late March, but the bulk of migrants are seen from mid-April to mid-May. Fall migrants are found from early August to late October, but a few linger to mid-November. **Taxonomy**: The subspecies found in Texas is *C. a. texensis* Lawrence.

COMMON NIGHTHAWK *Chordeiles minor* (Forster)

Uncommon to common migrant and summer resident throughout the state, except in the Pineywoods, where it is a common migrant and rare summer resident. This species routinely lingers into the early winter, particularly along the coast and in large urban areas in the southern half of the state. There are numerous reports from the winter, when Common Nighthawks should be in South America. Documentation of sightings between 20 December and 15 March are needed to resolve the status of this species in winter. **Timing of occurrence**: Spring migrants arrive in the eastern two-thirds of the state from early April to mid-May. In the western third of the state they arrive later, with the primary migration period from late April to late May. Fall migrants are found from late August through late October, but small numbers straggle into mid-November and, exceptionally, into mid-December. **Taxonomy**: Seven subspecies of Common Nighthawk are known to occur in the state.

C. m. minor (Forster)

Common migrant through the eastern two-thirds of the state.

C. m. hesperis Grinnell

Status uncertain. Reported to be a migrant through the western half of the state, with specimens from as far east as Dallas County.

C. m. sennetti Coues

Uncommon migrant through the western two-thirds of the state.

C. m. howelli Oberholser

Common summer resident from the Panhandle south through the Rolling Plains and northern portion of the Edwards Plateau. Common migrant through the western half of the state.

C. m. henryi Cassin

Uncommon summer resident and migrant in the Trans-Pecos east to the southwestern part of the South Plains and the extreme western Edwards Plateau.

C. m. aserriensis Cherrie

Common summer resident in the South Texas Brush Country, ranging north across the southern edge of the Edwards Plateau and west to Terrell County.

C. m. chapmani Coues

Uncommon summer resident in the eastern third of the state west to the edge of the Rolling Plains and Edwards Plateau and south roughly to Calhoun County. Uncommon migrant through the eastern half of the state.

COMMON PAURAQUE *Nyctidromus albicollis* (Gmelin)

Common resident in the South Texas Brush Country north to southern Maverick County in the west and Atascosa and Wilson Counties in the east. Common Pauraques are rare north along the coast to Calhoun County and west to southern Val Verde and Uvalde Counties, although their status there is unclear. This species is casual north to Bastrop, Grimes, and Guadalupe Counties, with most of these records from April through July. **Taxonomy**: The subspecies found in Texas is *N. a. merrilli* Sennett.

COMMON POORWILL

Phalaenoptilus nuttallii (Audubon)

Common to uncommon migrant and summer resident through the Trans-Pecos, Edwards Plateau, and South Texas Brush Country. Common Poorwills are uncommon migrants and uncommon to rare summer residents on the High Plains and Rolling Plains. The eastern edge of this species' range is poorly defined, but birds are present in low numbers east to Bosque, Travis, Goliad, and Hidalgo Counties. This nightjar has been reported, primarily in the spring, east to Dallas, Galveston, and Navarro Counties. Common Poorwills are rare to locally uncommon migrants and winter residents in the southern third of the summer range and have been recorded in late fall and early winter north to the Panhandle and west to El Paso County. Determining the winter status is made difficult by the fact that they are active only on warm nights. **Timing of occurrence:** Spring migrants have been detected from mid-March to mid-May. Fall migrants are found from early September to late October, although a few straggle into early November. **Taxonomy:** The subspecies found in Texas is *P. n. nuttallii* (Audubon).

CHUCK-WILL'S-WIDOW

Antrostomus carolinensis (Gmelin)

Uncommon to common migrant and summer resident in the eastern two-thirds of the state, west to the eastern Rolling Plains and southern Edwards Plateau, and west to Kinney and Edwards Counties. Summer residents are found south to DeWitt and Karnes Counties and rarely south to Refugio County. Chuck-will's-widow is also rare and local in summer in the eastern Panhandle, although the extent of its range there is poorly understood. Migrants have been found as far west as Val Verde County and are rare but regular through the High Plains. Very unexpectedly, one spent three summers in the upper elevations of the Davis Mountains from 2007 to 2009. **Timing of occurrence:** Spring migrants are found from late March to mid-May, very rarely straggling into late May. Fall migrants are found from early August to late October, with a few straggling into early November. The peak of fall migration is during late August and September. **Taxonomy:** Monotypic.

EASTERN WHIP-POOR-WILL

Antrostomus vociferus (Wilson)

Rare to uncommon migrant in the eastern half of the state west to the central Edwards Plateau and casual farther west to the High Plains and western Edwards Plateau. Eastern Whip-poor-wills are very rare to casual winter visitors along the coast, with one documented record inland from Trinity County. Eastern Whip-poor-will was split from Mexican Whip-poor-will in 2010 (Chesser et al. 2010), and both taxa occur in the state. **Timing of occurrence**: Spring migrants are present from mid-March to mid-May, and migration peaks from late March through late April. They are quite vocal at dusk and dawn during spring migration. Fall migrants are much more difficult to detect and appear to pass through the state from mid-August to early November. **Taxonomy**: Monotypic.

MEXICAN WHIP-POOR-WILL

Antrostomus arizonae Brewster

Common summer resident at upper elevations in the Guadalupe, Davis, and Chisos Mountains of the Trans-Pecos. While presumably Mexican Whip-poor-will is a migrant through the western Trans-Pecos, its status away from breeding areas is virtually unknown. There have been only two instances of individuals noted at lower elevations: one in Alpine, Brewster County, on 25–27 July 2007; and another thought to be this species (rather than Eastern Whip-poor-will) at Lubbock, Lubbock County, on 27 April 2012. **Timing of occurrence**: Mexican Whip-poor-wills arrive on the breeding grounds in mid-April, becoming common by the end of the month. Summer residents appear to depart between late August and mid-September, although a few linger through the end of the month. **Taxonomy**: The subspecies found in Texas is *A. a. arizonae* Brewster.

ORDER APODIFORMES

Family Apodidae: Swifts

TBRC Review Species **WHITE-COLLARED SWIFT**
Streptoprocne zonaris (Shaw)

Accidental. Texas has six documented records of this tropical swift found throughout the Neotropics and Caribbean
and occurring as close as southern Tamaulipas, Mexico.
The second of these provided the second specimen for the
United States (Lasley 1984). Since then two others have
been supported with photographs, the first ever taken of
this species in the United States (Eubanks and Morgan
1989). All Texas records are from along or near the coast.
Considering the wide variety of habitats this species uses in
northeastern Mexico, observers should be aware of the potential for this large swift to occur in other areas of the
state. **Taxonomy**: The subspecies involved with the Texas
records is not conclusively known. The lone Texas specimen is consistent with the subspecies *S. z. mexicana*
Ridgway, but that specimen has not been compared to a
wide series that includes other subspecies.

CHIMNEY SWIFT *Chaetura pelagica* (Linnaeus)

Common migrant and summer resident through the eastern three-quarters of the state. In the Trans-Pecos, Chimney Swifts are rare spring migrants and early summer visitors to Pecos and Reeves Counties and are casual migrants
and summer visitors in Alpine, Brewster County, and Fort
Davis, Jeff Davis County (Peterson and Zimmer 1998).
Chimney Swifts are localized, but increasing, during the
summer months in the Lower Rio Grande Valley. They
winter in South America, and *Chaetura* swifts found from
December through February should be carefully documented since there are no accepted records for Texas of
Vaux's Swift (*C. vauxi*), although it has been recorded in
Louisiana during this season. **Timing of occurrence**: Spring
migrants appear in Texas in early March, although they
are not common until late March or early April for most of
the state. Fall migrants are seen from early September to
mid-October. They become less common through late October and are rarely present into mid-November. **Taxonomy**: Monotypic.

WHITE-THROATED SWIFT
Aeronautes saxatalis (Woodhouse)

Common summer resident and locally uncommon winter resident in the Trans-Pecos, east to central Val Verde County (Maxwell 1980). This species is a casual winter visitor to Palo Duro Canyon, Randall County, and at Caprock Canyons SP, Briscoe County, in the Panhandle (Seyffert 1984, 2001b). White-throated Swifts are very rare to casual migrants in the western Panhandle and have wandered eastward during the fall to the Edwards Plateau, Coastal Prairies, and Lower Rio Grande Valley. There is a single report of them wintering in Brazos County. **Timing of occurrence**: Spring migrants arrive in the state from early March to mid-April. Fall migrants are seen from late September through mid-November. **Taxonomy**: The subspecies found in Texas is *A. s. saxatalis* (Woodhouse).

Family Trochilidae: Hummingbirds

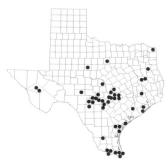

TBRC Review Species   **GREEN VIOLETEAR** *Colibri thalassinus* (Swainson)

Very rare to casual spring and summer visitor to the eastern two-thirds of the state and accidental in these seasons in the Trans-Pecos. Most of the Green Violetear records for Texas are from the eastern half of the Edwards Plateau, from the Lower Rio Grande Valley, and along the coast. There are six records from the Pineywoods and a single record from north-central Texas. This species is an annual visitor to the state and appears to be increasing in occurrence. There are 75 documented records for Texas, 31 of which have been since spring 2005. **Timing of occurrence**: The records for this species are all from 14 April to 3 October, but the primary season of occurrence is from mid-May through July. **Taxonomy**: The subspecies found in Texas is *C. t. thalassinus* (Swainson).

TBRC Review Species   **GREEN-BREASTED MANGO**
Anthracothorax prevostii (Lesson)

Casual visitor to the Lower Rio Grande Valley and the central coast, with one spring record from the upper coast in Jefferson County. The first record for the state occurred at Brownsville, Cameron County, from 14 to 23 September 1988 (TBRC 1988-272; TPRF 773). Since then there

have been 19 others, most involving immature birds. Separating immature-plumaged Green-breasted Mango from the primarily South American Black-throated Mango (*A. nigricollis*) is extremely difficult. Careful in-hand examination of one at Corpus Christi, Nueces County, which was present from 6 to 27 January 1992, confirmed that it was a Green-breasted Mango. The TBRC considers all records of *Anthracothorax* species to be of Green-breasted Mango until such time as another species is documented in or near Texas. With records from South Carolina, Georgia, and Wisconsin, this species could occur anywhere in the state. **Timing of occurrence:** This species has occurred in all seasons, although fall is the primary season. Half of the state records were found during August and September. **Taxonomy:** The subspecies found in Texas is not conclusively documented, but *A. p. prevostii* (Lesson) occurs in northeastern Mexico and is the most likely to occur in the state.

MAGNIFICENT HUMMINGBIRD
Eugenes fulgens (Swainson)

Uncommon and local summer resident in the Davis Mountains and rare in the Guadalupe Mountains. Magnificent Hummingbird is also a rare summer visitor to the Chisos Mountains but is not known to breed. This species does not often stray outside its breeding areas. Magnificent Hummingbirds are casual visitors to the El Paso area, primarily in the fall. There are four documented records from east of the Pecos River, all from the fall, with single individuals from Brazoria, Jim Hogg, Wilson, and Cameron Counties. Another report of Magnificent Hummingbird from Bexar County from 24 to 26 May 1959 has been widely cited (Oberholser 1974; TOS 1995). Examination of the material provided to the editor of *Audubon Field Notes* in 1959 appears to point toward Green Violetear. **Timing of occurrence:** The first spring migrants arrive at the breeding areas in mid-March, but the bulk of the population arrives in mid-April. Fall migrants depart breeding areas beginning in mid-September, and a few remain into mid-October. There is an exceptional late record from the Davis Mountains on 10 December 2009. There are also winter records from El Paso County. **Taxonomy:** The subspecies found in Texas is *E. f. fulgens* (Swainson).

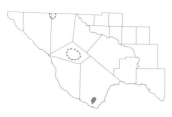

BLUE-THROATED HUMMINGBIRD

Lampornis clemenciae (Lesson)

Locally uncommon summer resident in the Chisos Mountains. This species is a rare to very rare summer visitor to the Davis and Guadalupe Mountains. There is a breeding record from the Guadalupe Mountains but none from the Davis Mountains. This hummingbird is a casual spring and fall migrant elsewhere in the Trans-Pecos. Blue-throated Hummingbirds have been reported during late summer and early fall at scattered locations from the South Plains south to the Lower Rio Grande Valley and east to the Brazos Valley and the upper coast. There are winter records at Mission, Hidalgo County, from 2 to 13 February 1996 (TPRF 1349) and Lake Jackson, Brazoria County, from 13 November 2006 to 27 February 2007. **Timing of occurrence**: Spring migrants arrive in the Chisos Mountains in mid-April and are present through early October. **Taxonomy**: The subspecies found in Texas is *L. c. phasmorus* Oberholser.

LUCIFER HUMMINGBIRD

Calothorax lucifer (Swainson)

Locally common summer resident in the foothills of the Chisos and Christmas Mountains. Lucifer Hummingbirds are rare and local during the summer in the Chinati and Davis Mountains as part of postbreeding dispersal. There are records falling within this pattern from the Glass Mountains and the Sierra Vieja as well. Elsewhere in the Trans-Pecos, this species has been found in the Guadalupe Mountains and in El Paso County. Lucifer Hummingbird is a casual visitor to the Edwards Plateau, primarily as a postbreeding wanderer. Five different individuals present in the summer of 2011 were associated with drought-related dispersal. There are three records from the South Texas Brush Country: a male in the Lower Rio Grande Valley on 27 April 2008 and two others that fall within the pattern of postbreeding dispersal. **Timing of occurrence**: Breeding birds arrive in early March, with the early arrival date of 23 February. Summer residents are present to early October, but a few linger as late as early November. Dispersal of individuals away from the breeding range occurs between late May and early September. **Taxonomy**: Monotypic.

RUBY-THROATED HUMMINGBIRD
Archilochus colubris (Linnaeus)

Common summer resident in the eastern third of Texas, west to the eastern edge of the Edwards Plateau, and south to Victoria County on the Coastal Prairies. Gravid females have been captured from locations on the Edwards Plateau and Concho Valley. Hybrids with Black-chinned Hummingbirds have been documented in these areas, clouding whether there are any true nesting records of Ruby-throated Hummingbirds. They are common to abundant migrants across the eastern half of the state, west through the eastern Edwards Plateau and eastern Rolling Plains, and south along the Coastal Prairies. Farther west through the remainder of the Edwards Plateau and South Texas Brush Country they are uncommon spring and common fall migrants. They are rare to uncommon migrants on the High Plains and eastern Trans-Pecos, although generally much more common in the fall than spring. This species is a very rare winter visitor along the coast and casual inland primarily in the eastern half of the state, although there are documented winter occurrences inland to Jeff Davis County. The majority of wintering birds are in female-type plumage. **Timing of occurrence**: Spring migrants are found from early March to late May. Fall migration occurs from early August through late October, although a few linger to mid-November. **Taxonomy**: Monotypic.

BLACK-CHINNED HUMMINGBIRD
Archilochus alexandri (Bourcier and Mulsant)

Common to locally abundant summer resident in the western two-thirds of the state. Black-chinned Hummingbirds are most common as breeding birds in the Trans-Pecos and Edwards Plateau. The eastern edge of the breeding range extends from the western edge of the Blackland Prairies in Dallas and Bell Counties, south and east to Goliad County. Black-chinned Hummingbirds are uncommon to rare summer residents south of the Balcones Escarpment to the Lower Rio Grande Valley. This species is generally absent from the High Plains, although there are nesting records from urban habitats within this region. In migration, this species is found much farther east, although it is rare in the Pineywoods. In winter, these hummingbirds are rare along the coast, mainly at feeders, and very rare to casual inland, as far inland as the Panhandle and the northern Piney-woods. **Timing of occurrence**: Spring migrants arrive in the state beginning in late February and are present until mid-May. Fall migrants are found as soon as early July, but the main passage of migrants occurs from mid-August through September. Very small numbers linger into early November. **Taxonomy**: Monotypic.

ANNA'S HUMMINGBIRD *Calypte anna* (Lesson)

Uncommon to rare fall migrant and winter resident in the western and central Trans-Pecos. This western hummingbird is a rare to very rare and irregular visitor, most often in late fall, to the remainder of the state. Anna's Hummingbirds are very rare winter residents outside the Trans-Pecos, mostly along the coast, where they appear to be increasing. Individuals have wintered as far north as Lubbock and Randall Counties on the High Plains. Texas has two breeding records: one from Jeff Davis County in 1976 (TPRF 101; Schmidt 1976) and another from El Paso County in 2000. **Timing of occurrence**: Fall migration extends from early September through early November, although a few very early individuals arrive in early July. Wintering birds generally depart the state by mid-February, but some remain through mid-March. An immature male spent the entire summer in the Davis Mountains during 2006. **Taxonomy**: Monotypic.

TBRC Review Species   ### COSTA'S HUMMINGBIRD *Calypte costae* (Bourcier)

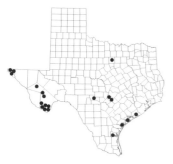

Very rare migrant and casual winter visitor to the Trans-Pecos and casual farther east. There are 36 documented records for Texas, 28 of which are from the Trans-Pecos. The documented occurrence of Costa's Hummingbirds in the state has increased dramatically during the past 10 years. A record of up to seven individuals present at a single location in El Paso County between 28 October and 12 December 1995 stands out as particularly noteworthy. All other records of Costa's Hummingbirds involve single individuals. There are seven records from east of the Pecos River, all but one of which are winter records and four of which are from the central Coastal Prairies. **Timing of occurrence**: Costa's Hummingbirds have been found from early fall through late spring. The earliest fall arrival is 20 August, and the latest spring date is 19 May. Notably, an adult male present near Fort Davis, Jeff Davis County, from 10 June to 1 September 2001 provides the only summer record. **Taxonomy**: Monotypic.

BROAD-TAILED HUMMINGBIRD
Selasphorus platycercus (Swainson)

Uncommon to locally common summer resident in the Chisos, Davis, and Guadalupe Mountains and common to uncommon migrant through the remainder of the Trans-Pecos. Broad-tailed Hummingbirds are rare spring and uncommon fall migrants on the High Plains and western Edwards Plateau, becoming less common farther east. They are rare and local winter residents in El Paso County and along the Coastal Prairies, and there are many instances of individuals lingering through the winter at feeders throughout the state. **Timing of occurrence**: Spring migrants arrive on the breeding grounds as early as late February, although the majority of birds pass through the state between late March and early May. Fall migrants are present primarily between mid-August and mid-October. **Taxonomy**: Monotypic.

RUFOUS HUMMINGBIRD *Selasphorus rufus* (Gmelin)

Very rare spring and common to abundant fall migrant in the Trans-Pecos. Rufous Hummingbird is an uncommon to common fall migrant on the High Plains and western Rolling Plains south through the Edwards Plateau, becoming locally uncommon to rare farther east. Rufous Hummingbirds follow an elliptical migratory path, passing through Texas in the fall and returning up the West Coast of North America in spring. Migrants found in Texas during the spring may be individuals that wintered along the Gulf Coast. They are rare to locally uncommon winter residents in the southern and western Trans-Pecos, on the Coastal Prairies, and in the Lower Rio Grande Valley and are found locally inland throughout the southern half of the state. There are numerous records of individuals lingering through the winter at feeders in virtually all areas of the state. **Timing of occurrence**: Fall migrants begin arriving in the state as soon as early July, and migration peaks from late July through mid-September. Fall migrants are still present through November, and small numbers linger as late as early January before departing. Wintering birds generally depart by mid-March, and spring migrants have been noted as late as early May. **Taxonomy**: Monotypic.

ALLEN'S HUMMINGBIRD *Selasphorus sasin* (Lesson)

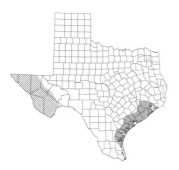

Rare to uncommon fall migrant through the Trans-Pecos. Allen's Hummingbirds are also rare winter residents on the Coastal Prairies and casual inland throughout the state. This species was removed from the Review List in 2004. The status of Allen's Hummingbirds has been elucidated by the extensive banding efforts along the upper and central coasts and in the central Trans-Pecos. Documenting Allen's Hummingbird requires examining the shape of retrix 2 and measuring retrix 5 (the outermost tail feather) to eliminate Rufous Hummingbird (Stiles 1972; McKenzie and Robbins 1999). These criteria require accurate aging and sexing, and the situation is further confused by molt-retarded, second-year male Rufous Hummingbirds that have retained largely green backs, but this is very rare compared to the proven rate of occurrence of Allen's Hummingbird in Texas. Recent studies suggest that adult male *Selasphorus* hummingbirds that appear to be Allen's Hummingbirds with fully developed gorgets and adult tail feathers are indeed Allen's Hummingbirds (B. Sargent, pers. comm.). **Timing of occurrence**: Fall migrants are found in the state between mid-August and early November, but a few linger as late as early December. Wintering birds generally depart by mid-March. **Taxonomy**: The subspecies that occurs in Texas is *S. s. sasin* (Lesson).

CALLIOPE HUMMINGBIRD
Selasphorus calliope (Gould)

Casual spring and uncommon fall migrant in the western half of the Trans-Pecos and rare fall migrant east through the High Plains and Concho Valley, becoming casual farther east to the Pineywoods. Most records of Calliope Hummingbird away from the Trans-Pecos are from the fall. There are scattered records of Calliope Hummingbirds lingering through the winter at feeders. These winter records occur almost annually on the upper coast and in El Paso County. **Timing of occurrence**: Males pass through the state between mid-July and early August, followed by females and young of the year from mid-August to mid-

September. Small numbers remain to early November. **Taxonomy:** Monotypic.

BROAD-BILLED HUMMINGBIRD
Cynanthus latirostris Swainson

Rare to very rare summer visitor to the Davis Mountains. Broad-billed Hummingbirds are casual to very rare winter visitors to El Paso County. Elsewhere in the state, this species is a casual visitor to virtually all areas, with records from all seasons. When this species was removed from the Review List in 2003, there were 54 documented records. Broad-billed Hummingbirds have occurred annually in the Davis Mountains since 2003, but clear evidence of nesting was observed only in 2010. Oberholser (1974) reported this species as a very rare summer resident in Brewster County during the 1930s. This is based on a report by Quillin (1935) of a nest with two eggs found on 17 May 1934. Van Tyne and Sutton (1937) knew of no other reports from Brewster County. **Timing of occurrence:** Summer residents in the Davis Mountains have arrived as early as late March and are generally present through mid-October. There are records from all seasons. **Taxonomy:** Monotypic.

TBRC Review Species

BERYLLINE HUMMINGBIRD
Amazilia beryllina (Deppe)

Accidental. The five documented records for the state are all from the Trans-Pecos. Four of these are from the central Davis Mountains, three of which are from late summer and probably involve postbreeding wanderers. The remaining records involved a male that spent the summer in the Davis Mountains. It was first seen coming to feeders in late May, and presumably the same individual returned on 4 June and 12 July. **Taxonomy:** The subspecies that occurs in Texas is *A. b. viola* (Miller).

18 AUG. 1991, CHISOS MOUNTAINS, BREWSTER CO.
  (TBRC 1991-121)
17 AUG.–4 SEPT. 1997, DAVIS MOUNTAINS, JEFF DAVIS CO.
  (TBRC 1997-137; TPRF 1661)
3 AND 8 AUG. 1999, DAVIS MOUNTAINS, JEFF DAVIS CO.
  (TBRC 1999-80; TPRF 1922)
25 MAY, 4 JUNE, AND 12 JULY 2000, DAVIS MOUNTAINS, JEFF
  DAVIS CO. (TBRC 2000-53; TPRF 1923)
25–28 AUG. 2007, DAVIS MOUNTAINS, JEFF DAVIS CO.
  (TBRC 2007–67; TPRF 2503)

BUFF-BELLIED HUMMINGBIRD
Amazilia yucatanensis (Cabot)

Uncommon to locally common summer resident in the Lower Rio Grande Valley and along the coast north to Victoria County. This species has expanded its range northward up the coast and inland into south-central Texas over the past decade. In recent years this species has become a rare spring and summer visitor, including a few nesting records, inland to Bastrop and Washington Counties, and is now casual during these seasons to the very southern edge of the Pineywoods. Most Buff-bellied Hummingbirds retreat southward during the winter and are rare to uncommon and local, but increasing, at feeders and ornamental plantings in the Lower Rio Grande Valley and along the coast as far north as Calhoun County. There are scattered winter records farther north along the coast and inland to Austin and Travis Counties. There is one record from the fall from the Trans-Pecos. **Timing of occurrence:** Migrants returning to the breeding grounds appear in mid-March, and most depart by late October. **Taxonomy:** The subspecies that occurs in Texas is *A. y. chalconota* Oberholser.

TBRC Review Species

VIOLET-CROWNED HUMMINGBIRD
Amazilia violiceps (Gould)

Very rare visitor to the Trans-Pecos and casual farther east. Violet-crowned Hummingbird has occurred with much greater frequency in the past decade, and 12 of the 19 documented records for the state have occurred since the spring of 2007. The first state record was at El Paso from 2 to 14 December 1987 (TBRC 1988-83; TPRF 594). It is a very rare visitor to the Trans-Pecos and accidental farther east and north. The most unexpected records of Violet-crowned Hummingbird in the state are birds at Lake Jackson, Brazoria County; Lubbock, Lubbock County; and Weslaco, Hidalgo County. The longest-staying individual was recorded in El Paso from 6 November 2001 to 16 February 2002 (TBRC 2001-135; TPRF 1996). **Timing of occurrence:** Violet-crowned Hummingbirds have been found between August and May, with the majority of records between 18 August and 8 January. **Taxonomy:** The subspecies that occurs in Texas is *A. v. ellioti* (Berlepsch).

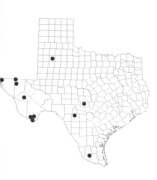

TBRC Review Species **WHITE-EARED HUMMINGBIRD**
Hylocharis leucotis (Vieillot)

Rare to very rare summer resident in the Davis Mountains. During summer and early fall they are casual visitors elsewhere in the Trans-Pecos and accidental on the Edwards Plateau and in the Lower Rio Grande Valley. There are 33 documented records from the state, most from the Davis Mountains. White-eared Hummingbirds were present in unexpectedly high numbers in the Davis Mountains Resort from 2005 to 2011, with as many as 15 documented in 2007. This trend reversed starting in 2011, with only three birds found from spring 2011 through June 2013. Other Trans-Pecos records exist from the Chisos and Guadalupe Mountains and from El Paso County. There are four documented records from outside the Trans-Pecos, all in July, which likely represent postbreeding dispersal. These include single birds present in Gillespie, Lubbock, Starr, and Uvalde Counties. **Timing of occurrence**: White-eared Hummingbirds have been found between early April and mid-September, and there are two records of birds present into mid-October. **Taxonomy**: The subspecies that occurs in Texas is *H.l. borealis* (Griscom).

ORDER TROGONIFORMES

Family Trogonidae: Trogons

TBRC Review Species **ELEGANT TROGON** *Trogon elegans* Gould

Casual. Texas has six documented records, including three well-documented occurrences from Hidalgo County. The remaining three are sight records from the Chisos Mountains. **Timing of occurrence**: The occurrences in the Lower Rio Grande Valley are from September and January, plus a long-staying individual present from 14 January to 12 May 2005 in Weslaco. The Trans-Pecos records are from April, June, and one from November into January. **Taxonomy**: The subspecies that has occurred in Hidalgo County is *T. e. ambiguous* Gould. The identification of the birds in the Trans-Pecos may be this subspecies or *T. e. canescens* van Rossem from northwestern Mexico.

ORDER CORACIIFORMES

Family Alcedinidae: Kingfishers

RINGED KINGFISHER *Megaceryle torquata* (Linnaeus)

Locally common resident from the Lower Rio Grande Valley west to Webb County. Ringed Kingfishers are uncommon and local north through the South Texas Brush Country. Beginning in the 1990s this species was reported with increasing frequency during all seasons farther north on the Edwards Plateau and southern Blackland Prairies, particularly along the Colorado, Guadalupe, and Nueces Rivers. Nesting was first documented for the Edwards Plateau along the Nueces River, Uvalde County, in 2001. Wandering Ringed Kingfishers are casual, primarily during spring and fall, north to Dallas, Randall, and Tarrant Counties and east on the Coastal Prairies to Chambers County. There are documented records from Oklahoma and Louisiana, demonstrating the potential for this species to appear anywhere in Texas. **Taxonomy:** The subspecies that occurs in Texas is *M. t. torquata* (Linnaeus).

BELTED KINGFISHER *Megaceryle alcyon* (Linnaeus)

Uncommon to locally common winter resident throughout the state. Belted Kingfishers are uncommon and local summer residents across the northern third of the state and south through the Pineywoods and Oaks and Prairies region. They are rare to very rare and local in summer in the remainder of the state, east of the Trans-Pecos. This is the most common and widespread kingfisher in Texas. **Taxonomy:** Monotypic.

TBRC Review Species   **AMAZON KINGFISHER** *Chloroceryle amazona* (Latham)

Accidental. There is one record for Texas and the United States. A female was present on a tributary close to the main channel of the Rio Grande in Webb County in 2010 (Wormington and Epstein 2010). This species is common and widespread from northern Mexico south to Argentina, and the closest known population of Amazon Kingfisher is in southern Tamaulipas, approximately 150 miles south of the Rio Grande. **Taxonomy:** Monotypic.

24 JAN.–3 FEB. 2010, LAREDO, WEBB CO.
(TBRC 2010–09; TPRF 2794)

GREEN KINGFISHER *Chloroceryle americana* (Gmelin)

Uncommon resident from the Edwards Plateau south to the Lower Rio Grande Valley and north along the Coastal Prairies to Victoria and Jackson Counties. Green Kingfishers are rare along the Rio Grande to Brewster County and casual west to Presidio County. They are also rare along the lower Devils and Pecos River drainages. Green Kingfishers are rare to locally uncommon east to Bastrop and Bell Counties, where nesting records exist. They have occurred as vagrants during the summer north to Somervell, Randall, and Wise Counties and east to Liberty County. Surprisingly, there are isolated nesting records from Montgomery and Washington Counties well away from the known breeding range. This species is very sensitive to cold weather. During colder-than-normal winters, northern populations retreat well to the south and often do not return for several years. **Taxonomy**: Two subspecies are known to occur in the state.

C. a. hachisukai (Laubmann)
Uncommon to rare resident north of the Lower Rio Grande Valley, as described above.

C. a. septentrionalis (Sharpe)
Uncommon resident in the Lower Rio Grande Valley north to Webb County. Range and possible contact zone with the previous subspecies require further study.

ORDER PICIFORMES

Family Picidae: Woodpeckers

LEWIS'S WOODPECKER *Melanerpes lewis* (Gray)

Irregular and very rare migrant and winter visitor to the western two-thirds of the state. This species was removed from the Review List in 2002 and had 59 documented records at that time. The majority are from the Trans-Pecos,

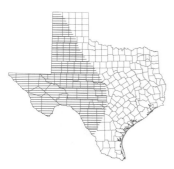

although there are scattered records eastward to Collin County. Lewis's Woodpeckers documented in the South Texas Brush Country during the winters of 2000–2001 and 2001–2002 are among the southernmost records known for this species. There is a single summer report of a bird in the Davis Mountains on 12 July 1901. These woodpeckers have shown an irruptive pattern of occurrence within the state. This was particularly evident between the fall of 2000 and the spring of 2004, when more than 35 individuals were reported. This level of occurrence declined sharply through the remainder of that decade. **Timing of occurrence:** Fall migrants generally arrive from mid-October through November, but there are records from as early as late August. Overwintering birds generally depart by late March, but in many instances individuals have lingered into mid-May. **Taxonomy:** Monotypic.

RED-HEADED WOODPECKER
Melanerpes erythrocephalus (Linnaeus)

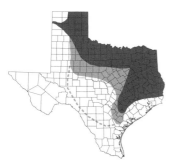

Locally common to rare resident in riparian habitats from the western Panhandle eastward across north-central Texas. Red-headed Woodpeckers are uncommon to locally common residents throughout the Pineywoods, becoming rare to locally uncommon westward through the Post Oak Savannah region. They are rare and irregular postbreeding wanderers and winter visitors west and south of the breeding range. Red-headed Woodpeckers are irruptive during winter on the southern Rolling Plains and Edwards Plateau, sometimes occurring in large numbers. This woodpecker is a casual winter visitor to the Trans-Pecos and South Texas Brush Country, but the numbers have declined significantly in recent decades through their entire range. **Taxonomy:** Monotypic.

ACORN WOODPECKER
Melanerpes formicivorus (Swainson)

Common resident in the Chisos, Davis, and Del Norte Mountains but uncommon and more local in the Guadalupe Mountains and the Sierra Diablo. A small remnant population is present in Bandera, Kerr, and Real Counties on the central Edwards Plateau, but these birds are found on private land and are only occasionally detected by birders. Away from the breeding range, Acorn Woodpeckers

are very rare and irregular visitors, primarily from fall through spring, to the remainder of the Trans-Pecos, with nesting documented in El Paso County in 2012. Farther east, Acorn Woodpeckers are irregular winter visitors to the western Edwards Plateau away from the small breeding area and north through the South Plains to the southern Panhandle. Wandering Acorn Woodpeckers have also been documented east to Bastrop County, along the Coastal Prairies to Aransas and Harris Counties, and south to the Lower Rio Grande Valley. **Taxonomy:** The subspecies that occurs in Texas is *M. f. formicivorus* (Swainson).

GOLDEN-FRONTED WOODPECKER

Melanerpes aurifrons (Wagler)

Common resident in the South Texas Brush Country north through the Edwards Plateau and western Rolling Plains to the south-central Panhandle. Golden-fronted Woodpeckers are also found in riparian habitats in the southern Trans-Pecos west to Presidio County. This species is a casual visitor, primarily in winter, eastward to Parker, Hill, and Bastrop Counties. It is a rare visitor during the winter on the High Plains as far west as Oldham County and on the northern Rolling Plains to Hemphill County. During the last century, this species has expanded its range in Texas west to the central Trans-Pecos and northward to the southern Panhandle (Husak and Maxwell 2000). **Taxonomy:** The subspecies that occurs in Texas is *M. a. aurifrons* (Wagler).

RED-BELLIED WOODPECKER

Melanerpes carolinus (Linnaeus)

Common to abundant resident in the eastern third of the state, west to the eastern edge of the Rolling Plains and Edwards Plateau, and south along the coast to Refugio County. Red-bellied Woodpeckers are fairly common residents in the eastern Panhandle and along the Canadian River drainage west to Oldham County, where they are rare. Red-bellieds are very rare to rare wanderers to the remainder of the Panhandle and to the eastern South Plains. On the Edwards Plateau, this species is a rare resident in riparian forests along the Colorado River drainage as far west as San Saba County. Vagrants have been noted as far west as Midland County and south along the coast to Cameron and Hidalgo Counties in the Lower Rio Grande Valley. **Taxonomy**: Monotypic.

WILLIAMSON'S SAPSUCKER
Sphyrapicus thyroideus (Cassin)

Rare to locally uncommon migrant and rare winter resident in the upper elevations of the Davis and Guadalupe Mountains in the Trans-Pecos. Williamson's Sapsucker is also a rare to very rare migrant and winter visitor to the remainder of the western half of the Trans-Pecos and very rare to casual eastward onto the High Plains. This species is a casual winter visitor to the western Edwards Plateau (Lockwood 2001) and is accidental farther eastward, having been documented east to Bastrop, Brazoria, Brazos, Kleberg, and Travis Counties. **Timing of occurrence**: Fall migrants begin passing through the state in mid-September, with an increase in occurrence from early October through mid-November. Spring migrants have been detected between late February and early April. **Taxonomy**: The subspecies that occurs in Texas is *S. t. nataliae* (Malherbe).

YELLOW-BELLIED SAPSUCKER
Sphyrapicus varius (Linnaeus)

Uncommon to locally common migrant and winter resident throughout most of the state, except west of the central mountain ranges of the Trans-Pecos, where it is rare. This species often shares habitat with the closely related Red-naped Sapsucker where their ranges overlap. **Timing of occurrence**: Fall migrants begin passing through the state in small numbers as early as mid-September. A notable increase occurs from early October through mid-November, by which time the wintering population is present. Spring migration begins in early February, and most of the wintering population departs the state by early April. Very small numbers linger as late as early May in some years. **Taxonomy**: Monotypic.

RED-NAPED SAPSUCKER *Sphyrapicus nuchalis* Baird

Rare to locally common migrant and winter resident in the Trans-Pecos. Red-naped Sapsucker is a rare migrant

through the High Plains and western Rolling Plains south to the Edwards Plateau. It is very rare to casual farther east to the Blackland Prairies and south through the South Texas Brush Country. Red-naped Sapsuckers are casual to accidental on the upper coast. They may be very rare and irregular summer residents in the Guadalupe Mountains of the Trans-Pecos, where Newman (1974) collected an adult female in breeding condition on 3 June 1971. No evidence of breeding activity has been noted before or since that record. **Timing of occurrence**: The first fall migrants reach Texas in very late August, and the majority of birds arrive by mid-September. Fall migration continues through November, and spring migration begins in early February. Most of the wintering population departs the state by mid-March. Small numbers linger as late as early May. **Taxonomy**: Monotypic.

TBRC Review Species

RED-BREASTED SAPSUCKER

Sphyrapicus ruber (Gmelin)

Accidental. Texas has three records of Red-breasted Sapsucker. The first record was an adult found in Central Texas, which remains the easternmost record for this species. There have also been a number of birds documented that appear to be hybrids with Red-naped Sapsucker (Lockwood and Shackelford 1998). The majority of the hybrids have been intermediate, but at least one photographed in Gonzales County was more like a Red-breasted Sapsucker. This species is found primarily along the Pacific slope from southeastern Alaska south to southern California but is found regularly in winter in southern Arizona. **Taxonomy**: The subspecies that has occurred in Texas is *S. r. daggetti* Grinnell.

27 FEB. 1996, MCGREGOR, MCLENNAN CO.
   (TBRC 1996-175; TPRF 1586)
3 DEC. 2000, BIG BEND RANCH SP, PRESIDIO CO. (TBRC 2001-30)
11–28 MAR. 2005, DAVIS MOUNTAINS, JEFF DAVIS CO.
   (TBRC 2005–56; TPRF 2302)

LADDER-BACKED WOODPECKER

Picoides scalaris (Wagler)

Common to uncommon resident throughout the western two-thirds of the state. Ladder-backed Woodpeckers can

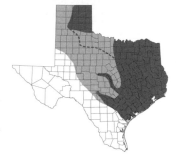

be found east into the western half of the Oaks and Prairies region, where they are locally uncommon to rare. This species has been found in small numbers in Brazoria and Galveston Counties, suggesting the species is expanding its range northeastward along the Coastal Prairies. There are no documented records from northeast Texas or the Pineywoods. **Taxonomy:** The subspecies that occurs in Texas is *P. s. cactophilus* (Oberholser).

DOWNY WOODPECKER *Picoides pubescens* (Linnaeus)

Common to uncommon resident in the eastern half of the state as far west as the eastern edge of the Rolling Plains and Edwards Plateau and south to the Guadalupe River delta. On the Edwards Plateau, Downy Woodpeckers are rare to locally uncommon residents only along the Colorado, Guadalupe, and Medina River drainages. These small woodpeckers are rare to uncommon residents in the Panhandle as far west as Potter and Randall Counties and have been found irregularly south through the northern portion of the Rolling Plains. Downy Woodpeckers are rare migrants and winter residents in the remainder of the Panhandle, Rolling Plains, and eastern Edwards Plateau. This species is a casual visitor farther west across the Edwards Plateau to the Trans-Pecos, with records from the Guadalupe and Davis Mountains and El Paso County. Downy Woodpecker is a casual visitor south to the lower Nueces River drainage, while farther south there is a single winter record from the Lower Rio Grande Valley in Starr County. **Timing of occurrence:** Winter occurrences away from breeding records generally are between early November and late March. **Taxonomy:** Three subspecies are found in Texas.

P. p. medianus (Swainson)
> Rare to uncommon resident in the eastern Panhandle. Specimens document the occurrence of this subspecies in winter south to Crockett County and east to Coleman and Tarrant Counties.

P. p. leucurus (Hartlaub)
> Casual visitor to the Davis and Guadalupe Mountains and El Paso County.

P. p. pubescens (Linnaeus)
> Common to uncommon resident in the eastern half of the state, including riparian areas in the eastern Edwards Plateau.

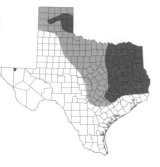

HAIRY WOODPECKER *Picoides villosus* (Linnaeus)

Uncommon resident in the Pineywoods and rare and local resident west to Tarrant and Bastrop Counties. This species is an uncommon and local resident at higher elevations in the Guadalupe Mountains of the Trans-Pecos and in the eastern Panhandle and along the Canadian River drainage as far west as Potter County. Hairy Woodpecker is casual to accidental in the northern Trans-Pecos west to El Paso County. These woodpeckers are very rare and irregular migrants and winter visitors elsewhere on the High Plains and eastern Edwards Plateau. They are rare winter visitors to the upper Coastal Prairies, although casual at coastal woodlots. Hairy Woodpeckers are casual winter visitors as far south as the Nueces River drainage along the Coastal Prairies. **Timing of occurrence:** Winter occurrences away from breeding areas generally are between late November and early March. **Taxonomy:** Three subspecies are found in Texas.

P. v. villosus (Linnaeus)

Uncommon resident in the eastern Panhandle. Specimens document the occurrence of this subspecies in winter south through the Rolling Plains.

P. v. orius (Oberholser)

Uncommon resident in the Guadalupe Mountains; casual to accidental visitor elsewhere in the Trans-Pecos.

P. v. auduboni (Swainson)

Uncommon resident in the eastern third of Texas.

RED-COCKADED WOODPECKER

Picoides borealis (Vieillot)

Rare to locally uncommon resident in open pine forests in the southern half of the Pineywoods. This species used to range considerably farther north and somewhat farther west in the Pineywoods until the first widespread logging began in Texas, which peaked in the early 1900s. Eggs were reported to have been collected in Lavaca (in 1912) and Lee (in 1883) Counties (Oberholser 1974); however, these may be of other species since appropriate habitat for this species was not widespread. Red-cockaded Woodpecker is a habitat specialist and requires intensive fire management of the open, mature pine forests that it occupies (Conner, Saenz, and Rudolph 2006). The US Fish and

Wildlife Service lists this species as Endangered, and the population present in the Sam Houston National Forest is one of the largest remaining in the country. As of 2012 there were fewer than 50 known family groups outside federal lands. **Taxonomy**: Monotypic.

NORTHERN FLICKER *Colaptes auratus* (Linnaeus)

Common to uncommon summer resident in the eastern Panhandle. Northern Flickers are rare to locally uncommon through the remainder of the Panhandle and south through the South Plains and northwestern Rolling Plains. These birds are fairly localized in urban habitats, woodlots, and riparian areas. They are also uncommon summer residents at mid- and upper elevations in the mountains of the Trans-Pecos and uncommon to rare summer residents in the Pineywoods. They are rare to very rare summer residents in woodland habitats on the Coastal Prairies along the upper coast. This species is a common winter resident throughout the northern two-thirds of the state, becoming rare in the southern South Texas Brush Country and very rare in the Lower Rio Grande Valley. **Taxonomy**: Four subspecies are found in Texas. These taxa can be divided into two distinct groups, Red-shafted and Yellow-shafted Flickers, which were formerly considered distinct species. Note that intergrades between Red-shafted and Yellow-shafted Flickers are rare to uncommon winter visitors, mostly in the central part of the state. Such birds can sometimes resemble Gilded Flicker, *C. chrysoides* (Malherbe), which has not been documented in Texas.

C. a. collaris Vigors
> Part of the Red-shafted Flicker group. Uncommon to locally common resident in the Chinati, Guadalupe, and Davis Mountains of the Trans-Pecos. Rare to common migrant and winter resident in the western two-thirds of the state and very rare in the eastern third.

C. a. nanus Griscom
> Part of the Red-shafted Flicker group. Uncommon resident in the Chisos Mountains and rare winter resident in the western Trans-Pecos.

C. a. luteus Bangs
> Part of the Yellow-shafted Flicker group. Common to uncommon resident in the Panhandle and northwestern Rolling

Plains. Uncommon migrant and winter resident through the eastern half of the state. This subspecies is very rare to locally uncommon in the northwestern part of the state and in the South Texas Brush Country to the Lower Rio Grande Valley. In winter, these flickers are also very rare west in the eastern half of the Trans-Pecos.

C. a. auratus (Linnaeus)

Part of the Yellow-shafted Flicker group. Uncommon to rare resident in the eastern third of the state. Common migrant and winter resident in the eastern half of the state south to the northeastern South Texas Brush Country and becoming very rare south to the Lower Rio Grande Valley. These flickers are rare to very rare winter visitors west to the Trans-Pecos.

PILEATED WOODPECKER
Dryocopus pileatus (Linnaeus)

Uncommon to locally common resident in the Pineywoods and Post Oak Savannah. Pileated Woodpeckers are rare residents west to Karnes, Travis, and Wise Counties. Vagrants have been found west to San Saba and Kerr Counties on the Edwards Plateau and to Eastland and Young Counties in north-central Texas. There are a few scattered records from along the central coast south of Lavaca Bay to Aransas County. In recent years, they have been found regularly in very small numbers in the Guadalupe River delta of Calhoun and Refugio Counties. **Taxonomy:** The subspecies that occurs in Texas is *D. p. pileatus* (Linnaeus).

TBRC Review Species ## IVORY-BILLED WOODPECKER
Campephilus principalis (Linnaeus)

The Ivory-billed Woodpecker has been extirpated from Texas and very likely from the United States. The last fully documented record for this species in the United States is from 1945 in northeast Louisiana. Ivory-billed Woodpeckers were formerly residents in the eastern third of the state, west to the bottomlands of the Trinity River, and along the coast to the Brazos River. Only three well-documented records exist for Texas, although there have been 45 other published reports from the state (Shackelford 1998). **Taxonomy:** The subspecies that occurs in Texas is *C. p. principalis* (Linnaeus).

3 MAY 1885, NECHES RIVER, JASPER CO.
  (MILWAUKEE PUBLIC MUSEUM 338)
CA. 1900, BOIS D'ARC ISLAND (TRINITY RIVER), DALLAS CO.
  (DALLAS MNH 6,216)
26 NOV. 1904, TARKINGTON, LIBERTY CO.
  (*USNM 195,199 AND 195,200; TPRF 2614 AND 2615)

ORDER FALCONIFORMES

Family Falconidae: Falcons

TBRC Review Species

COLLARED FOREST-FALCON
Micrastur semitorquatus (Vieillot)

Accidental. Texas has the only record of this tropical species for the United States. A light morph adult was discovered at Bentsen–Rio Grande Valley SP, Hidalgo County, during the winter of 1994. Collared Forest-Falcons are widespread in the Neotropics and occur within 150 miles of Brownsville in central Tamaulipas, Mexico. **Taxonomy:** The subspecies that has occurred in Texas is *M. s. naso* (Lesson).

22 JAN.–24 FEB. 1994, BENTSEN–RIO GRANDE VALLEY SP,
  HIDALGO CO. (TBRC 1994-40; TPRF 1227)

CRESTED CARACARA *Caracara cheriway* (Jacquin)

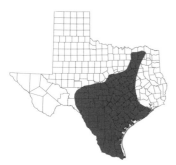

Uncommon to common resident in the South Texas Brush Country and Coastal Prairies, including associated barrier islands. This species occurs up the coast irregularly as far north as Chambers County. Crested Caracaras are locally common to uncommon residents in the Oaks and Prairies region north to Kaufman County and are rare north to Delta, Hunt, and Tarrant Counties. They have expanded their range over the past 20 years to include all of the Edwards Plateau, where they are rare to locally uncommon residents. Nesting has been documented well away from known populations, such as in western Crockett County and in Kent County. Crested Caracara is casual to the Trans-Pecos at all seasons. Some seasonal movements occur, as birds retreat somewhat from the northernmost portions of the range during cooler winters. **Taxonomy:** Monotypic.

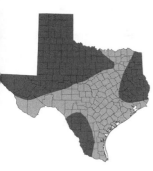

AMERICAN KESTREL *Falco sparverius* Linnaeus

Common to abundant migrant and winter resident throughout the state. American Kestrels are uncommon and local summer residents in the High Plains and Trans-Pecos. They are also rare to uncommon summer residents across the Rolling Plains to Dallas and Tarrant Counties, in the Pineywoods, and in the western South Texas Brush Country south to Starr County. There are isolated breeding records for other areas, but these kestrels are generally absent from the remainder of Texas during the summer. **Timing of occurrence:** Fall migrants arrive in the state in late August and become more common by mid-September through October. Winter residents begin to depart in late February, and most nonbreeding birds are gone by late April. **Taxonomy:** Two subspecies occur in the state.

F. s. sparverius Linnaeus

Common migrant and winter resident throughout the state. Uncommon and local summer resident in the western two-thirds of the state.

F. s. paulus (Howe & King)

Rare to uncommon resident in the eastern third of the state.

MERLIN *Falco columbarius* Linnaeus

Rare to uncommon migrant and winter resident throughout the state. Merlins are low-density winter residents throughout most of the state and are most common on the Coastal Prairies. **Timing of occurrence:** Fall migrants begin to arrive in mid-August but are rare until mid-September. Winter residents begin to depart in early February, and most are gone by late April. **Taxonomy:** Three subspecies have occurred in the state.

F. c. columbarius Linnaeus

As described above.

F. c. suckleyi Ridgway

Accidental. There are two photographically documented records: one from Bee County (TPRF 1959) and the other from Galveston County (Wheeler and Clark 1995).

F. c. richardsonii Ridgway

Rare migrant and winter resident primarily in the western two-thirds of the state.

APLOMADO FALCON *Falco femoralis* Temminck

Formerly a resident from the central Trans-Pecos east to Midland and in the South Texas Brush Country. There were scattered records from Cameron County and the western Trans-Pecos until the early 1950s, when the species was extirpated from Texas. Beginning in 1989, the Peregrine Fund, Inc. and the US Fish and Wildlife Service began a reintroduction/introduction program in South Texas. Aplomado Falcons are now rare residents along the Coastal Prairies from western Matagorda County southward, including associated barrier islands. In 1995 a pair successfully nested in Cameron County, followed by other nestings from Cameron to Calhoun Counties. Population estimates in 2012 included up to 35 nesting pairs. Aplomado Falcons from this project have wandered as far up the coast as Jefferson County. Birds are still being released as of 2012; thus, this population is still being augmented, which suggests the conservation groups involved do not consider the population to be self-sustaining. In the Trans-Pecos, this species may be a very rare to casual natural visitor to mid-elevation grasslands. The most recent records are from Culberson County in 1996 and Presidio/Jeff Davis Counties in 1991–92. These sightings, along with many others from southern New Mexico, are believed to be of individuals originating from a natural population in northern Chihuahua, Mexico, which has declined sharply in the past 10 years and appears on the verge of extirpation. Aplomado Falcons were released in Jeff Davis and Presidio Counties beginning in the summer of 2002, but this attempt has not been as successful as the releases on the Coastal Prairies, despite some successful nesting attempts. **Taxonomy**: The subspecies that occurs in Texas is *F. f. septentrionalis* Todd.

TBRC Review Species

GYRFALCON *Falco rusticolus* Linnaeus

Accidental. There is only one record of this Arctic falcon for Texas. An immature gray morph remained in Lubbock, Lubbock County, for more two months during the late winter and early spring of 2002. This individual faithfully roosted on a water tower within the city limits during its stay. This is the southernmost record for this species in the Americas. **Taxonomy**: Monotypic.

21 JAN.–7 APR. 2002, LUBBOCK, LUBBOCK CO.
  (TBRC 2002-16; TPRF 1982)

PEREGRINE FALCON *Falco peregrinus* Tunstall

Uncommon to rare migrant throughout the state. Peregrine Falcons are locally uncommon winter residents on the Coastal Prairies and can be common at times along the immediate coast, particularly near bays and estuaries. This species is a rare to very rare winter resident, most often in urban areas and around reservoirs, inland through the eastern half of the state. Formerly considered casual during winter in the western half of Texas, this species appears to be increasing and is now considered very rare. Peregrines are very local summer residents in the Trans-Pecos. Breeding populations are confined to the Guadalupe and Chisos Mountains and the cliffs that line the Rio Grande in southern Brewster and Presidio Counties. This falcon, once listed as Endangered, has made a remarkable comeback and was delisted in 1999. **Timing of occurrence:** Fall migrants are noted in the state as early as early July, with a notable increase in abundance during September and October. Spring migrants are found from late March through early May, but small numbers linger through the month. **Taxonomy:** Two subspecies occur in the state.

F. p. tundrius White
Uncommon to rare migrant and winter resident throughout the state.

F. p. anatum Bonaparte
Rare summer resident in the Trans-Pecos.

PRAIRIE FALCON *Falco mexicanus* Schlegel

Rare and local summer resident in the mountains of the western Trans-Pecos. Formerly a rare and local summer resident in the Panhandle (Seyffert 2001b) and South Plains; however, no nests have been found in the past decade even though there have been summer sightings that suggest the possibility of nesting. Prairie Falcons are rare to uncommon migrants and winter residents in the High Plains, western Rolling Plains, and Trans-Pecos. This species is a rare and local winter visitor across the eastern Rolling Plains and south to the South Texas Brush Country. Prairie Falcons are casual winter visitors farther east and south but have not been documented from the Pineywoods. **Timing of occurrence:** Fall migrants start to arrive in the state in early September, and the wintering population departs by early May. **Taxonomy**: Monotypic.

ORDER PSITTACIFORMES

Family Psittacidae: Parakeets and Parrots

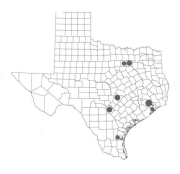

MONK PARAKEET *Myiopsitta monachus* (Boddaert)

This relatively drab parakeet is native to southern South America and has been introduced to the United States. Monk Parakeets are very locally common residents in a number of metropolitan areas, including Austin, Dallas, Galveston, Houston, and San Antonio. There are small colonies in Andrews, Cameron, and Hidalgo Counties, but the colonies are too small for there to be confidence that they will persist. The origin of most of these populations is presumed to be escaped caged birds, although a small number of Monk Parakeets were intentionally released in Austin in the early 1980s. Small groups have been found in other areas of the state but did not persist to become established. This is the only member of its family to construct communal stick nests. Monk Parakeets are common in the pet trade, and sightings away from known populations are likely local escapees. This hardy species naturally occurs in southern South America, where seasonal weather changes are very similar to those found in Texas. **Taxonomy**: The subspecies of the Monk Parakeets present in Texas is unknown. The subspecies most common in the pet trade is *M. m. monachus* (Boddaert), but two others can also be found for sale.

CAROLINA PARAKEET

Conuropsis carolinensis (Linnaeus)

Extinct. Apparently this species was once resident along the Red River Valley west to Montague County. The only report of breeding for Texas comes from Red River County (Oberholser 1974). Carolina Parakeets were irregular nonbreeding summer wanderers southwest to Brown County in the Rolling Plains (Bent 1940). During the fall and winter, these native parakeets were reported from across the eastern third of the state, west to Colorado and Nueces Counties. The last report of the species in the state was a bird killed in Bowie County around 1897 (Oberholser 1974). **Taxonomy**: Monotypic.

GREEN PARAKEET *Aratinga holochlora* (Sclater)

Uncommon to locally common resident of the Lower Rio Grande Valley. Green Parakeets are found primarily in urban areas and were first reported in the state in 1960 (Oberholser 1974). Populations are now well established and appear to be increasing. There has been considerable debate about the provenance of these birds, and it is plausible that some arrived as a result of displacement due to extensive habitat loss in adjacent northeastern Mexico. Escapees cannot be ruled out, although the species has been regarded as an undesirable cage bird (Oberholser 1974). Green Parakeets of unknown provenance have been reported north to Corpus Christi, Nueces County, and Kingsville, Kleberg County. **Taxonomy**: The subspecies that occurs in Texas is *A. h. holochlora* (Sclater).

RED-CROWNED PARROT

Amazona viridigenalis (Cassin)

Red-crowned Parrots have been in the Lower Rio Grande Valley since at least the early 1970s (Neck 1986; J. Arvin, pers. comm.). They are now common in urban areas, where flocks containing up to 250–300 birds have been reported. They are most common in Brownsville, Harlingen, McAllen, and Weslaco, with smaller populations in other metropolitan areas. Red-crowned Parrots are critically endangered in their range in northeastern Mexico, where habitat loss has been extensive. The population in Texas may include some birds that arrived as a result of displacement due to habitat loss. As a result, the Texas population is very important to the continued global existence of this species. Local escapees are very likely responsible for some of the Red-crowned Parrots found in the Lower Rio Grande Valley as well. The provenance of any individual or flock is indeterminable. **Taxonomy**: Monotypic.

ORDER PASSERIFORMES

Family Thamnophilidae: Antbirds

BARRED ANTSHRIKE
Thamnophilus doliatus (Linnaeus)

Accidental. There is only one record for Texas and the United States, and it is the only reported occurrence of any member of this Neotropical family. This record also has the unusual distinction of being the only species on the state list based on an audio recording as the voucher. The bird was recorded near a bright security light in a suburban yard after dark and was never observed. Antbirds are generally not known to vocalize after dark, making this occurrence more unusual. The ABA Checklist Committee chose not to include Barred Antshrike on its list (Pranty et al. 2007). **Taxonomy:** The audio recording was analyzed and determined to represent the subspecies found in northeastern Mexico, *T. d. intermedius* Ridgway.

1 SEPT. 2006, HARLINGEN, CAMERON CO. (TBRC 2006–95)

Family Tyrannidae: Tyrant Flycatchers

NORTHERN BEARDLESS-TYRANNULET
Camptostoma imberbe Sclater

Rare to locally uncommon resident in the Lower Rio Grande Valley, northward to Zapata County and through the Coastal Sand Plain. Northern Beardless-Tyrannulet appears to be increasing in abundance in the Lower Rio Grande Valley. There are records from north of the expected range on the Coastal Prairies from Calhoun and Goliad Counties. There have been two unexpected records from Presidio County in the Trans-Pecos: one in the foothills of the Chinati Mountains on 28 July 2001 (TPRF 1963) and another observed near Ruidoso on 21 January 2002. **Taxonomy:** The subspecies that occurs in South Texas is *C. i. imberbe* Sclater; however, it is likely that the vagrant records from the Trans-Pecos refer to *C. i. ridgwayi* (Brewster), the form found in southern Arizona and western Mexico that is partially migratory.

TBRC Review Species

GREENISH ELAENIA *Myiopagis viridicata* (Vieillot)

Accidental. There is one record for the United States of this tropical flycatcher (Morgan and Feltner 1985). Interestingly, the occurrence of this individual coincided with the presence of the state's lone record of Yucatan Vireo. **Taxonomy:** The subspecies for this record is unknown, but there are two possible source populations. The northern population, *M. v. placens* (Sclater), ranges from southern Tamaulipas, Mexico, to Honduras and appears to be the most likely to occur in Texas. However, *M. v. viridicata* (Vieillot) of south-central South America is an austral migrant and could potentially occur in the state.

20–23 MAY 1984, HIGH ISLAND, GALVESTON CO.
(TBRC 1998-289; TPRF 330)

TBRC Review Species

WHITE-CRESTED ELAENIA
Elaenia albiceps (d'Orbigny & Lafresnaye)

Accidental. The occurrence of this South American flycatcher was one of the more astounding ornithological discoveries made in Texas. The species had never been found outside South America, but the careful documentation of the record (including voice recordings) allowed for a subspecific identification of the migratory southernmost population (Reid and Jones 2009). White-crested Elaenias are found in the Andes from Colombia south through southern Chile. Some researchers have suggested that the southernmost populations are best treated as a separate species, which would include the Texas record. **Taxonomy:** The subspecies for this record is the austral migrant *E. a. chilensis* Hellmayr.

9–10 FEB. 2008, SOUTH PADRE ISLAND, CAMERON CO.
(TBRC 2008-09; TPRF 2535)

TBRC Review Species

TUFTED FLYCATCHER
Mitrephanes phaeocercus (Sclater)

Accidental. There are three records of this tropical flycatcher for the state. The first was discovered at Big Bend NP, Brewster County, in 1991 (Zimmer and Bryan 1993). There are also three records from Arizona. **Timing of occurrence:** When all six US records are considered, there is no established pattern of occurrence. The Texas records might suggest that late fall and early winter are the most likely times to find one of these birds, but the Arizona records are from February, May, and July. **Taxonomy:** The subspecies for the Texas occurrences is unknown. The subspecies found in the mountains of northeastern Mexico is *M. p. phaeocercus* (Sclater), and those from northwestern Mexico are *M. p. tenuirostris* Brewster. Both are possible sources for the Texas records; the western population may be slightly more likely considering the occurrences in Arizona.

3 NOV. 1991–17 JAN. 1992, RIO GRANDE VILLAGE, BREWSTER CO. (TBRC 1991-132; TPRF 1000)

2–6 APR. 1993, 25 MI. WEST OF FORT STOCKTON, PECOS CO. (TBRC 1993-41; TPRF 1149)

21 NOV. 2010–4 JAN. 2011, RIO GRANDE VILLAGE, BREWSTER CO. (TBRC 2010–76; TPRF 2917)

OLIVE-SIDED FLYCATCHER
Contopus cooperi (Nuttall)

Uncommon to rare spring and fall migrant throughout the state. Olive-sided Flycatchers are uncommon summer residents at upper elevations of the Guadalupe Mountains, which is the southernmost extension of the breeding population found in the southern Rocky Mountains. They are rare and irregular during the summer in the Davis Mountains, but nesting has not been observed (Peterson et al. 1991). **Timing of occurrence:** Spring migrants arrive in late April and can be found through early June. The earliest documented spring record is from 13 April, and the latest spring migrant is from 18 June. Fall migrants arrive in the state in early August and are seen through mid-October. Fall migrants have been found as early as 24 July, and an exceptionally late migrant was present 10–15 November. **Taxonomy:** Monotypic.

TBRC Review Species   **GREATER PEWEE** *Contopus pertinax* Cabanis and Heine

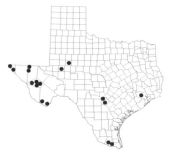

Very rare visitor to the Trans-Pecos and casual farther east. There are 24 documented records for the state, the first quite unexpectedly found in Big Spring, Howard County, from 22 October to 17 November 1984 (TPRF 329). The Trans-Pecos records are primarily from the central mountain ranges, with the majority, nine records, from the Davis Mountains. Three Greater Pewees were present at one location in the Davis Mountains during the summer of 2002, where nesting was documented, providing the only breeding record for the state (TBRC 2002-76). There have been seven additional records from east of the Pecos River: two from Bexar County, one from Midland County, one from Harris County, and three from the Lower Rio Grande Valley. **Timing of occurrence**: There are records from all seasons. Spring migrants have been noted from 7 April to 30 May, and postbreeding wanderers and fall migrants from 4 August to 17 November. The occurrences in the Lower Rio Grande Valley are all from winter, as are two records from the Trans-Pecos, one from Bexar County, and one from Harris County. Summer occurrences have been restricted to the Davis Mountains. **Taxonomy**: The subspecies that occurs in Texas is *C. p. pallidiventris* Chapman.

WESTERN WOOD-PEWEE *Contopus sordidulus* Sclater

Rare to locally common summer resident in the Davis, Chinati, and Guadalupe Mountains. Numerous summer sightings have come from the northern Panhandle, although nesting has not been observed (Seyffert 2001b). Western Wood-Pewee is a common to uncommon migrant through the High Plains, western Rolling Plains, and Trans-Pecos. Curiously, this species has rarely been detected on the Edwards Plateau but may be overlooked in this region because of the presence of Eastern Wood-Pewees. The two species can best be separated by voice, not appearance. Migrants have been documented east to north-central Texas and the upper coast, usually in the fall. **Timing of occurrence**: Spring migrants are found from late April through early June and on rare occasions linger into late June. Fall migrants have been noted from late July through early October. **Taxonomy**: Two subspecies are known to occur in the state.

C. s. veliei Coues
    As described above.

C. s. saturatus Bishop
    Status unknown. Possibly a rare migrant through the Trans-Pecos. There is one specimen from Midland County on 4 September 1965 (*TCWC 8,982).

EASTERN WOOD-PEWEE *Contopus virens* (Linnaeus)

Uncommon and declining summer resident in the eastern half of Texas, west to the eastern edge of the Rolling Plains

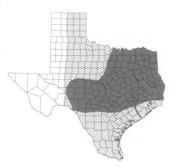

and throughout the Edwards Plateau. This species is a local breeder as far west as the Pecos River and its tributaries. Eastern Wood-Pewees are uncommon migrants in the eastern two-thirds of the state and rare migrants on the High Plains, the western edge of the Edwards Plateau, and the Trans-Pecos. The difficulty in separating these migrants from the more common Western Wood-Pewee in the western third of the state clouds the status of this species. **Timing of occurrence:** Spring migrants begin arriving in mid-April. The primary migration period and arrival of the breeding population is from late April through mid-May. A few lingering migrants have been found as late as early June. Fall migrants have been noted as early as late July and become more common between mid-August and early October. Lingering migrants have been found along the coast as late as early December. **Taxonomy:** Monotypic.

YELLOW-BELLIED FLYCATCHER

Empidonax flaviventris (Baird and Baird)

Uncommon to common migrant in the eastern half of the state. Yellow-bellied Flycatchers are rare migrants through the eastern Rolling Plains and eastern Edwards Plateau. This species is a casual migrant farther west, with records from the western Edwards Plateau (Kinney County), Concho Valley (Tom Green County), and Trans-Pecos (Brewster, Jeff Davis, and Reeves Counties). They are accidental in early winter to the Lower Rio Grande Valley. **Timing of occurrence:** Spring migrants pass through the state primarily in May, with small numbers present during the last week of April. They rarely linger into early June. Fall migrants are found between mid-August and mid-October. There are two reports of birds lingering into December from Cameron County, including one at Brownsville on 31 December 1971. **Taxonomy:** Monotypic.

ACADIAN FLYCATCHER *Empidonax virescens* (Vieillot)

Common to uncommon migrant and summer resident in the eastern third of the state, south in the summer to Bexar and Harris Counties. Acadian Flycatchers are also locally uncommon summer residents on the Edwards Plateau west to the Nueces River, Uvalde County. They have been found sporadically along the Devils River, Val Verde County, but nesting has not been confirmed. Migrants west of the Balcones Escarpment are very rarely encountered away from known breeding areas. **Timing of occurrence:** The first spring migrants arrive in late March and become more common between mid-April and mid-May when the breeding population is present. Fall migrants are found between mid-August and late September, but a few linger to mid-October. **Taxonomy:** Monotypic.

ALDER FLYCATCHER *Empidonax alnorum* Brewster

Uncommon to common migrant in the eastern third of the state. Alder Flycatchers are known to occur west to the eastern Edwards Plateau. Two reports from the South Plains are the westernmost sightings. The status of Alder Flycatcher in the western half of the state is obscured by the presence of the closely related Willow Flycatcher. These two species were formerly considered conspecific under the name Traill's Flycatcher. Alder Flycatchers can only be safely identified from Willow Flycatcher in the field by voice. **Timing of occurrence:** Spring migrants are found between very late April and early June, and fall migrants are present between late July and early October. **Taxonomy:** Monotypic.

WILLOW FLYCATCHER *Empidonax traillii* (Audubon)

Uncommon to rare migrant throughout the state. Willow Flycatcher was formerly considered conspecific with the Alder Flycatcher (the two forms together known as Traill's Flycatcher). Since the two taxa were recognized, Willow Flycatcher was initially thought to be the more common in Texas. However, recent evidence based on birds positively identified by voice shows that Willow Flycatcher is less common than Alder in the eastern half of the state. Willow Flycatcher was formerly a rare summer resident in the central and western Trans-Pecos, with the last nesting record from the 1890s in Brewster County. There is a breeding population in southeastern Oklahoma along the Red River, and these flycatchers may occur as a very rare breeder in northern Bowie and Red River Counties. Willow Flycatchers can only be safely identified from Alder Flycatcher in the field by voice. **Timing of occurrence:** Spring migrants are found between very late April and early June, and fall migrants are present between late July and early October. **Taxonomy:** Three subspecies are known to have occurred in the state; however, the extent of the range of each taxon is poorly understood.

E. t. adastus Oberholser

Uncommon to rare migrant through the Trans-Pecos east through at least the western parts of the High Plains south through the western Edwards Plateau to the western South Texas Brush Country.

E. t. extimus Phillips

Formerly a rare summer resident from El Paso County east to Brewster and Jeff Davis Counties, with the last reported records from the 1890s. These flycatchers are very likely very rare migrants through the western third of the Trans-Pecos. The Southwestern Willow Flycatcher is listed by the US Fish and Wildlife Service as Endangered.

E. t. traillii (Audubon)

Uncommon to rare migrant through at least the eastern two-thirds of the state.

LEAST FLYCATCHER
Empidonax minimus (Baird and Baird)

Common to uncommon migrant throughout the eastern three-quarters of the state and likely our state's most abundant *Empidonax* during migration. Least Flycatchers are generally uncommon migrants in the eastern third of the Trans-Pecos, becoming very rare to casual in the western third of the region. In the Trans-Pecos, this species occurs with much greater frequency in the fall, when these flycatchers can be common as far west as the Davis Mountains. Least Flycatchers are rare to very rare winter residents in the Lower Rio Grande Valley and along the coast, with a few scattered winter records from inland locations. **Timing of occurrence**: Spring migrants are present between mid-April and early June. Fall migrants have been found in early July, but the primary migration period is from late July through late October. Straggling birds have been present until late November. **Taxonomy**: Monotypic.

HAMMOND'S FLYCATCHER
Empidonax hammondii (Xántus de Vesey)

Uncommon migrant through the western half of the Trans-Pecos, becoming rare farther east in that region. Ham-

mond's Flycatcher is a casual to very rare migrant through the western High Plains south to the Concho Valley. This species is generally a more common fall migrant and has been recorded casually east to the central Coastal Prairies and accidentally farther east to Harris County. Hammond's Flycatcher is casual as a winter visitor in the western Trans-Pecos, and there are scattered winter records elsewhere, north to the South Plains, east to Bastrop County, and south to Cameron County. **Timing of occurrence:** Spring migrants are present from early April to late May, and migration peaks from late April through mid-May. Fall migrants have been found in early August, although the primary migration period is from late August through mid-September, and a few linger until early October. **Taxonomy:** Monotypic.

GRAY FLYCATCHER *Empidonax wrightii* Baird

Locally common summer resident in the Davis Mountains and very locally in the Guadalupe Mountains. The population in the Davis Mountains was discovered in 1989, and breeding was documented in 1991 (Peterson et al. 1991), but not until 2012 were they found nesting in the Guadalupe Mountains (Lockwood 2012). Gray Flycatchers are an uncommon to rare migrant in the western Trans-Pecos and a rare to casual migrant farther east in the region and on the western High Plains. There are very few records farther east, with an exceptional record from near Sam Rayburn Reservoir, San Augustine County, on 6 September 2004. Gray Flycatchers are rare to locally uncommon winter residents along the Rio Grande in Brewster and Presidio Counties. Winter records exist outside the Trans-Pecos north to Bailey County, east to Matagorda and Nueces Counties, and south to Hidalgo County. **Timing of occurrence:** Spring migrants are found from mid-April to late May, and the breeding populations in the Trans-Pecos also begin to arrive in mid-April. Fall migrants are found from mid-August through early October, although a few linger into early November. **Taxonomy:** Monotypic.

DUSKY FLYCATCHER *Empidonax oberholseri* Phillips

Common migrant through the western half of the Trans-Pecos and a rare migrant through the remainder of the Trans-Pecos. This flycatcher has been reported with in-

creasing frequency from the western High Plains during migration. Dusky Flycatchers are rare and very local summer residents at the highest elevations of the Davis Mountains (Bryan and Karges 2001). They probably nested in the Guadalupe Mountains in 1972 (Newman 1974) and possibly at other times, but the first documented nesting record for Texas came from the Davis Mountains in 2000 (TPRF 1809). Dusky Flycatchers are rare to locally uncommon winter residents along the Rio Grande in the southern Trans-Pecos. **Timing of occurrence:** Spring migrants are present from mid-April to late May. The breeding population in the Davis Mountains does not appear to be present until mid-May. Fall migrants are found from late July through early October, but a few linger into early November. **Taxonomy:** Monotypic.

CORDILLERAN FLYCATCHER
Empidonax occidentalis Nelson

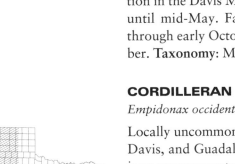

Locally uncommon to rare summer resident in the Chisos, Davis, and Guadalupe Mountains. Cordilleran Flycatcher is an uncommon to rare migrant throughout the remainder of the Trans-Pecos and on the western High Plains. This species may be a very rare migrant through the western Rolling Plains and western Edwards Plateau. Cordilleran Flycatcher was formerly considered conspecific with the Pacific-slope Flycatcher (*E. difficilis* Baird) under the name Western Flycatcher. These two species are impossible to visually differentiate in the field; therefore, the actual status of these species as migrants through Texas is unknown. There are vocal differences, but the diagnostic calls are very rarely given during migration. There are many reports of "Western" Flycatchers from the eastern two-thirds of the state, but few include documentation. There is one documented winter record from near Pearsall, Frio County. **Timing of occurrence:** Spring migrants are found from early April through late May, and a few linger into early June. Fall migrants pass through the state from late July through early October, and some linger until late November. **Taxonomy:** The subspecies that occurs in Texas is *E. o. hellmayri* Brodkorb.

TBRC Review Species

BUFF-BREASTED FLYCATCHER

Empidonax fulvifrons (Giraud)

Although there are 27 accepted summer records for this species, the status of Buff-breasted Flycatcher in Texas is difficult to assess. The first record for the state was a male discovered on the Davis Mountains Preserve, Jeff Davis County, on 3 May 1999 (Horvath and Karges 2000). A single pair of these birds was present within the Davis Mountains Preserve from 1999 to 2004 and in most years successfully fledged young. In 2005 a second territory was discovered, and a high count of five territories existed in 2006; however, in 2010 only a single territorial male could be found, and that continued in 2011 and 2012. Wildfires in 2011 and 2012 heavily impacted the open ponderosa pine (*Pinus ponderosa*) glades these birds occupy, making the continued occurrence of the species in the state uncertain. **Timing of occurrence:** The earliest spring arrival date is 8 April, and the latest fall date is 28 September. **Taxonomy:** The subspecies that occurs in Texas is *E. f. pygmaeus* Coues.

BLACK PHOEBE *Sayornis nigricans* (Swainson)

Locally uncommon to rare resident in the Trans-Pecos, the western two-thirds of the Edwards Plateau, and southward near the Rio Grande to the central Lower Rio Grande Valley. Black Phoebes are closely tied to rivers and other bodies of water and thus have a very discontinuous range. This species is a rare to very rare winter visitor to the remainder of the Edwards Plateau and to the South Plains. Populations in the Trans-Pecos fluctuate during the winter and become less common in the mountains and more common along the Rio Grande. Black Phoebes are casual visitors to the Panhandle from April to December and to the upper and central Coastal Prairies and South Texas Brush Country away from the Rio Grande during the fall and winter. Vagrants have been documented from north-central Texas. **Taxonomy:** The subspecies that occurs in Texas is *S. n. semiatra* (Vigors).

EASTERN PHOEBE *Sayornis phoebe* (Latham)

Uncommon to common summer resident in the northern two-thirds of the state, except in the eastern half of the

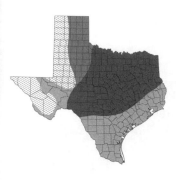

Panhandle. Eastern Phoebe is generally absent from the Trans-Pecos during the summer. These flycatchers are common migrants in the eastern two-thirds of the state, becoming increasingly less common westward to the extreme western Trans-Pecos, where they are rare. They are common winter residents over the eastern two-thirds of the state but generally absent from the High Plains south to the northwestern Edwards Plateau. This flycatcher is also a rare to locally uncommon winter resident in the eastern half of the Trans-Pecos, primarily along the Rio Grande and Pecos River drainage. **Taxonomy:** Monotypic.

SAY'S PHOEBE *Sayornis saya* (Bonaparte)

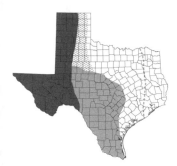

Common to uncommon resident in the Trans-Pecos and east to the western edge of the Edwards Plateau. Say's Phoebes are rare and local summer residents on the High Plains of the western Panhandle south through the South Plains to the northwestern edge of the Edwards Plateau. They are uncommon to rare winter residents in the western Edwards Plateau. They are also rare to very rare migrants and winter visitors through the Rolling Plains and uncommon during these seasons in the South Texas Brush Country. This species is a rare, but regular, winter visitor east to the central coast and the Blackland Prairies region from north-central Texas south to Bastrop County. They are casual visitors to the eastern quarter of the state, mostly in fall and winter. **Timing of occurrence:** Fall migrants arrive in wintering areas from mid-September in the north to mid-October in the south. These birds depart between early March and mid-May. **Taxonomy:** The subspecies that occurs in Texas is *S. s. saya* (Bonaparte).

VERMILION FLYCATCHER
Pyrocephalus rubinus (Boddaert)

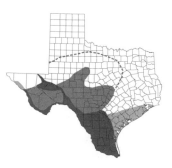

Uncommon to common summer resident across the southern Trans-Pecos east through the Edwards Plateau and South Texas Brush Country. Vermilion Flycatchers are rare summer visitors throughout the High Plains and Rolling Plains east through north-central Texas, with a few isolated breeding records (Sexton 1999). This species is an uncommon winter resident along the Rio Grande from eastern Hudspeth County in the Trans-Pecos eastward through the South Texas Brush Country and north along the Coastal Prairies. These flycatchers regularly wander eastward during the fall and winter as far as the Pineywoods. **Timing of occurrence**: Spring migrants reach breeding areas in early March, and most are present by early April. Summer residents are present into early October, which coincides with the arrival of wintering individuals farther south and east. **Taxonomy**: Two subspecies occur in the state.

P. r. mexicanus Sclater
Uncommon to common summer resident from the southern Rolling Plains through the Edwards Plateau and resident in the South Texas Brush Country. This subspecies is found in winter east and north of the breeding range.

P. r. flammeus van Rossem
Uncommon to common summer resident in the Trans-Pecos, where it is also uncommon in winter along the Rio Grande.

DUSKY-CAPPED FLYCATCHER
Myiarchus tuberculifer (d'Orbigny and Lafresnaye)

Rare and local summer resident in the Davis Mountains and very rare and irregular in summer in the Chisos Mountains. Prior to 1991 Texas had only three documented records of this species, but since the late 1990s Dusky-capped Flycatchers have colonized the upper elevations of the Davis Mountains and have been known to nest there annually since 2003. The first breeding record for Texas was documented in the Chisos Mountains during the summer of 2000. In West Texas away from these two mountain ranges there are single records from El Paso (11 May 1891) and Midland (3 September 1994) Counties. Dusky-capped Flycatchers are very rare to rare winter visitors to the Lower Rio Grande Valley. The first occurrence in this area was in the winter of 2000–2001, and they are now almost or actually annual in very small numbers. There were 51 documented records when this species was removed from the Review List in 2010; however, details are still requested for sightings from South Texas in order to better document their true abundance in this region. **Timing of occurrence**: The nesting population in the Trans-Pecos is present from late April through late July. Wintering birds in the Lower Rio Grande Valley have been found between 5 November and 3 April. **Taxonomy**: Two subspecies occur in the state.

M. t. olivascens Ridgway

Rare summer resident in the Davis and Chisos Mountains of the Trans-Pecos.

M. t. lawrenceii (Giraud)

Very rare to rare winter visitor to the Lower Rio Grande Valley. The TBRC requests documentation for all sightings of this taxon.

ASH-THROATED FLYCATCHER

Myiarchus cinerascens (Lawrence)

Common to uncommon summer resident in the western half of the state. Ash-throated Flycatchers are found in summer east through the Rolling Plains and Edwards Plateau and south through the South Texas Brush Country. They are rare migrants farther east in the easternmost Oaks and Prairies region. This species is rare to locally uncommon in winter along the Rio Grande in the southern Trans-Pecos, in the South Texas Brush Country, and along the Coastal Prairies. Ash-throated Flycatchers are casual migrants and winter visitors in the Pineywoods, and there are isolated winter records north to north-central Texas. **Timing of occurrence:** They begin to arrive in the state in mid-March, and migrants are present away from breeding areas through early May. The breeding population begins to disperse in late July, and fall migrants are present through mid-October, although a few straggle into early November. **Taxonomy:** The subspecies that occurs in Texas is *M. c. cinerascens* (Lawrence).

TBRC Review Species

NUTTING'S FLYCATCHER *Myiarchus nuttingi* Ridgway

Accidental. There is one record for the state, and it is by far the easternmost for the United States. Nutting's Flycatchers occur from western Chihuahua south along the west coast of Mexico to northwestern Costa Rica. The distinctive call notes of this species alerted observers to its presence and eliminated the similar Ash-throated Flycatcher. The winter occurrence of this bird fits the pattern of occurrences from elsewhere in the United States, of which there are five records from Arizona and one from California. **Taxonomy:** The subspecies that has occurred in Texas is *M. n. inquietus* Salvin & Godman.

31 DEC. 2011–11 JAN. 2012, SANTA ELENA CANYON, BIG BEND
NP, BREWSTER CO. (TBRC 2012–001; TPRF 2971)

GREAT CRESTED FLYCATCHER
Myiarchus crinitus (Linnaeus)

Common to uncommon summer resident in the eastern
half of the state westward through the eastern Panhandle
and eastern Rolling Plains and across the Edwards Plateau
to the Nueces River drainage. The southern edge of the
breeding range in Texas is bounded by the Nueces River
drainage on the central coast. Great Crested Flycatchers
are common migrants in the eastern half of the state, west
to the central Edwards Plateau. Farther west, this species
becomes increasingly rare as a migrant and is casual during
fall in the Trans-Pecos. Great Crested Flycatcher is casual
in winter along the Coastal Prairies and in the South Texas
Brush Country. **Timing of occurrence:** Spring migrants arrive as early as mid-March, but the majority of migrants
are present from early April to mid-May. Summer residents
begin to depart in early September. Fall migrants are present through mid-October, although a few linger as late as
December. Most winter records are from December and
are presumed to represent lingering migrants, but there is
one documented record from Live Oak County on 23 January 2010. **Taxonomy:** Monotypic.

BROWN-CRESTED FLYCATCHER
Myiarchus tyrannulus (Statius Müller)

Common summer resident in the Lower Rio Grande
Valley. Brown-crested Flycatchers are uncommon to rare
summer residents north through the South Texas Brush
Country and western Edwards Plateau to the Concho
Valley. They are very rare and local summer residents
along the Rio Grande to western Brewster County and are
rare along riparian corridors in Terrell and Val Verde
Counties. They are locally uncommon summer residents as
far north as Bexar, Gonzales, and Calhoun Counties, with
summer records east into Matagorda County. This species
appears to be expanding its range northward, with summer
records often involving pairs found north to Travis and
Bastrop Counties. Brown-crested Flycatcher is also a very
rare to casual summer visitor to El Paso County in far West

Texas. In winter, the species is very rare to casual on the Coastal Prairies north to Brazoria and Galveston Counties and accidental in the Lower Rio Grande Valley north to Webb County. **Timing of occurrence**: Summering birds arrive on breeding grounds from mid-April to early May. Summer residents begin to depart between mid-August and early October. **Taxonomy**: Two subspecies occur in the state.

M. t. cooperi Baird
   As described above.

M. t. magister Ridgway
   Very rare to casual summer visitor to El Paso County and possibly elsewhere in the western Trans-Pecos.

GREAT KISKADEE *Pitangus sulphuratus* (Linnaeus)

Locally common resident in the Lower Rio Grande Valley. Great Kiskadee is an uncommon and local resident north along the Rio Grande and its tributaries to southern Val Verde County. This species is also an uncommon and local resident north along the coast to Nueces County and locally north to Calhoun County. Great Kiskadees nested farther up the coast in Baytown, Chambers County, in 2002. They have wandered during all seasons as far north as the Panhandle and north-central Texas, east to the Pineywoods, and west to Big Bend NP, Brewster County, and Imperial Reservoir, Pecos County. They are occurring with increasing frequency in all seasons on the Coastal Prairies north to Jefferson County and inland to Bexar, Bastrop, and Travis Counties, suggesting that the species is undergoing a northward range expansion. **Taxonomy**: The subspecies that occurs in Texas is *P. s. texanus* van Rossem.

TBRC Review Species   **SOCIAL FLYCATCHER** *Myiozetetes similis* (von Spix)

Accidental. There are three accepted records of this tropical flycatcher from the Lower Rio Grande Valley. The first is a specimen from Cameron County housed in the British Museum that was recently brought to the TBRC's attention. Of the two modern records, only the most recent one allowed for thorough documentation (Arvin and Lockwood 2006). **Taxonomy**: The subspecies that has occurred in Texas is *M. s. texensis* (Giraud).

15 FEB. 1895, CAMERON CO.
 (*BMNH 98-7-12-88; TBRC 2009–39; TPRF 2721)
17 MAR.–5 APR. 1990, ANZALDUAS COUNTY PARK, HIDALGO CO.
 (TBRC 1990–83)
7–14 JAN. 2005, BENTSEN–RIO GRANDE VALLEY SP, HIDALGO CO.
 (TBRC 2005–8; TPRF 2261)

TBRC Review Species   ## SULPHUR-BELLIED FLYCATCHER
Myiodynastes luteiventris Sclater

Very rare visitor to the state, with 22 documented records. The first record for the state was a bird at Anahuac NWR, Chambers County, on 2 September 1965. Oddly, the second occurrence involved a nesting pair on the Santa Margarita Ranch, Starr County, in 1975. The majority of the documented occurrences are of spring migrants (11) and postbreeding dispersers (5) that found their way to the Coastal Prairies. There are two summer records from the Trans-Pecos and one from the Lower Rio Grande Valley of a bird that engaged in nesting activity. Reports of Sulphur-bellied Flycatcher require detailed notes and preferably close photography to eliminate the similar Streaked Flycatcher, *M. maculates* (Statius Müller). Streaked Flycatcher has not been documented in the United States but is a potential vagrant to Texas due to its migratory nature and the close proximity of its northernmost summer range in northern Mexico. There is one accepted state record of "Sulphur-bellied/Streaked Flycatcher" where the documentation established that it was one of these two species but was not sufficient to identify it to species. There are 15 reports of Sulphur-bellied Flycatchers from prior to the development of the Review List in 1988 for which there is no documentation on file. **Timing of occurrence**: Records pertaining to spring migrants are from 18 April to 22 May, with summering birds present from 7 May to 20 August. The records from the fall are from 27 August to 22 September. **Taxonomy**: Monotypic.

TBRC Review Species   ## PIRATIC FLYCATCHER *Legatus leucophaius* (Vieillot)

Accidental. Texas has five documented records of this tropical flycatcher. The only other records of Piratic Flycatcher for the United States come from Florida and New Mexico. The very similar Variegated Flycatcher, *Empidonomus varius* (Vieillot), has also been documented in the United States and Canada, and care must be taken in identification

to eliminate this species. There is one additional sighting, including photographs, that the TBRC accepted as a Piratic or Variegated Flycatcher. **Taxonomy:** The single specimen from Texas is from the subspecies found in Mexico, *L.l. variegatus* (Sclater). It has been postulated that some of the Piratic Flycatchers found in the United States could be from the migratory population in South America, *L.l. leucophaius* (Vieillot).

4 APR. 1998, RIO GRANDE VILLAGE, BREWSTER CO.
(TBRC 1998-60; TPRF 1685)
21–22 OCT. 2000, OFF PADRE ISLAND, WILLACY CO.
(TBRC 2000-126; TPRF 1933)
20–28 MAR. 2006, BENTSEN–RIO GRANDE VALLEY SP, HIDALGO
CO. (TBRC 2006–32)
28 SEPT. 2007, PASADENA, HARRIS CO.
(*LSUMNS; TBRC 2007–99; TPRF 2530)
3–6 MAY 2008, CORPUS CHRISTI, NUECES CO.
(TBRC 2008–32; TPRF 2549)

TROPICAL KINGBIRD *Tyrannus melancholicus* Vieillot

Uncommon and local resident in the Lower Rio Grande Valley. The status of Tropical Kingbird in Texas has changed dramatically during the past 20 years. Prior to 1991 only a single documented record existed for the state, a specimen collected on 5 December 1909 near Brownsville, Cameron County. Since 1991 this species has become a permanent resident in Cameron and Hidalgo Counties. Tropical Kingbirds have been found occasionally farther up the Rio Grande in Starr and Webb Counties, where they are still rare. Since 1997 Tropical Kingbirds have nested annually at Cottonwood Campground in Big Bend NP, Brewster County. Interestingly, there are summer records from elsewhere in the southern Trans-Pecos, but nesting has not been found. Since 2010, pairs have been discovered on the Coastal Prairies in Aransas, Galveston, and Nueces Counties, which may be part of a continuing range expansion. There are additional records of single Tropical Kingbirds from all seasons that have been conclusively identified by voice from along the Coastal Prairies, but this is still a very rare occurrence. There are only four sightings from farther inland and away from the known breeding site in Brewster County. Two are of nesting pairs: one near Marathon, Brewster County, in 1999; and the other in Midland County in 2001. The other two records were a single individual along the Nueces River in Uvalde County and one or two birds in Laredo, Webb County, both in 2012. In addition, sightings of silent kingbirds that are either this species or Couch's Kingbird (*T. couchii*) are increasing in regularity along the coast and inland through the southeastern quarter of the state. **Taxonomy:** The subspecies that occurs in Texas is *T. m. satrapa* (Cabanis & Heine).

COUCH'S KINGBIRD *Tyrannus couchii* Baird

Common to uncommon summer resident in the Lower Rio Grande Valley. Couch's Kingbirds are locally uncommon northward through the South Texas Brush Country to Val Verde, Uvalde, and Bexar Counties and along the coast to Calhoun County. The breeding range of this species appears to be slowly expanding northward. There are isolated breeding records up the Coastal Prairies to Chambers County and as far inland as Kimble, San Saba, and Williamson Counties. Couch's Kingbirds are very rare visitors to the upper Coastal Prairies and Pineywoods in fall and winter, although the frequency has increased in recent years. Vocalizing individuals definitively identified as Couch's have been found as far north as Lubbock and Tarrant Counties. Couch's Kingbirds withdraw southward from the northern portions of their range during winter and are uncommon to rare, though irregular, during this season in the Lower Rio Grande Valley. Increasingly, they are wintering along the immediate coast in urban and riparian habitats where they can be locally uncommon as far north as the central coast. **Timing of occurrence:** Spring migrants begin to arrive in mid-March, and summer residents north of the Lower Rio Grande Valley depart by mid-November. **Taxonomy:** Monotypic.

CASSIN'S KINGBIRD *Tyrannus vociferans* Swainson

Uncommon to locally common summer resident in mid- and upper elevations of the central Trans-Pecos from the Guadalupe Mountains south through the Davis and Chinati Mountains. Elsewhere in the state, a pair of Cassin's Kingbirds was discovered breeding in the northwestern corner of the Panhandle in 1983. There have been other summer sightings from Dallam County, but additional nestings have not been confirmed. Cassin's Kingbird is an uncommon to rare migrant throughout the remainder of the Trans-Pecos and through the western High Plains. Migrants are casual to accidental east to Chambers, Nolan, Travis, and Lee Counties. Cassin's Kingbird is a casual winter visitor to the South Texas Brush Country and the southernmost portion of the Trans-Pecos. **Timing of occurrence:** Spring migrants arrive in very late March, and migration peaks between mid-April and mid-May. Fall migration occurs between late August and early October, but small numbers linger into early November. **Taxonomy:** The subspecies that occurs in Texas is *T. v. vociferans* Swainson.

TBRC Review Species

THICK-BILLED KINGBIRD
Tyrannus crassirostris Swainson

Casual to accidental visitor to the state, with 18 documented records, 11 of which pertain to returning pairs in Big Bend NP more than 20 years ago. Thick-billed Kingbird was first documented in Texas in the Chisos Basin of Big Bend NP, Brewster County, on 21 June 1967 (Wauer 1967). One was again encountered at Big Bend during the summer of 1985. Between 1985 and 1991, Thick-billed Kingbirds were summer residents at Rio Grande Village (seven documented records) and Cottonwood Campground (four documented records). They bred successfully at Cottonwood Campground from 1988 through 1991. Since then there have been only four records in the southern Trans-Pecos. There are also four records well away from the Trans-Pecos: one from Palo Duro Canyon, Randall County; and an individual present for three consecutive winters at Selkirk Island, Matagorda County, from December 2002 to December 2004. **Timing of occurrence:** The records in the Trans-Pecos have been between 1 April and 24 September. **Taxonomy:** The subspecies that has occurred in Texas is *T. c. pompalis* Bangs & Peters.

WESTERN KINGBIRD *Tyrannus verticalis* Say

Common to uncommon summer resident in the western two-thirds of the state. In the eastern third, Western Kingbirds are uncommon to locally common summer residents in the Post Oak Savannah but generally absent from the Pineywoods. In the Trans-Pecos, Western Kingbirds are generally absent during summer from the mid- and upper elevations of the three major mountain ranges. These kingbirds are common to uncommon migrants throughout the state west to the Pineywoods, where they are rare in spring and very rare in fall. **Timing of occurrence:** Spring migrants start to arrive in very late March, and migration peaks between mid-April and mid-May. In the eastern half of their range the breeding population disperses by late July and early August, but small numbers of migrants can be encountered through October. On rare occasions late migrants can be encountered into December along the Coastal Prairies and are casual as late as early January. In the Trans-Pecos and east to the western High Plains, fall migration is more pronounced from September through mid-October, and a few straggle into early November. **Taxonomy:** Monotypic.

EASTERN KINGBIRD *Tyrannus tyrannus* (Linnaeus)

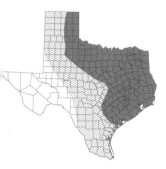

Common to locally uncommon summer resident in the eastern half of the state as far west as the central Panhandle and the eastern edge of the Edwards Plateau. This species is a common to uncommon migrant through the eastern two-thirds of the state, becoming abundant along the upper and central coasts. During migration these kingbirds frequently congregate along the coast in large flocks, sometimes numbering in the thousands. Eastern Kingbird is a rare migrant in the eastern Trans-Pecos and very rare to casual in the western half. There are well-documented winter records from Bexar and Nueces Counties, both from late January. **Timing of occurrence:** Spring migrants start to arrive in late March, and migration peaks between mid-April and mid-May. Fall migration is between mid-August and early October, although a few linger into early November. **Taxonomy:** Monotypic.

TBRC Review Species ### GRAY KINGBIRD *Tyrannus dominicensis* (Gmelin)

Casual visitor, with all records in close proximity to the coast. There are 10 documented records of this kingbird for the state. The birds have been found from Jefferson County close to the Louisiana border all the way to South Padre Island, Cameron County. Interestingly, half of them are from locations on the central coast, with two different individuals at the same site in successive years. Gray Kingbirds are summer residents along the eastern Gulf Coast to southern Alabama and islands off the coast of Mississippi. **Timing of occurrence:** Six of the records are during spring migration, with dates ranging from 24 April to 2 June. Three are from fall migration between 31 August and 29 October, and the latest record is from 6 to 21 November. **Taxonomy:** The subspecies that has occurred in Texas is *T. d. dominicensis* (Gmelin).

SCISSOR-TAILED FLYCATCHER

Tyrannus forficatus (Gmelin)

Common to locally abundant summer resident throughout the eastern three-quarters of the state. Scissor-tailed Flycatchers are locally uncommon in the western Panhandle south through the eastern Trans-Pecos. In the rest of the Trans-Pecos they are rare and very local migrants and summer visitors west through Reeves County and eastern Brewster County. They are casual migrants and summer visitors farther west. There are many records of birds overwintering in the state, but they are most commonly encountered along the coast and the southernmost parts of the state, where they are rare to very rare north to the vicinity of Choke Canyon Reservoir in McMullen and Live Oak Counties. **Timing of occurrence:** Spring migrants start to arrive in southernmost Texas in late February, and summering birds are present throughout the state by early April. Fall migration occurs from early September through October, with numbers present well into November. In fall this species often gathers in flocks that sometimes exceed more than 200 birds. Late migrants can be found through December in the southern half of the state and particularly along the coast. **Taxonomy:** Monotypic.

TBRC Review Species

FORK-TAILED FLYCATCHER *Tyrannus savana* Vieillot

Very rare visitor to the state, with records from the fall, winter, and spring. There are 24 documented records for Texas, and all but six are from along the Coastal Prairies. Those six records include four from Travis County (one of which involved two birds), one on a petroleum platform off North Padre Island, Kenedy County, one from El Paso County, and one from Reeves County. Fork-tailed Flycatchers are annual visitors to the United States, with the large majority of records from along the east coast (Lockwood 1999). These birds are austral migrants that overshoot their normal wintering grounds in northern South America or have undergone a reverse migration. **Timing of**

occurrence: Spring migrants have been found between 15 March and 17 May, fall migrants between 10 September and 10 November, and winter records between 4 December and 4 February. **Taxonomy:** Two subspecies are thought to have occurred in the state.

T. s. monachus Hartlaub
Status uncertain. There are two, or possibly three, records from the lower coast that, based on documentation and photographs, appear to belong to this subspecies, which is typically resident in northern Central America north to southern Mexico but is prone to nomadic wandering.

T. s. savana Vieillot
There are two specimens of this subspecies for the state. There have also been two recent records for which photos were obtained of the spread primaries that strongly suggest this taxon. Records for the United States are generally believed to belong to this highly migratory subspecies from South America.

Family Tityridae: Becards, Tityras, and Allies

TBRC Review Species

MASKED TITYRA *Tityra semifasciata* (Spix)

Accidental. The one record for the state is also the only record for the United States. Masked Tityra occurs as close as central Tamaulipas in northeastern Mexico, where this species occupies humid and semiarid forest edges in lowlands up to 5,000 feet in elevation (Howell and Webb 1995). **Taxonomy:** The subspecies of the Texas record is not known with certainty. *T. s. personata* Jardine & Selby occurs in northeastern Mexico and is the most likely to occur in Texas.

17 FEB.–10 MAR. 1990, BENTSEN–RIO GRANDE VALLEY SP, HIDALGO CO. (TBRC 1990-33; TPRF 860)

TBRC Review Species

ROSE-THROATED BECARD
Pachyramphus aglaiae (Lafresnaye)

Very rare and irregular visitor to the Lower Rio Grande Valley, primarily Hidalgo County, although annual since 2008. Rose-throated Becards were formerly a rare and local resident in the Lower Rio Grande Valley, limited to Cameron and Hidalgo Counties. Since the mid-1970s there have been 48 documented records for the state, 28 of

which were from Hidalgo County. There have been six summer records since 1980 and four nesting attempts, all unsuccessful and two where only a female was present. Three records come from outside the Lower Rio Grande Valley. Single individuals were documented near Fort Davis, Jeff Davis County, on 18 July 1973 (Runnels 1975); near Rockport, Aransas County, on 10 January 1990; and on the King Ranch, Kenedy County, on 20 January 1992. The 27 reports of Rose-throated Becards from prior to the development of the Review List in 1988 have no documentation on file. **Timing of occurrence:** There are records from all seasons, but the majority are between late November and mid-March. **Taxonomy:** The subspecies that has occurred in southern Texas is *P. a. gravis* (van Rossem). The subspecies of the record from Jeff Davis County is unknown, but could be *P. a. albiventris* (Lawrence) from southeastern Arizona and western Mexico.

Family Laniidae: Shrikes

LOGGERHEAD SHRIKE *Lanius ludovicianus* Linnaeus

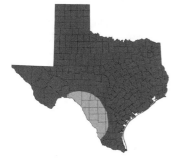

Rare to locally common resident throughout the state, except the northwestern South Texas Brush Country and southwestern Edwards Plateau, where these birds are largely absent during the summer. Loggerhead Shrikes are generally more common as migrants and winter residents throughout the state. In many areas the local resident populations increase with the influx of migrants that overwinter, and this is particularly true on the Coastal Prairies. Loggerhead Shrike populations have declined precipitously, especially in the eastern half of the United States, from which Texas derives much of its wintering population. In contrast, Loggerhead Shrikes in the southern South Texas Brush Country are increasing as nesting birds. **Timing of occurrence:** Fall migrants begin to arrive in the state in early August and are arriving through October, when the winter population becomes established. Spring migrants begin departing in mid-March, and most are gone by mid-April, leaving the resident or summering population. **Taxonomy:** The taxonomy of this species is complex, with weakly differentiated subspecies. Three subspecies are included here, following A. Phillips (1986).

L.l. excubitorides (Swainson)

Uncommon to common resident in the western half of the state, east to the eastern edge of the Rolling Plains, and south to the central coast. Includes *L.l. sonoriensis* Miller and *L.l. gambeli* Ridgway.

L.l. migrans Palmer

Rare resident in the northeastern portion of the state. This subspecies has declined significantly and is considered of conservation concern. These shrikes were formerly common winter residents, with specimens from as far south as the Lower Rio Grande Valley from the early 1900s.

L.l. ludovicianus Linnaeus

Locally common to rare resident in the eastern half of the state. Intergrades with *L.l. excubitorides* to the west.

NORTHERN SHRIKE *Lanius excubitor* Linnaeus

Rare to very rare winter resident in the Panhandle and very rare on the South Plains. Northern Shrikes are casual to accidental winter visitors away from the Panhandle and South Plains. There are six records from the northeastern Rolling Plains and northwestern Oaks and Prairies region, one from Irion County in the Concho Valley, and three from the Trans-Pecos. The primary wintering area in Texas is the northern Panhandle, an area that does not have consistent coverage by birders, which allows for uncertainty about the status of this species in the state. The winter of 2012–13 was an invasion year in the Panhandle and South Plains and provided the most sightings for a single season in the last 25 years. **Timing of occurrence:** The first wintering individuals are rarely found before mid-November, and most winter residents depart by early March. **Taxonomy:** The subspecies that occurs in Texas is *L. e. borealis* Vieillot.

Family Vireonidae: Vireos

WHITE-EYED VIREO *Vireo griseus* (Boddaert)

Common to uncommon migrant throughout the eastern two-thirds of the state. Farther west, White-eyed Vireo is a very rare spring and casual fall migrant to the eastern Panhandle and South Plains south through the eastern Trans-Pecos. These vireos are common to locally abundant sum-

mer residents through the eastern two-thirds of the state
west to the central Edwards Plateau. Farther west they are
uncommon and somewhat local summer residents through
the Concho Valley and western Edwards Plateau to the
Devils River drainage, Val Verde County, and more local
along the Rio Grande and Pecos River in Terrell County.
This species is a locally uncommon to rare winter resident
from the southern Pineywoods westward to the Balcones
Escarpment and an uncommon to common winter resident
south of the Edwards Plateau. The winter range has moved
northward into the southern Edwards Plateau in the past
10 years, and they are now locally uncommon to rare winter residents in the southern third of that region, including
along the Devils River in Val Verde County. White-eyed
Vireos are also rare winter visitors along the Rio Grande
west to eastern Presidio County. **Timing of occurrence**:
Spring migrants are found from mid-March to early May.
Fall migrants pass through the state between mid-August
and mid-October. **Taxonomy**: Two subspecies are known
to occur in the state.

V. g. griseus (Boddaert)
>   Locally abundant to uncommon migrant and summer resident
>   throughout the northern three-quarters of the state, west to
>   Val Verde County.

V. g. micrus Nelson
>   Common resident in the Lower Rio Grande Valley and north
>   through the South Texas Brush Country to Maverick, McMul-
>   len, and Nueces Counties.

BELL'S VIREO *Vireo bellii* Audubon

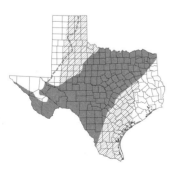

Locally common to uncommon summer resident in the
Trans-Pecos away from the arid northwest and eastward to
the eastern Edwards Plateau. Bell's Vireo is a rare to very
rare and local summer resident eastward through the
northeastern Pineywoods, north through the Rolling
Plains, and south through the western South Texas Brush
Country. There are scattered summer records from the
eastern Panhandle, but evidence of nesting is generally
lacking (Seyffert 2001b). Bell's Vireos are uncommon to
rare migrants from the central Panhandle and South Plains
east through the Oaks and Prairies region and very rare
east and west of this corridor. They are very rare to casual

winter visitors in the Lower Rio Grande Valley, with scattered records northward on the Coastal Prairies, and one documented winter record from El Paso County. The population of Bell's Vireos in Texas has declined over recent decades, which is particularly evident in the eastern half of the state. Brood parasitism by Brown-headed Cowbirds (*Molothrus ater*) likely plays an important role in this decline. **Timing of occurrence:** Spring migrants arrive in the state in late March, and migration peaks from early April to early May. The breeding population in the western portions of the state is in place by mid-April, while those breeding farther east are in place by early May. Fall migrants pass through the state between mid-August and early October. **Taxonomy:** Two subspecies are known to occur in the state.

V. b. medius Oberholser
Locally common summer resident in the western half of the Trans-Pecos.

V. b. bellii Audubon
Common to very rare summer resident in the eastern three-quarters of Texas, west to the eastern Panhandle, and south through the Edwards Plateau and eastern Trans-Pecos to the western South Texas Brush Country. Uncommon to very rare migrant through the eastern three-quarters of the state. Very rare winter visitor in the Lower Rio Grande Valley.

BLACK-CAPPED VIREO *Vireo atricapilla* Woodhouse

Rare to locally uncommon summer resident from Brewster County eastward across the Edwards Plateau and north to Taylor and Palo Pinto Counties. A small population of Black-capped Vireos has recently been discovered in Montague County. The stronghold for the species in Texas appears to be on the southwestern Edwards Plateau and locally elsewhere with intensive management. Black-capped Vireos are rarely reported during migration, but a few reports have come from outside the breeding range. Most notable are individuals found in Cameron, Hidalgo, Jeff Davis, Kleberg, Midland, Presidio, and Starr Counties. This species is a habitat specialist that requires open shrublands where foliage is present down to ground level. Heavy browsing by livestock or deer changes the structure of the vegetation, making habitat unsuitable for nesting. Due to significant brood parasitism, the management of Brown-headed Cowbirds plays an important role in the protection of this state and federally listed Endangered Species. **Timing of occurrence**: Spring migrants arrive on the breeding grounds in late March and early April, and most depart by early September. **Taxonomy**: Monotypic.

GRAY VIREO *Vireo vicinior* Coues

Locally uncommon to rare summer resident from the southern Trans-Pecos east to the western Edwards Plateau. Gray Vireos are rare and local nesters in the Guadalupe Mountains but are absent from the Davis Mountains except as a very rare migrant. They are rare winter residents in southern Brewster and Presidio Counties and may occur in southern Terrell County as well. Vagrants have been found in Coke, El Paso, Kerr, and Randall Counties. This species appears to have greatly expanded its range in Texas in the past 40 years (Bryan and Lockwood 2000). The first record from east of the Pecos River was obtained in 1974 (Pulich and Parrot 1977). Not until the mid-1980s were Gray Vireos discovered to be locally uncommon on the southwestern Edwards Plateau. **Timing of occurrence**: Spring migrants arrive on the breeding grounds in late March and early April, and most depart by early September. **Taxonomy**: Monotypic.

YELLOW-THROATED VIREO *Vireo flavifrons* Vieillot

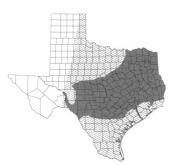

Common to uncommon summer resident in the eastern third of the state north of the Coastal Prairies and westward through the Oaks and Prairies region to the western Edwards Plateau, including very locally along the Pecos River drainage. Yellow-throated Vireos are also rare summer residents in riparian habitats along the Concho River as far west as Tom Green County (Maxwell 1979). They are uncommon summer residents south and east of the Edwards Plateau in the riparian woodlands of the Nueces, Frio, Sabinal, and San Antonio Rivers south to Zavala, Frio, and Karnes Counties. This species is a common to uncommon migrant through the eastern half of the state, becoming increasingly rare farther west in the state. These vireos are casual in early winter along the Coastal Prairies and the Lower Rio Grande Valley. **Timing of occurrence**: Spring migrants begin arriving on the breeding grounds in late March, and there is a peak in birds passing through the state from early April through mid-May. Fall migrants are recorded from late August through early October and on rare occasions linger into early January along the coast and the Lower Rio Grande Valley. **Taxonomy:** Monotypic.

PLUMBEOUS VIREO *Vireo plumbeus* Coues

Common summer resident in the Davis and Guadalupe Mountains of the central Trans-Pecos. Plumbeous Vireo is a common migrant throughout the Trans-Pecos and an uncommon to rare migrant east through the western High Plains and western Edwards Plateau. This species is a rare winter visitor throughout the Trans-Pecos. These vireos are very rare to rare in late fall and winter to the South Texas Brush Country and along the Coastal Prairies and casual inland, including in the Pineywoods. Plumbeous Vireo is one of three species that were formerly treated as subspecies of the Solitary Vireo (*V. solitarius*). **Timing of occurrence:** Spring migrants begin to arrive in the state in early March, and the primary migration period is from late March through early May. Fall migration occurs between late August and early October. **Taxonomy:** The subspecies that occurs in Texas is *V. p. plumbeus* Coues.

CASSIN'S VIREO *Vireo cassinii* Xántus de Vesey

Uncommon to rare migrant through the western half of the Trans-Pecos, becoming very rare to casual east through the western Edwards Plateau. Cassin's Vireo is more common during fall migration and has been found with regularity east to the western High Plains. This species is apparently a very rare winter visitor to the state, with reports from the Trans-Pecos, South Texas Brush Country, and Lower Rio Grande Valley and along the Coastal Prairies. Cassin's Vireo is part of the Solitary Vireo complex. Observers should be aware of the similarity between Cassin's Vireo and other members of this complex and the identification challenge these similar species represent (Heindel 1996). **Timing of occurrence:** Spring migrants are present from early March through late April, although a few linger through mid-May. Fall migrants pass through the state between late August and early November, and a few linger to mid-December. **Taxonomy:** The subspecies that occurs in Texas is *V. c. cassinii* Xántus de Vesey.

BLUE-HEADED VIREO *Vireo solitarius* (Wilson)

Common to uncommon migrant east of the Pecos River, becoming a rare to casual migrant in the Trans-Pecos. Blue-headed Vireo is an uncommon winter resident in much of the eastern two-thirds of the state. Formerly considered conspecific with Plumbeous and Cassin's Vireos, it is the most likely member of the Solitary Vireo complex to be encountered east of the Pecos River. **Timing of occurrence:** Spring migrants are present from late March to mid-May, and a few linger to late May. Returning fall migrants arrive as early as late August, and migration peaks from mid-September to early November. **Taxonomy:** Two subspecies are known to occur in the state.

V. s. solitarius (Wilson)
   As described above.

V. s. alticola Brewster
   Status uncertain. This subspecies is reported to have been collected near Silsbee, Hardin County, on 16 January 1917.

HUTTON'S VIREO *Vireo huttoni* Cassin

Locally common summer and uncommon winter resident in the Davis and Chisos Mountains. Hutton's Vireo is very rare in the Trans-Pecos away from the breeding range and is a casual visitor to El Paso County. Scattered summer and winter records exist for the Guadalupe Mountains, where nesting has been confirmed. A pair of Hutton's Vireos was discovered nesting in Real County in 1990 (Lasley and Gee 1991), providing the first record for the Edwards Plateau. Since then this species has colonized the Edwards Plateau region, with the current range expansion beginning in the spring of 1999. Hutton's Vireos are now locally uncommon residents in the southwestern Hill Country and rare but increasing east to the Balcones Escarpment. They have been found east of the Balcones Escarpment in Bastrop, Bell, and Guadalupe Counties, with a nesting record for the latter (Eitniear and Schaezler 2012). **Timing of occurrence:** Up to half of the summering population departs the state in the fall, most from late September to early November. Spring migrants return from late March through late April. **Taxonomy:** Two subspecies are known to occur in the state.

V. h. stephensi Brewster

> Casual visitor, primarily in winter, to El Paso County east to the Guadalupe Mountains.

V. h. carolinae Brandt

> Locally common to uncommon resident in the Davis and Chisos Mountains. The population on the Edwards Plateau is likely this subspecies, but verification is needed.

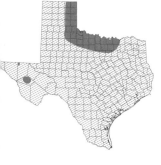

WARBLING VIREO *Vireo gilvus* (Vieillot)

Uncommon to rare summer resident in the Davis and Guadalupe Mountains and sporadic nester in the Chisos Mountains. This vireo is also an uncommon to rare summer resident from the eastern Panhandle east, very locally, across north-central Texas. Warbling Vireos are very rare and local breeders along the lower Trinity River in the Pineywoods, and there are isolated breeding records east of the Balcones Escarpment south to Travis County. They are uncommon to common migrants throughout the state. **Timing of occurrence:** Spring migrants pass through the state between early April and early June. Fall migrants are found from early August to mid-October. **Taxonomy:** Two subspecies are known to occur in the state.

V. g. swainsoni Baird

> Uncommon to rare summer resident in the Davis and Guadalupe Mountains. Uncommon migrant through the western third of the state. Extent of occurrence during migration is poorly known and requires further study.

V. g. gilvus (Vieillot)

> Uncommon to rare summer resident in the eastern Panhandle and eastward across north-central Texas. Uncommon to common migrant through the eastern two-thirds of the state.

PHILADELPHIA VIREO *Vireo philadelphicus* (Cassin)

Uncommon to rare spring and rare to very rare fall migrant in the eastern third of the state. This species is most commonly encountered along the coast. Migrants have been reported from all areas of the state but are rare to casual west of the Blackland Prairies. The only midwinter record is one photographed at the Tejano-Formosa Wetland, Jackson County, on 5 February 2013. **Timing of occur-**

rence: Spring migrants are found between early April and late May, and migration peaks from late April to early May. Fall migrants are seen from late August through late October and very rarely straggle into mid-December. **Taxonomy**: Monotypic.

RED-EYED VIREO *Vireo olivaceus* (Linnaeus)

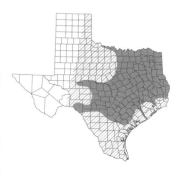

Locally abundant to uncommon summer resident in the eastern half of the state. This species reaches the western edge of its breeding range in Texas in the riparian woodlands that line the rivers of the Edwards Plateau. The San Antonio River in Victoria County on the central coast marks the southern extent of the breeding range. In the north it typically breeds only as far west as Parker County. During migration, Red-eyed Vireos can be found in all areas of the state but are very rare in the western half. **Timing of occurrence**: Spring migrants first appear in mid-March. Migration peaks from early April through late May, and a few linger into early June. Fall migrants pass through the state from late July through mid-October, rarely straggling to late November. **Taxonomy**: The subspecies that occurs in Texas is *V. o. olivaceus* (Linnaeus).

YELLOW-GREEN VIREO *Vireo flavoviridis* (Cassin)

Very rare to locally rare spring migrant along the coast and rare summer resident in the Lower Rio Grande Valley. There were 42 documented records for the state when this species was removed from the Review List in 2004. The summer records are primarily from Cameron and Hidalgo Counties, with a few reports of successful nesting. There are isolated summer and fall records westward from Big Bend NP, Brewster County, and northward to Frio and Travis Counties. Spring migrants that presumably have overshot their intended nesting areas have been found nearly annually at patches of stopover habitat along the coast as far north as Jefferson County. In some years there have been more occurrences of this type than of territorial birds in the Lower Rio Grande Valley. **Timing of occurrence**: The breeding population in the Lower Rio Grande Valley arrives in mid-April and has departed by mid-September. Migrants along the coast have been found between 15 April and 28 May. **Taxonomy**: The subspecies that occurs in Texas is *V. f. flavoviridis* (Cassin).

TBRC Review Species **BLACK-WHISKERED VIREO** *Vireo altiloquus* (Vieillot)

Very rare spring and accidental fall visitor along the coast. Texas has 36 documented records, with 21 from Galveston and Jefferson Counties. Most records are from the spring and presumably represent migrants that overshoot the western edge of their regular breeding range along the eastern Gulf Coast of Florida. The majority of records are of birds that remained for only one or two days, but a small number indicate that they remained as long as eight days. There is one summer record from Galveston where a single individual was present from 26 June to 3 July 2001. Two accepted fall records from High Island, Galveston County, report two individuals present from 20 August to 2 October 1989 (TBRC 1989-195) and a single bird there from 23 to 24 August 1991 (TBRC 1991-112). **Timing of occurrence**: Spring records fall between 4 April and 25 May. **Taxonomy**: The subspecies that occurs in Texas is *V. a. barbatulus* (Cabanis).

TBRC Review Species **YUCATAN VIREO** *Vireo magister* (Lawrence)

Accidental. The only record for Texas was a single bird near Gilchrist, Galveston County, discovered in the spring of 1984. This unexpected first record for the United States (Morgan et al. 1985) remains the only record from north of the Yucatán Peninsula, Mexico. **Taxonomy**: The subspecies of the Texas individual is unknown, but it is most likely from the Yucatán Peninsula population, *V. m. magister* (Lawrence).

28 APR.–27 MAY 1984, GILCHRIST, GALVESTON CO. (TPRF 318)

Family Corvidae: Jays and Crows

TBRC Review Species **BROWN JAY** *Psilorhinus morio* (Wagler)

Formerly uncommon and local resident along the Rio Grande in Starr County but now a very rare visitor to Starr and Zapata Counties. A precipitous decline in the Texas population occurred during the late 1990s, and by 2003 there were fewer than 12 individuals, which were found around the communities of Chapeño and Salineño, Starr County. This decline continued until 2006, when there was no longer a year-round presence in the state. Brown Jay

was added to the Review List in 2007. Since then there have only been five documented occurrences (see map), including three birds at San Ygnacio, Zapata County, from 22 January to 17 April 2010. Immature Brown Jays have been observed since 2007, suggesting that the species still nests either at an unknown location in Texas or within dispersal distance in Mexico. There is also a historic specimen and egg set from Cameron County (Hubbard and Niles 1975). **Timing of occurrence:** Since the species was added to the Review List, there have been three records of birds discovered in midwinter that remained into spring and two short-stay summer records. **Taxonomy:** The subspecies that occurs in Texas is *P. m. palliatus* (van Rossem).

GREEN JAY *Cyanocorax yncas* (Boddaert)

Common to uncommon resident from the Lower Rio Grande Valley north to Live Oak, Bee, and southern Maverick Counties. Green Jays are rare residents north to southern Val Verde and Kinney Counties. They are rare winter visitors to Uvalde County, and wandering individuals have been reported just north of the normal range to Bexar, Victoria, and Calhoun Counties. Vagrants have been found farther north to Brazos, Johnson, and Midland Counties. During the winter of 2007–8 there was a major influx of birds northward, including onto the Edwards Plateau in northern Uvalde, Real, and Bandera Counties. Small numbers returned to northern Uvalde County in subsequent winters and lingered into the summer in 2012. This northward shift may indicate a range expansion. **Taxonomy:** The subspecies taxonomy in Texas is unclear, with A. Phillips (1986) recognizing two subspecies, *C. y. luxuosus* (Lesson) and *C. y. glaucescens* (Ridgway). Some authorities have followed this treatment, but others have recognized only *C. y. luxuosus,* following the AOU (1957). Following Phillips, the Green Jays in the Lower Rio Grande Valley are *C. y. luxuosus,* while those to the north in the South Texas Brush Country are *C. y. glaucescens.*

TBRC Review Species   PINYON JAY *Gymnorhinus cyanocephalus* Wied

Casual visitor to the state. Pinyon Jay is a highly irruptive species to Texas, typically during the fall and winter. Such incursions occurred regularly through the 1970s and 1980s but have become increasingly rare events in recent decades,

with the last irruption taking place in the winter of 2002–3. Since then Pinyon Jays have been reported in the state on only four occasions. As a result, the species was added to the Review List in 2011. These irruptions often involve large flocks and are usually confined to the Trans-Pecos, but some have included the High Plains. A few individuals may linger into the summer, but nesting activities have never been reported. **Timing of occurrence**: Historically Pinyon Jays were found in the state between late August and early June, with the majority of occurrences from early October through late March. **Taxonomy**: Monotypic.

STELLER'S JAY *Cyanocitta stelleri* (Gmelin)

Locally common resident of the Davis and Guadalupe Mountains of the Trans-Pecos and an irregular winter visitor to El Paso County. Winter invasions occasionally occur in the remainder of the Trans-Pecos and, on very rare occasions, include the South Plains and Panhandle. Stray Steller's Jays reached the western Edwards Plateau during such an invasion in the winter of 1972–73. Breeding populations in the Trans-Pecos are found above 6,000 feet, although they occasionally descend to lower elevations in the fall and winter. **Timing of occurrence**: Winter incursions have taken place as early as mid-September, and birds have lingered into early April. **Taxonomy**: The subspecies that occurs in Texas is usually considered *C. s. macrolopha* Baird. However, some authors include the breeding population under *C. s. diademata* (Bonaparte) of northwestern Mexico.

BLUE JAY *Cyanocitta cristata* (Linnaeus)

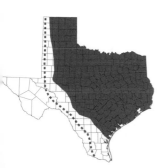

Common to locally uncommon resident in the eastern three-quarters of the state, west to the western Edwards Plateau, Permian Basin, and eastern High Plains, along the Canadian River drainage to Hartley County, and south to northern Nueces County. In the western third of this range, particularly in the High Plains and Edwards Plateau, Blue Jays are largely restricted to urban areas and are found elsewhere primarily during migration and winter irruptions. This species is a regular winter visitor in the northwestern Trans-Pecos to the town of Pecos, Reeves County, and occurs irregularly to Brewster, El Paso, and Jeff Davis Counties. Blue Jays are very rare south of the Nueces River, although a nesting pair was discovered in Hidalgo County in 1999 (Brush 2000), and very small numbers have persisted. In some years, massive migrations are noted in Texas, with loose flocks of up to 4,000 birds reported from the eastern third of the state. **Taxonomy:** Two subspecies are known to occur in the state. During winters when there have been large influxes into Texas, other subspecies may be involved.

C. c. cristata (Linnaeus)

Common resident in the eastern third of the state, west to about Hunt County in the north and through the southern Edwards Plateau and northeastern South Texas Brush Country. Intergrades with the following subspecies in north-central Texas and the Edwards Plateau.

C. c. cyanotephra Sutton

Rare to common resident from the Panhandle southward to the northern Edwards Plateau and east through the Rolling Plains and Blackland Prairies.

WESTERN SCRUB-JAY

Aphelocoma californica (Vigors)

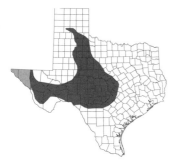

Common resident on the Edwards Plateau west through the mountains of the central Trans-Pecos north of the Big Bend region. Western Scrub-Jays are rare to locally uncommon residents in the canyonlands of the southern Panhandle, south through the western Rolling Plains. They are rare to locally uncommon winter visitors to El Paso and Hudspeth Counties. Elsewhere in the Trans-Pecos, Rolling Plains, and High Plains they are found only during fall and winter irruptions. Western Scrub-Jays are very rare to casual visitors during fall and winter east of the Balcones Escarpment. In recent decades they appear to be declining in portions of the easternmost Edwards Plateau near Austin and San Antonio, presumably due to habitat loss by urbanization and competition with the expanding Blue Jay population. An incursion into the northern South Texas Brush Country in the fall of 1994 was unprecedented. **Timing of occurrence:** Winter incursions have taken place as early as mid-September, and birds have lingered into early April. **Taxonomy:** Two subspecies are known to occur in the state.

A. *c. woodhouseii* (Baird)

Common resident in the Davis, Guadalupe, and other mountains of the Trans-Pecos.

A. *c. texana* Ridgway

Common resident in the Edwards Plateau and rare to locally uncommon north through the western Rolling Plains. The contact zone, if one exists, with the previous taxon in the extreme eastern Trans-Pecos is poorly understood.

MEXICAN JAY *Aphelocoma wollweberi* Kaup

Common resident of the Chisos Mountains. Mexican Jays very rarely wander into the surrounding lowlands, and only two documented records exist away from Big Bend NP. One was collected near Alpine, Brewster County, on 24 March 1935, and another was discovered in El Paso County on 24–25 January 2001. This species was formerly known as Gray-breasted Jay. **Taxonomy:** Two subspecies are known to occur in the state.

A. *w. arizonae* (Ridgway)

Accidental. The record from El Paso pertains to this subspecies.

A. *w. couchii* Ridgway

As described above.

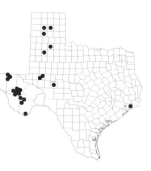

TBRC Review Species

CLARK'S NUTCRACKER

Nucifraga columbiana (Wilson)

Very rare winter visitor to the Trans-Pecos, High Plains, and western Rolling Plains. There are 23 documented records for the state, and most are part of winter incursions that typically coincided with influxes of Steller's or Pinyon Jays. One such event took place during the fall and winter of 2000–2001 when small numbers of Clark's Nutcrackers were present in the Trans-Pecos. There is also a remarkable record from the Upper Texas Coast from the fall of 2000. The largest incursion took place during the fall and winter of 1972–73 in the Panhandle, South Plains, and Trans-Pecos. During this invasion, more than 30 individuals were reported from the Panhandle alone (Seyffert 2001b). There are two summer records: the first involved two specimens collected on 1 June 1969 in the Guadalupe Mountains,

Culberson County; the more recent record involved birds from a winter incursion into the Davis Mountains that remained from 13 December 2002 to 19 June 2003. There are 20 reports of Clark's Nutcrackers from prior to the development of the Review List in 1988 for which there is no documentation on file. **Timing of occurrence:** The documented records for the state are all from 27 August to 19 June, but most are between 1 October and 6 April. **Taxonomy:** Monotypic.

TBRC Review Species

BLACK-BILLED MAGPIE *Pica hudsonia* (Sabine)

Accidental. Texas has but four documented records, even though magpies occur regularly in the Oklahoma Panhandle and northeastern New Mexico. Three of these records are from the extreme northern Panhandle, while the fourth is from El Paso County. Black-billed Magpies have also been found well to the east in Dallas, Hays, and Travis Counties, but questions of provenance have precluded them from being accepted by the TBRC. There are more than a dozen undocumented reports, primarily from the Panhandle, from prior to the formation of the TBRC. **Timing of occurrence:** All are from early winter to early spring, with dates ranging from 5 December to 4 April. **Taxonomy:** Monotypic.

27 DEC. 1983, TEXLINE, DALLAM CO. (TBRC 1984-11)
4–6 AND 17 FEB. 1990, EL PASO, EL PASO CO.
(TBRC 1990-28; TPRF 890)
5 DEC. 1997–16 JAN. 1998, NEAR GRUVER, HANSFORD CO.
(TBRC 1998-5)
5 DEC. 1999–4 APR. 2000, NEAR GRUVER, HANSFORD CO.
(TBRC 1999-106; TPRF 1855)

AMERICAN CROW *Corvus brachyrhynchos* Brehm

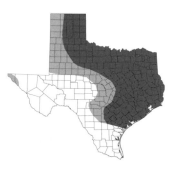

Common to abundant resident in the eastern half of the state, west to the central Panhandle and to the eastern edge of the Edwards Plateau, and south on the central coast south to Victoria and Refugio Counties. American Crows are rare to locally uncommon migrants and winter visitors to the remainder of the High Plains, western Rolling Plains, eastern Edwards Plateau, and northeastern parts of the South Texas Brush Country. They are also common to uncommon and local winter visitors in El Paso County and rare in Hudspeth County. During winter, large numbers of American Crows move into Texas from other states, increasing the overall population in the state. Formerly, flocks of many thousands were noted in agricultural areas, especially moving to and from roost sites; however, such flocks are seldom noted today. **Timing of occurrence:** Fall migrants begin to arrive in mid-October, with larger numbers present by early November. Wintering birds depart for more northern breeding areas by late March. **Taxonomy:** Three subspecies are known to occur in the state.

C. b. hesperis Ridgway
> Common to uncommon winter resident in El Paso County and rare in Hudspeth County.

C. b. paulus Howell
> Common to abundant resident in the Pineywoods and the upper coastal Prairies west to the Balcones Escarpment and south to Refugio County. This taxon is often included in *C. b. hesperis*.

C. b. brachyrhynchos Brehm
> Common to uncommon resident from the eastern Panhandle east through the Rolling Plains to the western edge of the Pineywoods.

TBRC Review Species

TAMAULIPAS CROW *Corvus imparatus* Peters

Casual visitor to the Brownsville area in southern Cameron County. Formerly a common winter resident and very rare summer resident in Cameron County, with the winter population found at the Brownsville Sanitary Landfill. This species was first reported in 1968, and the largest reported concentration was more than 2,300 individuals at Laguna Atascosa NWR on 30 January 1970 from which two specimens were obtained. Currently there is no wintering population, at least partially due to changes in landfill management that resulted in the garbage being covered more quickly. All of the occurrences since the winter of 1999–2000 are from spring and summer, most from subdivisions with palm trees close to the Brownsville airport. The last nesting attempt was in 2007, and the species has not been documented in the state since spring 2010. In the fall of 2000, the TBRC voted to place Tamaulipas Crow on the Review List in an effort to maintain a record of its presence in the state. Documented records obtained since 2000 are included on the range map. **Timing of occurrence:** There are nine records since the species was added to the Review

List, all between 13 March and 26 July. **Taxonomy**: Monotypic. This species was formerly called Mexican Crow and was considered conspecific with Sinaloa Crow (*C. sinaloae* Davis).

FISH CROW *Corvus ossifragus* Wilson

Uncommon to locally common resident along the Sabine River north to Sabine County and in northeast Texas along the Red River drainage west to Grayson County. Fish Crows are expanding their range westward elsewhere in northeast Texas, including along the Sulphur River to Delta and Hopkins Counties. They are vagrants west to Smith and Rains Counties. In southeast Texas, they are common residents in Jefferson and Orange Counties, rare to locally uncommon residents west to Harris County, and very rare visitors to Brazoria County. Fish Crows were first recorded in Texas in Orange County in 1921 (Oberholser 1974). **Taxonomy**: Monotypic.

CHIHUAHUAN RAVEN *Corvus cryptoleucus* Couch

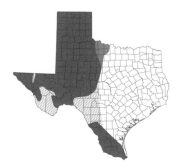

Uncommon to common resident in the western half of the state from the Panhandle south to the northwestern Edwards Plateau, and from Dimmit and La Salle Counties in the South Texas Brush Country south to the Lower Rio Grande Valley. In the Trans-Pecos, they are uncommon to common residents in the northern half of the region and primarily only migrants away from their grassland habitat. On the remainder of the western Edwards Plateau this raven is a common migrant, with a few scattered nesting records southeast to Kerr County. Chihuahuan Ravens have declined in recent decades from Wichita County southward through Throckmorton County, where they are now rare to locally uncommon summer residents. They can be common during the winter in the South Texas Brush Country and Lower Rio Grande Valley, although the population has declined during the last 10 years. This species is also a casual visitor east to Aransas, Caldwell, Hays, and Nueces Counties, with reports from the upper coast from Galveston and Jefferson Counties. In the fall, they typically stage in large concentrations prior to moving to their wintering grounds. **Taxonomy**: Monotypic.

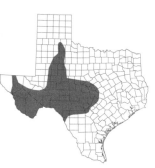

COMMON RAVEN *Corvus corax* Linnaeus

Uncommon to common resident in the mountains of the Trans-Pecos and east through the Edwards Plateau. This species is a rare to locally uncommon resident north of the Edwards Plateau in the western Rolling Plains north to Briscoe County. Vagrant Common Ravens have been noted in the eastern Rolling Plains, South Texas Brush Country, and Blackland Prairies. Perhaps the most unexpected record was one discovered at Galveston in November 1998. Almost all reports of the species east of its normal range are from the winter. Identification of out-of-range ravens poses a significant challenge. **Taxonomy**: The subspecies that occurs in Texas is *C. c. sinuatus* Wagler.

Family Alaudidae: Larks

HORNED LARK *Eremophila alpestris* (Linnaeus)

Locally common to uncommon resident in the Trans-Pecos, from the Panhandle south to the Concho Valley, and along the Coastal Prairies south through the Lower Rio Grande Valley, although much less local in winter. Horned Larks are also rare to locally uncommon residents across the eastern Rolling Plains to northeast Texas south to Williamson County, as well as on the northern Edwards Plateau. This species is a locally common migrant and winter resident in Central Texas, primarily in the Oaks and Prairies region, south to Guadalupe County, where these larks often associate with longspur flocks. Horned Larks have recently been confirmed as breeding in northeast Texas. In the Pineywoods, they are very rare to rare winter visitors and casual spring migrants. Overall distribution in the state is rather patchy, so breeding in many regions is localized. This is especially true in the central part of its resident range, including most of the Edwards Plateau. **Timing of occurrence**: Fall migrants begin to arrive in the state in early October, and spring migrants have generally left wintering areas in the state by early April. **Taxonomy**: Six subspecies have been reported in the state. The relative abundance is not known for many of these taxa, so the ranges given are general and additional research is needed.

E. a. leucolaema Coues

Common to uncommon resident in the Trans-Pecos, east through the northern Edwards Plateau, and north to the southern Panhandle. Uncommon winter resident away from breeding areas in this region and south through the western South Texas Brush Country.

E. a. enthymia (Oberholser)

Common to uncommon resident in the northern Panhandle. Common migrant and winter resident south to the northwestern Edwards Plateau and west through the Trans-Pecos.

E. a. praticola (Henshaw)

Migrant and winter resident to the Oaks and Prairies region south to Brazos County.

E. a. lamprochroma (Oberholser)

Status poorly known. This species nests in the northwestern Rocky Mountains and could be a very rare winter visitor in the western third of the state. Stevenson (1937a) reported a specimen collected in Randall County, 9 February 1936 (USNM).

E. a. occidentalis (McCall)

Migrant and winter resident in the western Trans-Pecos.

E. a. giraudi (Henshaw)

Common to uncommon resident on the Coastal Prairies south to include most of the South Texas Brush Country.

Family Hirundinidae: Martins and Swallows

PURPLE MARTIN *Progne subis* (Linnaeus)

Common to uncommon summer resident in most of the state east of the Pecos River. Purple Martins are rare and local summer residents on the western High Plains but absent from the westernmost portion of the Panhandle. Purple Martins are common to locally abundant migrants in the eastern two-thirds of the state, especially in the fall when they congregate in enormous roosts, often consisting of many thousands of birds. They are very rare to casual migrants in the Trans-Pecos. **Timing of occurrence:** Spring migrants begin arriving along the coast in early January. Numbers slowly build through February until the main push of migrants occurs between early March and mid-

April. Fall migrants are recorded beginning in early July and continuing through mid-September. They are rare through mid-October, with a very few straggling as late as mid-November. **Taxonomy:** Two subspecies are known to occur in the state.

P. s. subis (Linnaeus)
> Common to uncommon migrant and summer resident east of the Pecos River.

P. s. arboricola Behle
> Very rare to casual migrant through the Trans-Pecos.

TBRC Review Species

GRAY-BREASTED MARTIN *Progne chalybea* (Gmelin)

Accidental. Texas has two documented records of this species. Both refer to presumed spring migrants that overshot the breeding areas in northeastern Mexico. Gray-breasted Martins occur as close as within 150 miles of the Rio Grande in southern Tamaulipas and are widespread south through the Neotropics. The two Texas records are from the 1880s and are represented by specimens. Since the population of Gray-breasted Martins in northern Mexico is migratory, this species might be expected to occur again. However, the similarity to female Purple Martins makes identification in the field very difficult. **Taxonomy:** The subspecies that has occurred in Texas is *P. c. chalybea* (Gmelin).

25 APR. 1880, RIO GRANDE CITY, STARR CO.
  (*AMNH 89,806; TPRF 2611)
18 MAY 1889, HIDALGO, HIDALGO CO.
  (*AMNH 84,808; TPRF 2612)

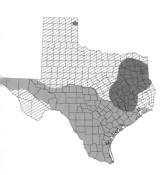

TREE SWALLOW *Tachycineta bicolor* (Vieillot)

Common migrant throughout the state, often locally abundant along the coast. Tree Swallows are rare to uncommon winter residents in the southern third of the state, especially along the coast. They are very rare in the Pineywoods in the winter, in contrast to their status on the upper coast. Likewise, in winter they are rare to very rare, and irregular, across the northern Trans-Pecos eastward through the Concho Valley and Edwards Plateau. This species is a rare to locally uncommon summer resident from northeast Texas and the northern Pineywoods south through the

Oaks and Prairies region to Harris County. Farther west, nesting has been documented in Hemphill County in the northeastern Panhandle. This swallow formerly bred as far south as Bexar and Jackson Counties. **Timing of occurrence:** Spring migrants begin arriving in the state in late February and are recorded through mid-May. Fall migrants are found between early August and mid-November. **Taxonomy:** Monotypic.

VIOLET-GREEN SWALLOW
Tachycineta thalassina (Swainson)

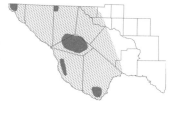

Uncommon to common summer resident in the mountains of the Trans-Pecos. Violet-green Swallows are common migrants in the central and western Trans-Pecos, becoming rare in the eastern third of the region. This species is a very rare migrant elsewhere in the state, with the majority of reports occurring in the fall. Violet-green Swallow is a very rare winter visitor in the Trans-Pecos, usually occurring along the Rio Grande. **Timing of occurrence:** Spring migrants begin arriving in the state in early March and are recorded through mid-May. Fall migrants are found between late July and early October. **Taxonomy:** The subspecies that occurs in Texas is *T. t. thalassina* (Swainson).

NORTHERN ROUGH-WINGED SWALLOW
Stelgidopteryx serripennis (Audubon)

Common to uncommon migrant throughout the state. Northern Rough-winged Swallows are rare to locally uncommon summer residents west to the Pecos River drainage and along the Rio Grande to the New Mexico border. This species is absent from the Coastal Prairies from eastern Nueces County northward to the Louisiana border during the summer. Northern Rough-winged Swallows are rare to uncommon winter residents in the South Texas Brush Country and extending west along the Rio Grande to southern Presidio County in the Trans-Pecos. They are very rare elsewhere during the winter, with most occurrences in the early winter pertaining to lingering migrants. They can be more common in these areas during particularly mild early winters. **Timing of occurrence:** Spring migrants begin arriving in the state in mid-February. The primary migration period is from mid-March through mid-May. Fall migrants are found from late July to late

October, with stragglers present until mid-December. **Taxonomy**: Two subspecies are known to occur in the state.

S. s. serripennis (Audubon)
> Locally uncommon to rare summer resident in the eastern half of the state. Common to uncommon migrant throughout the state and uncommon winter resident as described above.

S. s. psammochrous Griscom
> Locally uncommon to rare summer resident in the western half of the state and uncommon migrant through the same region.

BANK SWALLOW *Riparia riparia* (Linnaeus)

Locally uncommon summer resident along the Rio Grande from Cameron to Val Verde Counties. There are scattered breeding records from the Edwards Plateau, east to Hays County, and possibly northward. Bank Swallows are common to uncommon migrants throughout the state. There are winter reports from along the central and lower coasts and in the Lower Rio Grande Valley, but these require documentation. **Timing of occurrence**: Spring migrants begin arriving in the state in mid-March. The primary migration period is from early April through the end of May. Fall migrants are found from early July to late October, rarely to mid-November. **Taxonomy**: The subspecies that occurs in Texas is *R. r. riparia* (Linnaeus).

CLIFF SWALLOW *Petrochelidon pyrrhonota* (Vieillot)

Common to locally abundant summer resident throughout the state. Cliff Swallows may be the most common breeding swallows in Texas, now nesting commonly in areas where they were formerly rare or absent. Their adaptation to using human-made structures has allowed for the expansion of both overall range and population size. They are uncommon to abundant migrants throughout the state. There are undocumented reports of this species in winter (December and January), but great care must be taken to eliminate Cave Swallow (*P. fulva*), which is the default winterer in this genus. **Timing of occurrence**: Spring migrants begin to arrive in mid-February, with an increase in migrants from early April through mid-May. Fall migrants are found from early August to late October, although small numbers linger into November. **Taxonomy**: Three subspecies are known to occur in the state.

P. p. pyrrhonota (Vieillot)
> Common to locally abundant summer resident from the Panhandle south through the High Plains to Midland County and east across the northern Rolling Plains east to the edge of the Pineywoods. Common to abundant migrant throughout the state east of the Pecos River.

P. p. ganieri (Phillips)
> Common summer resident and migrant from the Pineywoods west through the central and southern Oaks and Prairie region to the Balcones Escarpment and along the Coastal Prairies to at least Calhoun County.

P. p. tachina Oberholser
> Common to locally abundant summer resident in the Trans-Pecos east through the Edwards Plateau and south through the South Texas Brush Country.

CAVE SWALLOW *Petrochelidon fulva* (Vieillot)

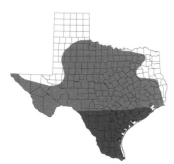

Cave Swallows have greatly expanded their range during the past 30 years. This species once nested only in limestone caves in the southwestern Edwards Plateau but now uses a variety of human-made structures. Cave Swallows are common to locally abundant migrants and summer residents in the southern half of the state, north through the southern Rolling Plains, and west through most of the South Plains and Trans-Pecos. They are uncommon to locally common migrants and summer residents eastward through the southern Post Oak Savannah. This species is a rare but increasing migrant and summer resident through north-central Texas north to near the Red River. These swallows are rare to uncommon summer residents on the upper coast to Jefferson County and northward through the southern Pineywoods. Cave Swallows are rare to locally common winter residents in the southern third of the state north to Bexar and Val Verde Counties. They are rare to very rare winter visitors to Bell and Williamson Counties, east to Galveston County. In winter they are now found with increasing frequency farther north, including in the southern Trans-Pecos. **Timing of occurrence:** Spring migrants begin to arrive in early February, and migration peaks between early March and mid-April. Breeding populations begin to disperse in early September, and migrants are found through October and increasingly into November, with lingerers found well into December. **Taxonomy:** The subspecies that occurs in Texas is *P. f. pallida* Nelson.

BARN SWALLOW *Hirundo rustica* Linnaeus

Rare to common summer resident in every region of the state. Barn Swallows are common to abundant migrants throughout the state. In the late fall and early winter they are rare to very rare along the coast and in the South Texas Brush Country and are very rare to casual north through the southern Pineywoods, southern Oaks and Prairies region, and southern Trans-Pecos. Overwintering birds are rare in the Lower Rio Grande Valley and casual north to the southern edge of the Edwards Plateau. **Timing of occurrence:** Spring migrants begin to arrive in mid-February, and migration peaks between mid-March and mid-May. Fall migrants are found from mid-July through November, with stragglers found well into December. **Taxonomy:** The subspecies that occurs in Texas is *H. r. erythrogaster* Boddaert.

Family Paridae: Chickadees and Titmice

CAROLINA CHICKADEE *Poecile carolinensis* (Audubon)

Common resident in the eastern half of the state west through the eastern Rolling Plains and west-central Edwards Plateau, south to northern Live Oak and Aransas Counties on the Coastal Bend. On the southern perimeter of their range they become localized and are found mostly along riparian drainages. They are uncommon to common residents in the eastern Panhandle and locally along the Canadian River valley as far west as Hartley County. Carolina Chickadees are very rare visitors south and west of their resident range, including as far west as Midland County. The southernmost records include one at Santa Ana NWR, Hidalgo County, from 25 October to 9 November 1992 and one at the Sabal Palm Sanctuary, Cameron County, from July 1997 through February 1998. The Carolina Chickadee is endemic to the southeastern United States and has never been documented in Mexico. **Taxonomy:** Two subspecies are known to occur in the state.

P. c. atricapilloides (Lunk)

> Common resident from the Panhandle south and east through the Rolling Plains to Fannin and Navarro Counties and south through the Edwards Plateau into the northeastern South Texas Brush Country. Intergrades with the following subspecies along the eastern edge of the Edwards Plateau.

P. c. agilis (Sennett)

> Common resident from northeast Texas south through the Pineywoods, west across the southern Oaks and Prairies region to the Balcones Escarpment, and south to the central Coastal Prairies.

TBRC Review Species

BLACK-CAPPED CHICKADEE
Poecile atricapillus (Linnaeus)

Accidental. Texas has only one documented record, a specimen from the Franklin Mountains, El Paso County, in 1881. Black-capped Chickadees were reported from the northeastern quarter of the state on several occasions between 1880 and 1920, and one was supposedly collected from Navarro County in 1880, but the location of this specimen is not known by the TBRC. The Black-capped Chickadee has been confirmed in the western Oklahoma Panhandle, and observers should be alert to its possible occurrence in the Texas Panhandle. **Taxonomy:** The subspecies that has occurred in Texas is *P. a. garrinus* (Behle).

10 APR. 1881, EL PASO CO.
  (*YALE PEABODY MUSEUM 9723; TBRC 1995-35; TPRF 1327)

MOUNTAIN CHICKADEE *Poecile gambeli* (Ridgway)

Locally common to uncommon resident in the higher elevations of the Davis and Guadalupe Mountains of the Trans-Pecos. Some Mountain Chickadees move to lower elevations within these ranges during the winter, including into desert scrublands surrounding the Guadalupe Mountains. This species is an uncommon and irregular fall and rare winter visitor to the western Trans-Pecos, particularly in El Paso County. Mountain Chickadee is a casual fall and winter visitor to the western Panhandle and northwestern South Plains. **Timing of occurrence:** During winter incursions to lower elevations in the Trans-Pecos and in the Panhandle and South Plains these birds arrive as early as early October and can remain through mid-March. **Taxonomy:**

The subspecies that occurs in Texas is *P. g. gambeli* (Ridgway).

JUNIPER TITMOUSE *Baeolophus ridgwayi* (Richmond)

Locally uncommon to rare resident in the foothills of the Guadalupe Mountains. Juniper Titmouse may be declining within its Texas range based on a perceived drop in sightings in appropriate habitat over the past decade. These birds are accidental in other areas of the western Trans-Pecos and western Panhandle. During the winter of 2000–2001, this species was found in El Paso, Dallam, Jeff Davis, and Potter Counties. Although Juniper Titmouse was known to occur in the Guadalupe Mountains for decades, breeding was not confirmed until 1973. **Taxonomy:** The subspecies that occurs in Texas is *B. r. ridgwayi* (Richmond). Juniper Titmouse was formerly known as Plain Titmouse and considered conspecific with the Oak Titmouse (*B. inornatus* Gambel).

TUFTED TITMOUSE *Baeolophus bicolor* (Linnaeus)

Common resident of the eastern third of the state. This species is found west to the eastern edge of the Rolling Plains, the Balcones Escarpment, and south to Refugio County on the central coast. It was formerly considered conspecific with Black-crested Titmouse (*B. atricristatus*). There is a narrow region where the ranges of these species overlap, forming a stable hybrid zone that extends from western north-central Texas southward, just east of the Balcones Escarpment to the central coast. In this region hybrid individuals dominate the population and interbreed with either parental species along the edges of this zone. The hybrid zone appears to be about 15–25 miles in width. The appearance of hybrid offspring is variable, but most have a chestnut patch on the forehead. **Taxonomy:** Monotypic.

BLACK-CRESTED TITMOUSE

Baeolophus atricristatus Cassin

Common resident in the western two-thirds of the state. Black-crested Titmouse is found from the Lower Rio Grande Valley northward through the Edwards Plateau to the northern Rolling Plains. The eastern edge of the range

extends from along the San Antonio River on the central coast northward to just east of the Balcones Escarpment and on to Clay County on the Red River. This species has a more localized distribution in the Trans-Pecos, including much of the Stockton Plateau and the Davis, Del Norte, and Chisos Mountains. This titmouse is a very rare visitor to the High Plains and the Guadalupe Mountains. There is a stable zone of hybridization with the Tufted Titmouse east of the Balcones Escarpment (see Tufted Titmouse account). **Taxonomy:** Three subspecies are known to occur in the state.

B. a. paloduro Stevenson
Common resident from the southeastern Panhandle south through the western Rolling Plains to the extreme western Edwards Plateau and west into the Trans-Pecos.

B. a. sennetti Ridgway
Common resident in western north-central Texas south through the Edwards Plateau to the northern part of the South Texas Brush Country.

B. a. atricristatus Cassin
Common resident in the Lower Rio Grande Valley and the southern South Texas Brush Country north to about Kleberg and Maverick Counties.

Family Remizidae: Penduline Tits and Verdins

VERDIN *Auriparus flaviceps* (Sundevall)

Uncommon to common resident in the western two-thirds of the state from the Lower Rio Grande Valley north to Briscoe County in the southeastern Panhandle and west through the Trans-Pecos. Verdins are found along the coast as far north as western Calhoun County. The eastern edge of their range roughly follows a line from western Calhoun County to Travis County north to Clay County. They are casual in winter east of the Balcones Escarpment to Bastrop County. Verdins were formerly residents in Palo Duro Canyon in Randall County. **Taxonomy:** The subspecies that occurs in Texas is *A. f. ornatus* (Lawrence).

Family Aegithalidae: Bushtits

BUSHTIT *Psaltriparus minimus* (Townsend)

Uncommon to common resident from the southern Panhandle and Rolling Plains south through the Edwards Plateau and west locally into the Trans-Pecos. This species formerly occurred east to Dallas County (Pulich 1988). In the Trans-Pecos, Bushtits can be locally abundant, occurring in all areas containing oak-juniper habitat, west to eastern Hudspeth County. There is one nesting record from Hemphill County in the northeastern Panhandle, and strays during all seasons have been found elsewhere in the Panhandle and South Plains. Populations in the Trans-Pecos and, to a lesser extent, the Edwards Plateau exhibit polymorphism. First-year males frequently have black auriculars and were formerly considered a separate species, the Black-eared Bushtit. **Taxonomy**: Two subspecies are known to occur in the state.

P. m. plumbeus (Baird)
    Uncommon resident from the southeastern Panhandle south through the Edwards Plateau.

P. m. dimorphicus van Rossem & Hachisuka
    Common resident in the mountains of the Trans-Pecos.

Family Sittidae: Nuthatches

RED-BREASTED NUTHATCH

Sitta canadensis Linnaeus

This species is highly irruptive in fall and winter throughout the state, although it seldom reaches the southern South Texas Brush Country. This species is present somewhere in Texas every year, but abundance varies locally from virtually absent to fairly common. Red-breasted Nuthatches are generally rare and irregular winter visitors to the Trans-Pecos, except at upper elevations in the Guadalupe Mountains, where they are uncommon. Nesting was documented in the Guadalupe Mountains in 1993, and a young juvenile was found in 2003, indicating possible low-density summer residency. **Timing of occurrence**: Fall migrants have arrived as early as late August but more commonly are first seen in late September and early October.

Wintering birds generally depart during March, but small numbers sometimes linger into late May. There are records of birds found during July and August in the Panhandle, South Plains, and Trans-Pecos, suggesting these are post-breeding wanderers. **Taxonomy**: Monotypic.

WHITE-BREASTED NUTHATCH
Sitta carolinensis Latham

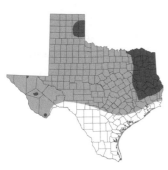

Common to uncommon resident in the Chisos, Davis, and Guadalupe Mountains. White-breasted Nuthatch is also a uncommon to locally common resident in northeast Texas, becoming rare to locally uncommon in the central and southern Pineywoods and the eastern Panhandle. In the Post Oak Savannah this species is a locally common resident in the northeast, becoming rare and local farther south to Gonzales and Travis Counties. White-breasted Nuthatches were formerly summer residents across north-central Texas, and there are isolated historical breeding records from Tom Green and Kerr Counties. This species is a rare to very rare migrant and winter visitor to the northern half of the state away from those areas with resident populations. **Timing of occurrence**: Fall migrants have been recorded from late August through early October. Wintering birds and spring migrants generally depart during March, although small numbers sometimes linger through May. **Taxonomy**: Three subspecies are known to occur in the state.

S. c. carolinensis Latham
Uncommon to locally common resident in the eastern third of the state, becoming rare in the southern Oaks and Prairies region. Uncommon resident in the eastern Panhandle. Rare migrant and winter visitor in the eastern half of the state, including very rarely on the Coastal Prairies.

S. c. nelsoni Mearns
Common to uncommon resident in the mountains of the Trans-Pecos north of the Chisos Mountains. Rare in winter in the western half of the state.

S. c. oberholseri Brandt
Uncommon resident in the Chisos Mountains.

PYGMY NUTHATCH *Sitta pygmaea* Vigors

Uncommon resident at upper elevations in the Guadalupe Mountains of the Trans-Pecos and very rare to accidental winter invader to nearby areas. This species was formerly a rare resident in the upper elevations of the Davis Mountains; however, the population there appears to have been sustained by birds from winter invasions that then remained to become temporarily resident. There has not been a significant invasion to the Davis Mountains since the mid-1990s, and this species was last reliably recorded there in 2005. Pygmy Nuthatches are casual winter visitors to the Chisos Mountains, El Paso County, and the western Panhandle and accidental in winter on the South Plains. An unexpected winter record was provided by a specimen collected from Dallas County on 31 December 1966 (Pulich 1988). **Timing of occurrence:** Winter invasions generally occur between mid-October and mid-April. **Taxonomy:** The subspecies that occurs in Texas is *S. p. melanotis* van Rossem.

BROWN-HEADED NUTHATCH *Sitta pusilla* Latham

Locally common to rare resident in the Pineywoods. This species is largely confined to mature pine and pine-hardwood forests. The Brown-headed Nuthatch's range extends from Bowie County in the extreme northeast, south to Chambers and Harris Counties, and east to Van Zandt County. These nuthatches are casual to very rare visitors, primarily in winter, to Grimes County and the Coastal Prairies west to Brazoria County. Local populations of this nuthatch are adversely impacted by the removal of older, large pines that serve as potential nesting sites. **Taxonomy:** The subspecies that occurs in Texas is *S. p. pusilla* Latham.

Family Certhiidae: Creepers

BROWN CREEPER *Certhia americana* Bonaparte

Generally uncommon to rare migrant and winter resident through the northern three-quarters of the state. Brown Creeper can be locally common in the Pineywoods and is rare to very rare in the southern half of the South Texas

Brush Country. The occurrence of Brown Creepers in Texas is at least somewhat irruptive, and they can be much more common during these incursions. This species is an uncommon summer resident in the Guadalupe Mountains. There has been at least one summer sighting in the Davis Mountains (Peterson et al. 1991), but no nesting activities have been observed. **Timing of occurrence:** Fall migrants begin to arrive in late October, and winter populations are generally in place by early December. Winter residents begin to move northward in late February, and most have departed by late March, with a few lingering through April. **Taxonomy:** Three subspecies are known to occur in the state.

C. a. montana Ridgway

Uncommon resident in the Guadalupe Mountains. Uncommon to rare in winter elsewhere in the Trans-Pecos and possibly farther east. There is a specimen reported from as far east as Nueces County from 29 January 1957 that is housed at the Welder Wildlife Foundation.

C. a. americana Bonaparte

Uncommon to rare migrant and winter resident east of the Pecos River. Very rare or casual in the Trans-Pecos, with a specimen reported from the Chisos Mountains, Brewster County (Van Tyne & Sutton 1937).

C. a. nigrescens Burleigh

Status uncertain. Texas is well west of the reported wintering range of this subspecies, but there is a specimen reported from Nueces County on 19 March 1964 that is housed at the Welder Wildlife Foundation.

Family Troglodytidae: Wrens

ROCK WREN *Salpinctes obsoletus* (Say)

Common resident in the western half of the state east through the Rolling Plains and locally on the Edwards Plateau. In the northern half of their range, Rock Wrens retreat during midwinter, becoming uncommon to locally rare. They are rare and very local in winter in the western South Texas Brush Country south to Falcon Dam, Starr County, and casual downstream to central Hidalgo County. This species is a vagrant across much of the remainder of the state, primarily in winter. Wandering individuals are often found near large, loose rock structures such as dams and have been reported as far east as Harris, Hopkins, Nacogdoches, and Van Zandt Counties. **Timing of occurrence**: Fall migrants begin moving into wintering areas in mid-October and can remain into late April. **Taxonomy**: The subspecies found in Texas is *S. o. obsoletus* (Say).

CANYON WREN *Catherpes mexicanus* (Swainson)

Uncommon to locally common resident in the Trans-Pecos eastward through the Edwards Plateau. Canyon Wrens are locally uncommon in the Concho Valley as well as the Lampasas Cut Plains north to southern Palo Pinto County. They are also common residents in the canyonlands of the southeastern Panhandle, although they are very rare to absent along the extreme western Rolling Plains along the Llano Estacado Escarpment. Vagrants have been reported from Denton, Gonzales, Parker, and Zapata Counties. This wren seldom strays far from its preferred canyon habitat. **Taxonomy**: The taxonomy of this species is unsettled, with three to eight subspecies recognized. The more conservative approach appears to be followed by most authorities and includes two subspecies that occur in the state.

C. m. consperus Ridgway
   Uncommon to locally common resident as given above except the western half of the Trans-Pecos.

C. m. albifrons (Giraud)
   Common resident in the western Trans-Pecos, east to the Guadalupe, Davis, and Chisos Mountains.

HOUSE WREN *Troglodytes aedon* Vieillot

Common to uncommon migrant throughout the state. House Wrens are common to rare winter residents throughout the state south of the Panhandle. They are most common in winter along the Coastal Prairies and in the South Texas Brush Country. As a summer resident, House Wrens are uncommon along the Canadian River in the eastern Panhandle, becoming uncommon to rare westward to Hartley County. In the Trans-Pecos, this species is a locally uncommon summer resident above 7,000 feet in the Davis Mountains and rare in the Guadalupe Mountains. **Timing of occurrence**: Fall migrants begin to arrive in early September, with a notable increase in migrants from late September through October. Spring migrants occur from late March and mid-May, and very few linger into late May. **Taxonomy**: Two subspecies occur in the state.

T. a. parkmanii Audubon

Rare to uncommon summer resident in the Guadalupe and Davis Mountains and presumably the same for those in the Panhandle. Uncommon migrant and winter resident in the western half of Texas, east through the Rolling Plains and Edwards Plateau, and south through the South Texas Brush Country.

T. a. aedon Vieillot

Common to uncommon migrant and winter resident in the eastern two-thirds of the state, south through the South Texas Brush Country to the Lower Rio Grande Valley. This taxon includes the sometimes recognized subspecies *T. a. baldwini* Oberholser.

WINTER WREN *Troglodytes hiemalis* Vieillot

Uncommon to common migrant and winter resident in the eastern two-thirds of the state, occurring most commonly in the eastern third. Winter Wrens are found south to about the Nueces River drainage during most winters. This species is a rare to casual migrant and winter visitor through much of the remainder of the state. These wrens frequent woodlots and riparian corridors and are quite vocal in the winter, especially at dawn, revealing their presence in places where they might otherwise be overlooked. The circumpolar Winter Wren was split into three species in 2011 (Chesser et al. 2011). The species in the eastern two-thirds of North America retained the common name, while the species in the west became Pacific Wren (*T. pacificus* Baird). The latter may be a casual winter visitor to the western portions of the state in riparian and other mesic habitats, but conclusive documentation of its presence has not been obtained. A specimen of Pacific Wren reported to have been collected in Big Bend NP, Brewster County, on 22 October 1937 (Oberholser 1974) has not been located. **Timing of occurrence**: Fall migrants begin to arrive in early October. Wintering populations are present through mid-

March, and small numbers linger as late as early April. **Taxonomy:** Two subspecies occur in the state.

T. h. hiemalis Vieillot
  As described above.

T. h. pullus (Burleigh)
  Rare migrant and winter resident in the Pineywoods and eastern Oaks and Prairies region.

SEDGE WREN *Cistothorus platensis* (Latham)

Common to uncommon migrant through the eastern half of the state, becoming rare to casual westward across the Trans-Pecos. Sedge Wrens are uncommon to locally common winter residents along the Coastal Prairies and farther inland throughout the eastern third of the state west to the Balcones Escarpment. This wren has also been found during the winter at scattered locations across the remainder of the state and, due to its secretive nature, may occur as a winter resident over a larger portion of the state than is currently known. **Timing of occurrence:** Fall migrants begin arriving in late September, with an increase in migrants beginning in mid-October. Spring migrants are found between late March and late May and have lingered into early June. There are three August records from northeast Texas (White 2002). **Taxonomy:** The subspecies found in Texas is *C. p. stellaris* (Naumann).

MARSH WREN *Cistothorus palustris* (Wilson)

Rare to locally common winter resident throughout the state. Marsh Wrens are uncommon to locally common in the summer in the marshes east of Galveston Bay and locally from central Matagorda County south to northern Aransas County. There are also isolated nesting records along the Red River east of the Panhandle and along the Rio Grande in the Trans-Pecos. There are no recent nesting records from the Trans-Pecos, and the channelization of the Rio Grande has eliminated potential breeding habitat in many areas. **Timing of occurrence:** Fall migrants begin arriving in early September, and the number increases from late September through October. Spring migrants are found between late March and mid-May. **Taxonomy:** Six subspecies occur in the state. Research suggests that eastern and western populations should be considered as separate species (Kroodsma 1989). The first three taxa listed are in the western group, and the remaining three are in the eastern group.

C. p. pulverius (Aldrich)

Rare migrant through the Trans-Pecos and possibly the western High Plains. This subspecies has been reported to be a winter resident in South Texas, but status is uncertain during that season.

C. p. plesius (Wilson)

Very rare and sporadic in summer in the western Trans-Pecos. Uncommon migrant through the western Trans-Pecos and rare eastward through the western half of the state. Reported east to Tarrant County (Pulich 1979).

C. p. laingi (Harper)

Rare to uncommon migrant through the western half of the state, with a specimen as far east as Bexar County. This subspecies has been reported to winter in the Lower Rio Grande Valley, but status is uncertain during that season.

C. p. iliacus (Ridgway)

Common migrant in the eastern two-thirds of the state, west to the eastern Panhandle. Common winter resident along the Coastal Prairies south to the Lower Rio Grande Valley.

C. p. dissaeptus Bangs

Rare to uncommon migrant in the eastern third of Texas and uncommon winter resident on the Coastal Prairies.

C. p. marianae Scott

Uncommon to locally common resident in coastal marshes east of Galveston Bay and along the central coast from Matagorda County to Aransas County.

CAROLINA WREN *Thryothorus ludovicianus* (Latham)

Common to abundant resident in the eastern two-thirds of the state. In the more arid portions of the western Rolling Plains and Edwards Plateau, this species is most often found along riparian corridors. In the Panhandle, Carolina Wrens occur westward to Oldham County in the Canadian River drainage, in the upper reaches of Palo Duro Canyon, and in Amarillo. They are rare visitors, in all seasons, to the South Plains. This wren has been found with increasing

regularity in the southern Trans-Pecos, where a small resident population is present along the Rio Grande in southern Brewster County. Elsewhere in the region, Carolina Wrens are very rare to casual visitors. **Taxonomy:** Two subspecies occur in the state.

T.l. ludovicianus (Latham)

As described above south through the northern and western South Texas Brush Country.

T.l. lomitensis Sennett

Uncommon to rare resident in the Lower Rio Grande Valley upriver through Starr County and northward through Kenedy County.

BEWICK'S WREN *Thryomanes bewickii* (Audubon)

Uncommon to common resident in the western two-thirds of the state, east through north-central Texas, and to the mouth of the Colorado River along the coast. Bewick's Wrens were formerly rare residents northward through the Coastal Prairies to San Jacinto and Jefferson Counties. Eastern populations have declined significantly beginning in the late 1960s (Kennedy and White 1997). It is estimated that populations in the eastern United States, including eastern Texas, have declined by 75 to 80 percent. **Timing of occurrence:** Fall migrants begin arriving in winter areas outside the breeding range in late September and are present through mid-March. **Taxonomy:** There is considerable debate over the validity of some races. Three subspecies are included here, following Brewer (2001).

T. b. bewickii (Audubon)

Rare to locally uncommon migrant and winter resident in the eastern third of the state and has been found as far west as the eastern edge of the Edwards Plateau.

T. b. eremophilus Oberholser

Uncommon to common resident in the Trans-Pecos. Rare to uncommon migrant and winter resident across the Edwards Plateau and south to at least San Patricio County on the central coast.

T. b. cryptus Oberholser

Uncommon to common resident from the Panhandle east through north-central Texas and the Edwards Plateau, south to the central coast and to the Rio Grande, and west to the Pecos River. Rare winter resident east to Brazos County.

CACTUS WREN
Campylorhynchus brunneicapillus (Lafresnaye)

Uncommon to locally common resident from the Lower Rio Grande Valley, north through the Edwards Plateau to the southeastern Panhandle, as well as throughout the Trans-Pecos. The eastern edge of this species' range generally follows a line from Refugio County on the central coast northwest to the Balcones Escarpment to Baylor and Foard Counties. Vagrants have been reported east to Calhoun and Tarrant Counties. Although fairly common below the Caprock Escarpment, Cactus Wrens are very rare visitors away from breeding areas in the remainder of the High Plains. **Taxonomy:** Two subspecies occur in the state.

C. b. couesi Sharpe
Uncommon to locally common in the Trans-Pecos and east through the northern Edwards Plateau into the Rolling Plains and southern Panhandle.

C. b. guttatus (Gould)
Uncommon to locally common resident in the southern Edwards Plateau south through the South Texas Brush Country.

Family Polioptilidae: Gnatcatchers

BLUE-GRAY GNATCATCHER *Polioptila caerulea*
(Linnaeus)

Rare to locally common summer resident in the eastern half of Texas, west to the Rolling Plains, and across the Edwards Plateau to the Pecos River. Blue-gray Gnatcatchers are generally absent as a breeding species from all but the northernmost portions of the Rolling Plains and from the South Texas Brush Country, except in the oak mottes of the Coastal Sand Plain and along the Rio Grande from Cameron to Starr Counties. They are also common to uncommon summer residents in the Guadalupe and Chisos Mountains but, oddly, are very rare in the Davis Mountains. They are common to abundant migrants throughout the eastern half of the state, becoming uncommon in the west. In winter, this species is uncommon to rare in the

76. The winter range of Williamson's Sapsucker (*Sphyrapicus thyroideus*) includes the southern Rocky Mountains through southern New Mexico and extends well into Mexico. This makes the wintering population in the Davis Mountains well to the east of the main range. Although the population there is small, they are present every winter. This male was photographed on 4 November 2011. *Photograph by Mark W. Lockwood.*

77. The Red-breasted Sapsucker (*Sphyrapicus ruber*) is a denizen of the mixed forests along the Pacific slope of North America. Its occurrence in Texas is rather unexpected. There have been three such records, which are among the farthest east for the species. This male was at the Lawrence Woods Picnic Area in the Davis Mountains, Jeff Davis County, from 11 to 28 March 2005. *Photograph by Mark W. Lockwood.*

78. The Endangered Red-cockaded Woodpecker (*Picoides borealis*) is a habitat specialist that requires large tracts of open pine forest with a grassy understory. Recent research suggests that the species continues to slowly decline despite careful habitat management. This adult was near Jasper, Jasper County, on 12 June 2008. *Photograph by Greg W. Lasley.*

79. Pileated Woodpeckers (*Dryocopus pileatus*) are widespread through the eastern half of the state and can be found in forest habitats from the Pineywoods west through the Oaks and Prairies region. This male was at Atlanta SP, Cass County, on 1 April 2011. *Photograph by Mark W. Lockwood.*

80. One of the more astounding finds for Texas is this White-crested Elaenia (*Elaenia albiceps*). It is from the migratory population that breeds in Chile and Argentina and winters in Peru and possibly Ecuador. There are several records of migrants from Colombia, but that is a long way from Texas. This bird was at South Padre Island, Cameron County, from 9 to 10 February 2008. *Photograph by Erik Breden, The Otter Side.*

81. There are six documented occurrences of Tufted Flycatcher (*Mitrephanes phaeocercus*) for the United States. Two are from Big Bend's Rio Grande Village. This attractive flycatcher is at least partially migratory in the northern portion of its range in Mexico. This adult was at Rio Grande Village, Brewster County, from 21 November 2010 to 4 January 2011. *Photograph by Mark W. Lockwood.*

82. The pattern of occurrence of Greater Pewee (*Contopus pertinax*) in Texas is unexpectedly complex. There are summer and winter records from the Trans-Pecos and winter records from the Lower Rio Grande Valley, the Coastal Prairies, and inland to Bexar and Howard Counties. This one was near Fort Davis, Jeff Davis County, from 19 December 2006 to 23 January 2007. *Photograph by Mark W. Lockwood.*

83. Gray Flycatcher (*Empidonax wrightii*) has been known to be a breeding bird in the pinyon-oak woodlands of the upper elevations of the Davis Mountains since 1991. It was still a surprise to find a small nesting population in Dog Canyon in the Guadalupe Mountains in 2012. This fresh-plumaged adult was at Big Bend's Cottonwood Campground, Brewster County, on 7 January 2012. *Photograph by Mark W. Lockwood.*

84. Despite an apparent small population, Buff-breasted Flycatchers (*Empidonax fulvifrons*) have maintained a foothold in the ponderosa pine glades of the Davis Mountains since first being discovered in 1999. The wildfires of 2011 and 2012 killed many trees and thus impacted their preferred habitat. Only time will tell if this species continues to occur in the state. This adult was photographed on The Nature Conservancy's Davis Mountains Preserve, Jeff Davis County, on 14 June 2008. *Photograph by Mark W. Lockwood.*

85. Dusky-capped Flycatcher (*Myiarchus tuberculifer*) is now known to be a rare summer resident in the Davis Mountains and sporadically in the Chisos Mountains. This range extension into Texas has not resulted in large numbers of breeding pairs. They are still restricted to canyons with extensive stands of large pines, unlike the wider variety of habitats these flycatchers use in other parts of their range. This adult was on The Nature Conservancy's Davis Mountains Preserve, Jeff Davis County, on 26 June 2010. *Photograph by Mark W. Lockwood.*

86. Nutting's Flycatcher (*Myiarchus nuttingi*) is a species some expected would be found in Texas and had even been reported from the Big Bend region a few times. The first fully documented record for the state was discovered at Santa Elena Canyon, Brewster County, on 31 December 2011. This image was taken on 7 January 2012. *Photograph by Mark W. Lockwood.*

87. Although there was an accepted sight record for Social Flycatcher (*Myiozetetes similis*) from spring 1990, the species was not officially added to the state list until this individual was discovered at Bentsen–Rio Grande Valley SP, Hidalgo County. This cooperative bird was there from 7 to 14 January 2005. *Photograph by Geoff Malosh.*

88. Piratic Flycatcher (*Legatus leucophaius*) has an interesting pattern of occurrence in the United States. There are scattered records from South Texas northwest to the Trans-Pecos and southeastern New Mexico. This individual was at Corpus Christi, Nueces County, on 3–6 May 2008. *Photograph by Dan Roberts.*

89. Thick-billed Kingbirds (*Tyrannus crassirostris*) are casual visitors to the state and even nested in the Big Bend region from 1989 to 1991. This individual was quite a surprise when it was discovered at Selkirk Island, Matagorda County, in December 2002. It returned to the same location for two more winters, and this photograph was obtained on 20 December 2003. *Photograph by Brush Freeman.*

90. Gray Kingbirds (*Tyrannus dominicensis*) have been found along the Texas coast on 10 occasions. These birds presumably have wandered west along the Gulf Coast from nesting areas in Florida. This one was at Corpus Christi, Nueces County, from 26 to 29 October 2006 and photographed there with two other species of *Tyrannus* on 27 October. *Photograph by Garrett Hodne.*

91. Fork-tailed Flycatcher (*Tyrannus savana*) is an austral migrant well known for overshooting the wintering grounds in northern South America and reaching the United States and Canada in the fall. A few birds also undergo a reverse migration and arrive here in the spring rather than on their breeding grounds. This one was at High Island, Galveston County, from 24 to 25 April 2010. *Photograph by Linda Gail Price.*

92. Rose-throated Becards (*Pachyramphus aglaiae*) have once again become an almost annual visitor to the Lower Rio Grande Valley since 2002. Despite this increase in occurrence, there have been no successful nesting attempts discovered. This first-year male was at Estero Llano Grande SP, Hidalgo County, from 6 November 2008 through 15 April 2009. *Photograph by David McDonald.*

93. The normal winter range of Northern Shrike (*Lanius excubitor*) just reaches into the Panhandle of Texas. The true status of this species in Texas can be debated because of the very limited coverage of the Panhandle by birders in winter. It seems clear that the number of birds present in a given winter fluctuates greatly. This first-winter bird was at Lake Meredith, Hutchinson County, on 3 December 2012. *Photograph by Martin Reid.*

94. The Black-capped Vireo (*Vireo atricapilla*) is one of two Endangered songbirds found in Texas. This vireo often nests in loose colonies and favors open shrubland habitats. Intensive cowbird management is under way in many areas to help protect this species. This male was at Balcones Canyonlands NWR, Williamson County, on 20 April 2007. *Photograph by Greg W. Lasley.*

95. Despite its rather drab plumage, the Gray Vireo (*Vireo vicinior*) is much sought after by birders. The arid juniper and oak habitat of this species is often in fairly rugged terrain that, combined with the very large areas the territorial males defend, can often make Gray Vireos difficult to find. This one was at Kickapoo Cavern SP, Kinney County, on 26 March 2007. *Photograph by Mark W. Lockwood.*

96. Cassin's Vireo (*Vireo cassinii*) is an uncommon migrant through the Trans-Pecos and rare farther east. Separating Cassin's Vireos from Blue-headed Vireos is an identification challenge that many observers face when looking at dull-plumaged "Solitary" Vireos in Texas. This adult Cassin's Vireo was at Big Bend's Cottonwood Campground, Brewster County, on 29 April 2011. *Photograph by Mark W. Lockwood.*

97. Hutton's Vireos (*Vireo huttoni*) have greatly expanded their range on the Edwards Plateau in the past decade. Although the species has been known to occur in the region since 1991, not until the early 2000s was a population established that continues to increase. This adult was in the Chisos Mountains, Brewster County, on 19 May 2012. *Photograph by Mark W. Lockwood.*

98. Black-whiskered Vireo (*Vireo altiloquus*) has become a near-annual visitor to the Texas coast and particularly Jefferson and Galveston Counties since 1995. This individual was farther down the coast at Port Aransas, Nueces County, on 10 April 2011. *Photograph by Martin Reid.*

99. The continued occurrence in the United States of Brown Jays (*Psilorhinus morio*) appears to be in jeopardy. As many as 200 individuals were found in Starr County in the 1980s, but in 2012 there were no known resident birds in the state. Brown Jays have been found on five occasions since 2007. This adult was at Salineño, Starr County, on 20 January 2012. *Photograph by Larry Ditto.*

100. Green Jays (*Cyanocorax yncas*) are common and colorful residents of the wooded areas of South Texas. The bright greens and blues of their plumage serve them well, as they blend into the foliage. This adult was at Bentsen–Rio Grande Valley SP, Hidalgo County, on 10 November 2012. *Photograph by Mark W. Lockwood.*

101. The population of Mexican Jay (*Aphelocoma wollweberi*) in Texas is restricted to the Chisos Mountains of the Big Bend region. There are very few instances of their being found elsewhere in the state. This adult was in the Chisos Mountains, Brewster County, on 12 December 2009. *Photograph by Mark W. Lockwood.*

102. Like Brown Jay, the Tamaulipas Crow (*Corvus imparatus*) has all but vanished from the state's avifauna. At one time several hundred wintered at the Brownsville Sanitary Landfill, but more recently the number had dwindled to a single nesting pair last documented in 2010. This adult was at the Brownsville Sanitary Landfill, Cameron County, on 29 February 1992. *Photograph by Mark W. Lockwood.*

103. Juniper Titmouse (*Baeolophus ridgwayi*) is typically found in Texas only in the foothills surrounding the Guadalupe Mountains. These birds are often difficult to find, leading to concerns about the health of that population. This adult was at Dog Canyon, Guadalupe Mountains NP, Culberson County, on 26 May 2012. *Photograph by Mark W. Lockwood.*

104. Brown-headed Nuthatches (*Sitta pusilla*) are found in the extensive pine forests of the southeastern United States. From a Texas perspective, they are one of the specialties of the Pineywoods of East Texas. This adult was at Angelina National Forest, Jasper County, on 14 June 2008. *Photograph by Greg W. Lasley.*

105. The recent split of Winter Wren (*Troglodytes hiemalis*) into two species has left questions about the status of the western species, known as the Pacific Wren, in Texas. Although there is some circumstantial evidence that Pacific Wrens have occurred, to date all documented occurrences refer to the eastern species. This Winter Wren was in Austin, Travis County, on 19 December 2010. *Photograph by Greg W. Lasley.*

106. There are two records of Northern Wheatear (*Oenanthe oenanthe*) for Texas. The second of these was a crowd pleaser, as it remained at a rural farmhouse near Olmos, Bee County, from 30 December 2009 through 29 March 2010. *Photograph by Matthew Matthiessen.*

107. The only record of Black-headed Nightingale-Thrush (*Catharus mexicanus*) for the United States is that of a very long-staying bird in Pharr, Hidalgo County, from 28 May through 29 October 2004. This image was obtained on 3 June. *Photograph by Larry Ditto.*

108. The olive-backed subspecies of Swainson's Thrush (*Catharus ustulatus*) is a common migrant through much of the eastern half of the state, but the russet-backed subspecies of the Pacific Northwest is a casual migrant through the western Trans-Pecos. The TBRC requests documentation for all sightings of russet-backed Swainson's Thrushes. This one was in El Paso, El Paso County, on 8 May 2011. *Photograph by Jim Paton.*

109. The White-throated Thrush (*Turdus assimilis*) is a tropical species that has visited the Lower Rio Grande Valley on 13 occasions. Surprisingly, seven of those records came from January through March 2005. This one was photographed at Estero Llano Grande SP, Hidalgo County, on 11 February 2011. *Photograph by Robert Epstein.*

110. Prior to the 1990s there were only two documented records of Rufous-backed Robin (*Turdus rufopalliatus*) for Texas. Since 1992 there have been 15 records. The easternmost was this adult near Utley, Bastrop County, from 7 January to 7 April 2006. *Photograph by Greg W. Lasley.*

111. Varied Thrush (*Ixoreus naevius*) is a casual winter visitor to the state from northwestern North America. Although it was first reported as early as 1935, the first documented record is from 1978. This male was in the Christmas Mountains, Brewster County, from 20 November 2012 to 19 March 2013. *Photograph by Mark W. Lockwood.*

112. Aztec Thrush (*Ridgwayia pinicola*) is a denizen of the mountains of western Mexico and has been documented in Texas on six occasions. Surprisingly, three of these records are from the central coast and Lower Rio Grande Valley, including this individual at Bentsen–Rio Grande Valley SP, Hidalgo County, on 16 February 2010. *Photograph by Rick Nirschl.*

113. There are three accepted records of Blue Mockingbird (*Melanotis caerulescens*) for Texas from the Lower Rio Grande Valley. Two of these records are presumed to refer to the same individual that was present during the winter of 2002–3. It then returned to the same location in Pharr, Hidalgo County, and remained there from mid-September 2003 through 26 March 2005. This image of that bird was taken on 29 May 2004. *Photograph by Robert Tizard.*

Coastal Prairies, southern Pineywoods, and southern Trans-Pecos and up the Pecos River drainage and common in the South Texas Brush Country. Blue-gray Gnatcatchers are very rare to casual in winter in many other parts of the state. **Timing of occurrence:** Spring migrants begin arriving in mid-March, and breeding populations are present by early May. Migrants are still passing through the state through late May. Fall migrants are recorded from early August to mid-October. Winter populations also arrive during this period. **Taxonomy:** Two subspecies occur in the state.

P. c. caerulea (Linnaeus)

Rare to locally common summer resident as described above east of the Pecos River and north of the Balcones Escarpment. Common migrant in all areas east of the Pecos River. Uncommon to rare in the Coastal Prairies, southern Pineywoods, and southern Trans-Pecos and up the Pecos River drainage. Common in the South Texas Brush Country.

P. c. amoenissima Grinnell

Common to uncommon summer resident in the Chisos and Guadalupe Mountains and common to uncommon migrant throughout the Trans-Pecos. Rare to locally uncommon winter resident in the southern Trans-Pecos and along the Pecos River drainage. This taxon is sometimes included under *P. c. obscura* Ridgway.

BLACK-TAILED GNATCATCHER

Polioptila melanura Lawrence

Common to uncommon resident in the western and southern Trans-Pecos and locally uncommon to rare on the southwestern Edwards Plateau, southward along the Rio Grande to the western half of the South Texas Brush Country. Black-tailed Gnatcatcher reaches the eastern limit of its range in Live Oak, Jim Wells, and Hidalgo Counties, although it is declining in the central Lower Rio Grande Valley. Recently these gnatcatchers have expanded their range eastward on the southwestern Edwards Plateau, where breeding was first noted in Edwards, Kinney, and Uvalde Counties in the early 1990s. Vagrants have been reported from Bexar, Cameron, Duval, Goliad, and San Patricio Counties. **Taxonomy:** The subspecies found in Texas is *P. m. melanura* Lawrence.

Family Cinclidae: Dippers

AMERICAN DIPPER *Cinclus mexicanus* Swainson

Casual visitor to the state, mostly in winter. There are eight documented records for Texas; the first came from Crosby County on 2 May 1969 (TPRF 37). Five of these records are from the western third of the state, with the remaining three involving single birds in Travis County on 5 March 1994, Dallas County on 23 December 2004, and Kerr County on 16 February 2010. American Dipper has been documented twice in successive years (1988, 1989) in McKittrick Canyon, Guadalupe Mountains NP, Culberson County. **Timing of occurrence:** The documented records range from 23 October to 3 May. **Taxonomy:** The subspecies that has occurred in Texas is *C. m. unicolor* Bonaparte. It is possible that the record from El Paso County, 8–16 November 1984, refers to *C. m. mexicanus* Swainson, which occurs as close as northern Chihuahua, Mexico.

Family Regulidae: Kinglets

GOLDEN-CROWNED KINGLET
Regulus satrapa Lichtenstein

Generally uncommon to locally common migrant and winter resident in the eastern two-thirds of the state, becoming uncommon to rare westward to El Paso. Golden-crowned Kinglets are normally casual to rare winter visitors from the southern South Texas Brush Country and southern Nueces County on the Coastal Plain southward to the Lower Rio Grande Valley. This species is somewhat irruptive: in some winters these birds may be scarce, and in others they may be fairly common. **Timing of occurrence:** Fall migrants begin to arrive in early October, with larger numbers recorded after mid-October. The wintering population is present through late March, and small numbers linger to mid-April and rarely to early May. **Taxonomy:** Two subspecies occur in the state.

R. s. apache Jenks
   Normally uncommon to rare migrant and winter resident in the Trans-Pecos.

R. s. satrapa Lichtenstein

> Generally uncommon to common migrant and winter resident in the eastern three-quarters of the state and rare in the western third.

RUBY-CROWNED KINGLET

Regulus calendula (Linnaeus)

Common to abundant migrant and winter resident throughout the state. Ruby-crowned Kinglets are less common in winter in the northernmost portions of the state. **Timing of occurrence**: Fall migrants can start to arrive as early as mid-August, although in most years the first migrants are found in mid-September, with numbers building through October. Spring migrants are recorded from mid-March through mid-May, and small numbers occasionally straggle into late May. **Taxonomy**: The subspecies found in Texas is *R. c. calendula* (Linnaeus).

Family Muscicapidae: Old World Robins, Flycatchers, and Allies

TBRC Review Species ## NORTHERN WHEATEAR

Oenanthe oenanthe (Linnaeus)

Accidental. There are two records for the state. Both individuals were in female-type plumage and occurred in the late fall and winter, which coincides with the pattern of occurrence in the southeastern United States. **Taxonomy**: The subspecies that has occurred in Texas is unknown. It is difficult, if not impossible, to determine sex or subspecies of fall birds. The very pale coloration of both Texas birds might suggest that they are from the population found in Alaska, *O. o. oenanthe* (Linnaeus). However, there is considerable variation within the northeast Canada and Greenland population, *O. o. leucorhoa* (Gmelin), and that taxon cannot be eliminated.

1–6 NOV. 1994, LAGUNA ATASCOSA NWR, CAMERON CO.
(TBRC 1994-165; TPRF 1310)
30 DEC. 2009–29 MAR. 2010, NEAR OLMOS, BEE CO.
(TBRC 2010–02; TPRF 2789)

Family Turdidae: Thrushes

EASTERN BLUEBIRD *Sialia sialis* (Linnaeus)

Uncommon to locally common summer resident in the eastern half of the state to the eastern Edwards Plateau, westward along the Concho River drainage, and south along the Coastal Prairies to northern Kenedy County. Eastern Bluebirds are also common summer residents in the eastern Panhandle westward along the Canadian River drainage to Hartley County. This species is a common to uncommon migrant and winter resident east of the Trans-Pecos and south through the South Texas Brush Country. In the Trans-Pecos and in the Lower Rio Grande Valley, Eastern Bluebirds are rare to locally uncommon migrants and winter residents. **Timing of occurrence:** Fall migrants begin to arrive in wintering areas in late September and October. Spring migrants are recorded from mid-February through March, and small numbers occasionally straggle into late April. **Taxonomy:** The subspecies found in Texas is *S. s. sialis* (Linnaeus).

WESTERN BLUEBIRD *Sialia mexicana* Swainson

Uncommon and local resident in the Davis and Guadalupe Mountains. During the breeding season Western Bluebirds are typically found above an elevation of 6,000 feet, descending to lower elevations at other seasons. Western Bluebird is an irruptive winter resident elsewhere in the Trans-Pecos, although numbers can vary from locally abundant to rare. This species is a rare to very rare winter visitor to the western Edwards Plateau and High Plains and casual to accidental farther east to the Blackland Prairies and south to Bexar County. **Timing of occurrence:** Fall migrants found away from breeding areas begin to arrive in mid-September and depart by early April. **Taxonomy:** The subspecies found in Texas is *S. m. bairdi* Ridgway. The breeding population in the Trans-Pecos is sometimes considered a separate subspecies, *S. m. jacoti* Phillips.

MOUNTAIN BLUEBIRD *Sialia currucoides* (Bechstein)

Common to uncommon migrant and winter resident in the Trans-Pecos and in the Panhandle south to the northwestern Edwards Plateau. Mountain Bluebirds are irruptive

winter visitors to Texas, sometimes in flocks of 50 or more birds, but occur with greater predictability in the Panhandle and Trans-Pecos than farther east. This species can be abundant during some years and virtually absent during others, even in the western third of the state. They are rare migrants and winter visitors to the eastern Edwards Plateau and eastern Rolling Plains during invasion years. They are very rare to casual migrants and winter visitors to all other areas of the state, including the Pineywoods and Lower Rio Grande Valley. There is a single breeding record for the state involving an adult with week-old fledglings discovered on 31 July 1995 in the Davis Mountains. **Timing of occurrence**: Fall migrants begin arriving in the state as early as mid-September, but the primary arrival period is from mid-October through mid-November. Wintering populations usually remain through mid-March, but large numbers have remained in the Trans-Pecos through early April and small numbers as late as early May. **Taxonomy:** Monotypic.

TOWNSEND'S SOLITAIRE

Myadestes townsendi (Audubon)

Uncommon to rare, although irregular, migrant and winter resident in the Panhandle and Trans-Pecos. Townsend's Solitaire is an irruptive winter visitor to Texas south of the Panhandle. These birds are present during most years but can be virtually absent some years. During invasion years, they are rare winter visitors to the South Plains, Edwards Plateau, and Rolling Plains south of the Panhandle. This species is casual east to Fannin and Hunt Counties and south to the central coast and Lower Rio Grande Valley. One photographed in Midland, Midland County, on 29 July 2005 is the lone midsummer record for the state. **Timing of occurrence**: Fall migrants begin to arrive in late September, with an increase in occurrence most years beginning in mid-October. Spring migrants are recorded from late March through April, with a few individuals routinely lingering to mid-May and exceptionally until 2 June. **Taxonomy**: The subspecies found in Texas is *M. t. townsendi* (Audubon). The subspecies found in northwestern Mexico, *M. t. calophonus* Moore, may occur as a vagrant to the western Trans-Pecos.

TBRC Review Species   **ORANGE-BILLED NIGHTINGALE-THRUSH**
Catharus aurantiirostris (Hartlaub)

Accidental. Texas has two records, with the first pertaining to an adult mist-netted at Laguna Atascosa NWR, Cameron County, in the spring of 1996 (Papish, Mays, and Brewer 1997). The population in northeastern Mexico is at least partially migratory, and the two Texas records would appear to be spring migrants that continued northward to the Lower Rio Grande Valley. They are found primarily in dry pine-oak forests in southwestern Tamaulipas, and any birds found in the thorn-scrub woodlands of South Texas are unlikely to remain for an extended period. Astoundingly, there is a third record for the United States of a territorial male in South Dakota during July 2010. **Taxonomy:** The Texas specimen has features consistent with the subspecies *C. a. melpomene* (Cabanis).

8 APR. 1996, LAGUNA ATASCOSA NWR, CAMERON CO.
(TBRC 1996-59; TPRF 1493)
28 MAY 2004, EDINBURG, HIDALGO CO.
(*LSUMNS 178,569; TBRC 2005–61; TPRF 2306)

TBRC Review Species   **BLACK-HEADED NIGHTINGALE-THRUSH**
Catharus mexicanus (Bonaparte)

Accidental. There is one record for the state and the United States. The northernmost population is migratory, and the lone occurrence is probably the result of an overshooting spring migrant (Lockwood and Bates 2005). This species is found in subtropical forests in southern Tamaulipas and western Nuevo León, Mexico, environments similar to that of the Lower Rio Grande Valley. Notably, this bird was found on the same day that the second state record for Orange-billed Nightingale-Thrush was discovered. **Taxonomy:** The subspecies that has occurred in Texas is *C. m. mexicanus* (Bonaparte).

28 MAY–29 OCT. 2004, PHARR, HIDALGO CO.
(TBRC 2004–50; TPRF 2213)

VEERY *Catharus fuscescens* (Stephens)

Uncommon to rare spring and very rare fall migrant in the eastern half of the state. Veery is most often encountered in woodland habitats on the Coastal Prairies. Away from the immediate coast this species is considered a rare to very rare spring migrant and a casual fall migrant. These birds

are casual in spring and fall farther west through the eastern Panhandle and South Plains. Banding data from the eastern Edwards Plateau show the species to be a rare spring migrant in Central Texas (Lockwood 2001). There is only one documented spring record from the Trans-Pecos, and observers in that region should be aware of the similarities between this species and the russet-backed form of Swainson's Thrush (*C. ustulatus*). This species winters in South America, and there are no documented winter records from Texas. **Timing of occurrence:** Spring migrants have been found as early as late March, but the primary migration period is from mid-April to mid-May. Very small numbers are found into late May and exceptionally into early June. Fall migrants have been recorded between early September and late October. **Taxonomy:** Two subspecies occur in the state. A third subspecies may occur as a casual migrant through the eastern half of the state, *C. f. fuliginosus* (Howe).

C. f. fuscescens (Stephens)

Uncommon to rare spring migrant and very rare fall migrant through the eastern half of the state.

C. f. salicicola (Ridgway)

Rare to casual spring migrant through the western half of the state, with specimen records from as far east as Cooke and Dallas Counties.

GRAY-CHEEKED THRUSH

Catharus minimus (Lafresnaye)

Uncommon to rare spring and very rare fall migrant in the eastern half of the state. Like the Veery, this species is most often encountered at stopover habitats on the Coastal Prairies. Away from the immediate coast this species is a rare to very rare spring migrant and a casual fall migrant west to the High Plains and western Edwards Plateau and accidental in the Trans-Pecos. There is an unexpected midsummer record from South Padre Island, Cameron County, from 30 June to 3 July 2011. This species winters in northern South America, and there are no documented winter records for Texas. **Timing of occurrence:** Spring migrants begin to arrive in the state in early April. The primary migration period is from mid-April to mid-May, and very small numbers are found into early June. Fall migrants have been recorded between early September and late October. **Tax-onomy**: The predominant subspecies that occurs in Texas is *C. m. aliciae* (Baird); however, there are a few smaller, shorter-winged, warmer-hued birds that have been found along the coast that seem to better match *C. m. minima* (Lafresnaye).

SWAINSON'S THRUSH *Catharus ustulatus* (Nuttall)

Common to uncommon spring and uncommon to rare fall migrant in the eastern half of Texas, becoming increasingly less common westward through the Trans-Pecos, where this species is an uncommon to rare spring and rare to very rare fall migrant. Swainson's Thrush is most frequently encountered in woodland habitats along the coast but is far more likely to be observed elsewhere inland than Veery or Gray-cheeked Thrush. Eastern populations of Swainson's Thrush winter in South America. The only documented winter records for Texas are one at Port Aransas, Nueces County, on 15 January 2009 and another at Port Aransas, Nueces County, from late November 2011 through until at least 15 January 2012. There is a single midsummer record from El Paso, El Paso County, on 30 June 2011. **Timing of occurrence**: Spring migrants have been found as early as late March, but the primary migration period is from early April into late May. Very small numbers are found in very early June. Fall migrants have been recorded between late August and late October, with records as late as early November. **Taxonomy**: Three subspecies occur in the state. This species consists of two distinctive groups of subspecies that were originally described as separate species. The eastern populations are part of the olive-backed group, and those from the Pacific Coast form the russet-backed group. There is a contact zone between these groups in western Canada (Ruegg 2007).

C. u. ustulatus (Nuttall)

Part of the russet-backed group. Casual migrant through El Paso County and accidental east to the Chisos Mountains. This subspecies winters in eastern Mexico and may occur more regularly than records indicate. Oberholser (1974) reports a specimen from Bexar County collected in spring 1890.

C. u. incanus (Godfrey)

Included in the olive-backed group, following Ruegg (2007).

Status uncertain, but appears to be a rare migrant through at least the Trans-Pecos.

C. u. swainsoni (Tschudi)

Part of the olive-backed group. Common to uncommon spring and uncommon to rare fall migrant in the eastern half of Texas, becoming increasingly less common westward. The winter and midsummer records pertain to this subspecies, which includes *C. u. clarescens* Burleigh and Peters.

HERMIT THRUSH *Catharus guttatus* (Pallas)

Common migrant and winter resident throughout most of the state, except in winter in the Trans-Pecos, Panhandle, and South Plains, where it is rare to uncommon. The Hermit Thrush is the only *Catharus* thrush expected to occur anywhere in Texas during the winter. These thrushes are uncommon summer residents in the upper elevations of the Davis and Guadalupe Mountains of the Trans-Pecos. There are three midsummer records from the Panhandle, plus a reported successful nesting at Muleshoe NWR, Bailey County, from 1980. **Timing of occurrence:** Fall migrants have been found as early as mid-August, but they generally begin arriving in late September, with an increase in abundance starting in mid-October through mid-November. Spring migrants are recorded from mid-March through mid-May. A few linger to the end of May in the western third of the state. **Taxonomy:** A. Phillips (1991) has proposed 13 subspecies of Hermit Thrush, and nine of those have been reported from Texas. There is no agreement on the validity of many of these taxa, and Collar (2005) advocated a conservative approach, which is followed here.

C. g. guttatus (Pallas)

Rare migrant and winter resident through the western two-thirds of the state, with specimens reported east to Dallas and Nueces Counties and south to Cameron County; includes *C. g. euborius* Oberholser.

C. g. slevini (Grinnell)

Rare to uncommon migrant and winter resident through the western half of the state, with specimens reported from east to Cooke and San Patricio Counties and south to Cameron and Hidalgo Counties; includes *C. g. oromelus* Oberholser and *C. g. jewetti* Phillips.

C. g. sequoiensis (Belding)

> Rare migrant and winter resident through the western half of the state, with specimens reported east and south to Kerr, Nueces, and Cameron Counties.

C. g. auduboni (Baird)

> Uncommon summer resident in the Davis and Guadalupe Mountains. Common to uncommon migrant and winter resident to the western two-thirds of the state, with specimens reported northeast to Cooke County and southeast to Kinney County; includes *C. g. munroi* Phillips.

C. g. faxoni (Bangs & Penard)

> Common migrant and winter resident to the eastern two-thirds of the state, including the South Texas Brush Country; includes *C. g. crymophilus* Burleigh and Peters.

WOOD THRUSH *Hylocichla mustelina* (Gmelin)

Uncommon to locally common summer resident in the Pineywoods and locally uncommon to rare west to Navarro County and the Lost Pines area of Bastrop County. There are also isolated breeding records from as far south as Victoria County. Wood Thrushes are uncommon to common spring migrants and uncommon to rare fall migrants in the eastern half of the state. They are rare to casual migrants in the western half of the state, with very few fall records. The largest concentrations of migrating birds are usually found at stopover habitats near the coast. They are casual to accidental in winter, and most of the reports are from the Coastal Prairies south to the Lower Rio Grande Valley. **Timing of occurrence**: Spring migrants have been recorded as early as late March and as late as early June, with the primary migration period between mid-April and mid-May. Fall migrants occur between mid-September and early November, but birds occasionally linger into December. **Taxonomy**: Monotypic.

CLAY-COLORED THRUSH *Turdus grayi* Bonaparte

Locally uncommon to rare resident in the Lower Rio Grande Valley. The population of Clay-colored Thrushes has increased steadily over the past 20 years, and now they are found in native thorn-scrub with emergent trees and in riparian woodlands along the Rio Grande, as well as in

urban habitats. They have expanded upriver to Webb County, where nesting was first documented in 2002. There has been an increase in occurrences north of the breeding range in the last 10 years, particularly so in the southern South Texas Brush Country and the Coastal Prairies north to Aransas County (Brush and Conway 2012). The dispersal of individuals farther north in the state has increased as well, with documented records from Brazoria, Gonzales, Bexar, and Uvalde Counties. Dispersal farther afield into the state has included records from Brewster, Schleicher, and Walker Counties. There were 53 documented records when this species was removed from the Review List in November 1998. **Timing of occurrence:** There have been two patterns of northward dispersal. The most common pattern is birds in spring moving up the Coastal Prairies and often staying into the summer. Individuals in Brazoria and Gonzales Counties remained for years. The second pattern includes records of birds discovered in the late fall that remained through the winter. **Taxonomy:** The subspecies that occurs in Texas is *T. g. tamaulipensis* (Nelson).

TBRC Review Species   **WHITE-THROATED THRUSH** *Turdus assimilis* Cabanis

Casual to very rare visitor to the Lower Rio Grande Valley. There are 14 documented records, all but one of which are from Cameron and Hidalgo Counties. The first was a single bird at Laguna Vista, Cameron County, discovered in 1990 (Lasley and Krzywonski 1991). There has been an increase in occurrence of this thrush during the past 10 years, which is highlighted by an astounding eight individuals in the state during 2005. A single bird was at the King Ranch Norias Division, Kenedy County, from 22 to 23 March 2011, providing the sole record outside the Lower Rio Grande Valley. White-throated Thrush is an uncommon resident in the mountains of southern Tamaulipas, Mexico (Howell and Webb 1995), where it is known to move to lower elevations in winter. **Timing of occurrence:** The Texas records have all occurred between 29 December and 12 April. **Taxonomy:** The subspecies that occurs in Texas is *T. a. assimilis* Cabanis. The population in southern Tamaulipas is sometimes considered to be a separate subspecies, *T. a. suttoni* Phillips.

TBRC Review Species

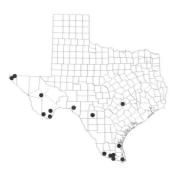

RUFOUS-BACKED ROBIN
Turdus rufopalliatus Lafresnaye

Very rare to casual winter visitor to the Trans-Pecos and the South Texas Brush Country. Texas has 17 documented records. Rufous-backed Robin is a rare fall and winter visitor to southern Arizona, less regularly to southern New Mexico, and might be expected to occur on a more regular basis in the western Trans-Pecos. More unexpected have been the five documented occurrences in the Lower Rio Grande Valley along with single records from Kenedy and Uvalde Counties. Most remarkable is one near Utley, Bastrop County, from 7 January through 7 April 2006. This species is undergoing a significant range expansion in Mexico (Martinez-Morales et al. 2010) that may be primarily a result of the release of caged birds or may be a natural expansion into human-modified habitats. More research is needed to determine if these populations are feral or part of a larger range expansion. **Timing of occurrence:** The Texas records have all occurred between 29 December and 12 April. **Taxonomy:** The subspecies that occurs in Texas is *T. r. rufopalliatus* Lafresnaye.

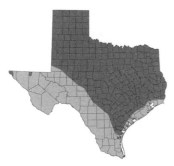

AMERICAN ROBIN *Turdus migratorius* Linnaeus

Common to locally abundant migrant and winter resident in the northern two-thirds of the state, becoming uncommon and more local south through the South Texas Brush Country and southern Coastal Plain. American Robins are common summer residents in the northern half of Texas east of the Pecos River, becoming locally rare to uncommon south to Nueces County. They are uncommon summer residents within the city of El Paso and are locally uncommon summer residents in the Guadalupe Mountains. This species was formerly a rare summer resident in the Davis Mountains, with no nesting birds detected since 2007. This species is increasing as a summer resident along the southern edge of the breeding range in the eastern half of the state, particularly in urban areas. **Timing of occurrence:** Fall migrants begin to arrive in late September. The numbers increase from mid-October through mid-November, when the bulk of the wintering population arrives. These fall migrants can become somewhat nomadic based on food availability. Spring migrants begin to move northward in mid-February, with a notable decline in abundance by late

March. A few birds linger in nonbreeding areas through April. **Taxonomy:** Three subspecies occur in the state.

T. m. migratorius Linnaeus

Common to abundant migrant and winter resident in the eastern half of the state, including the eastern South Texas Brush Country.

T. m. achrusterus (Batchelder)

Rare to uncommon resident in the eastern two-thirds of the state west through the Edwards Plateau. Uncommon in winter in the South Texas Brush Country.

T. m. propinquus Ridgway

Uncommon resident locally in the Panhandle, El Paso County, and the Guadalupe Mountains. Uncommon to rare winter resident elsewhere in the Trans-Pecos.

TBRC Review Species

VARIED THRUSH *Ixoreus naevius* (Gmelin)

Very rare winter visitor to the western half of the state and casual farther east. Of the 43 documented records of Varied Thrush in Texas, 29 are from the High Plains, western Rolling Plains, and Trans-Pecos. In addition, there are three records from the Lower Rio Grande Valley, five from the upper Coastal Prairies, and single records from Bexar, Kerr, Sherman, Tarrant, and Travis Counties. All of these records were of single individuals except one, when up to three Varied Thrushes were present in a city park in Lubbock, Lubbock County, during the winter of 1999–2000. **Timing of occurrence:** The documented records all occur between 2 October and 10 May, with the preponderance of birds found between late October and early January. **Taxonomy:** Monotypic.

TBRC Review Species

AZTEC THRUSH *Ridgwayia pinicola* (Sclater)

Casual. Texas has six documented records, including three from Big Bend NP. The remaining records include two from the central coast and one from the Lower Rio Grande Valley. The first record for the state and the United States was an immature bird at Boot Spring in the Chisos Mountains (D. Wolf 1978). Aztec Thrush is primarily found in the mountains of western Mexico but in the last decade has expanded its range eastward to the western edge of Nuevo León. **Timing of occurrence:** There is not a consistent pattern of occurrence in the state. Two of the Trans-Pecos records are from the late summer, suggesting postbreeding wandering, and the third is from spring. Two of the South Texas records are from the winter, and the third from the spring. **Taxonomy:** This species is often considered monotypic; however, the northernmost populations have been considered a distinct subspecies by some authorities under the name *R. p. maternalis* (Phillips).

Family Mimidae: Mockingbirds and Thrashers

TBRC Review Species   **BLUE MOCKINGBIRD** *Melanotis caerulescens* (Swainson)

Accidental. There are three documented records for presumably two individuals. The first of these was present at Weslaco, Hidalgo County, for almost three years starting in 1999. This bird appears to have departed for long periods of time during its stay, primarily during the summers, although details of its occurrence are scant at best. The second bird was discovered in fall 2002 and remained into the spring. It was in juvenile plumage when initially discovered, although it molted into adult plumage soon thereafter. It returned to the same location the following fall and remained for 18 months. These long-staying individuals do not fit the established pattern of occurrence for other montane species from northeastern Mexico that undertake altitudinal movements. **Taxonomy:** The subspecies that has occurred in Texas is *M. c. caerulescens* (Swainson).

9 MAY 1999–27 FEB. 2002, WESLACO, HIDALGO CO.
  (TBRC 1999–46; TPRF 1797; TBSL 230)
28 SEPT. 2002–26 MAY 2003, PHARR, HIDALGO CO.
  (TBRC 2002–110; TPRF 2117), RETURNING MID-SEPT. 2003–26
  MAR. 2005 (TBRC 2004–46; TPRF 2210)

TBRC Review Species   **BLACK CATBIRD** *Melanoptila glabrirostris* Sclater

Accidental. There is one record of Black Catbird for the state. This specimen was collected at Brownsville, Cameron County, by Frank B. Armstrong and is the only documented record away from the species' breeding range in the Yucatán, Mexico (AOU 1998). The TBRC's decision to accept the record was partially based on an examination of the known collections of Armstrong. There are no other species that Armstrong cataloged from Texas that have not been otherwise documented in the state. The ABA Checklist Committee accepted the record but placed it in the Origin Uncertain category, not on the main list (DeBenedictis et al. 1994). **Taxonomy**: Monotypic.

21 JUNE 1892, BROWNSVILLE, CAMERON CO.
(*ANSP 42,944; TPRF 2613)

GRAY CATBIRD *Dumetella carolinensis* (Linnaeus)

Uncommon to common migrant through the eastern two-thirds of the state west to the eastern High Plains and south through the Rolling Plains to the central Edwards Plateau, becoming increasingly less common farther west. This species is a locally uncommon to rare winter resident along the Coastal Prairies south to the Lower Rio Grande Valley. Gray Catbirds are very rare to casual winter visitors in the remainder of the state, including the Panhandle and Trans-Pecos. They are locally uncommon to rare summer residents across the northern portion of the state, rarely west to the Panhandle. Gray Catbirds are also rare and local breeders in the Pineywoods, mostly in urban settings, and very rare nesters in wooded habitats on the upper Coastal Prairies. **Timing of occurrence**: Spring migrants have been recorded from late March through late May, and a few have lingered into early June. Postbreeding wanderers can be found away from the breeding areas as early as late June, but true fall migrants occur from late August to late October. Gray Catbirds wintering away from the Coastal Prairies obscure the late date of fall migrants, as these individuals may be present from early October through late February. **Taxonomy**: Monotypic.

CURVE-BILLED THRASHER
Toxostoma curvirostre (Swainson)

Uncommon to common resident in the western half of the state. The range of Curve-billed Thrasher extends east through most of the Rolling Plains and Edwards Plateau and south through Gonzales County to Calhoun County on the Coastal Prairies. These thrashers are generally absent from the eastern quarter of the Panhandle. On several occasions, this species has strayed into eastern Texas (once to the Pineywoods), primarily between September and April. Recent studies have suggested that the two subspecies groups of Curve-billed Thrasher should be considered separate species (Zink and Blackwell-Rago 2000; Rojas-Soto 2003). **Taxonomy**: Two subspecies occur in the state, both in the eastern group of subspecies.

T. c. celsum Moore

Uncommon to common resident in the Trans-Pecos and the western half of the Edwards Plateau, north through the South Plains and Rolling Plains through the Panhandle.

T. c. oberholseri Law

Common resident from the Lower Rio Grande Valley north to the southern and eastern edges of the Edwards Plateau and to the Coastal Prairies to Calhoun County.

BROWN THRASHER *Toxostoma rufum* (Linnaeus)

Uncommon to locally common resident in the northeastern third of the state, west across north-central Texas to the eastern Panhandle, and south to the uppermost Coastal Prairies. This species is a very rare and local summer resident across the Oaks and Prairies region to the eastern edge of the Edwards Plateau. In the Panhandle, Brown Thrashers are fairly common in the eastern half and locally common along the major rivers farther west to Hartley County. They have expanded their range westward across the northern Rolling Plains over the past decade. This species is a common migrant and winter resident in the eastern half of the state west to the eastern Edwards Plateau, Concho Valley, and eastern High Plains. Brown Thrasher is a rare to very rare migrant and winter resident in the Trans-Pecos, western Edwards Plateau, and the South Texas Brush Country. **Timing of occurrence:** Fall migrants have been recorded from mid-September through early November. Spring migrants are found from early March through late April, with a few lingering into mid-May. **Taxonomy:** Two subspecies occur in the state.

T. r. rufum (Linnaeus)

As described above for the eastern three-quarters of the state.

T. r. longicauda (Baird)

Rare migrant and winter resident in the western third of the state. There are specimens from as far east as Brazos and San Patricio Counties.

LONG-BILLED THRASHER

Toxostoma longirostre (Lafresnaye)

Common to uncommon resident in the South Texas Brush Country and southern Coastal Plains. Along the coast

Long-billed Thrashers are found north to Matagorda County, where they are very local. This species is a rare and local resident along the southern edge of the Edwards Plateau from Val Verde County east to southern Bexar County. They also occur irregularly in Big Bend NP, suggesting that their range along the Rio Grande might extend farther west than is currently known. Long-billed Thrashers are rare to very rare in all seasons north of the normal range to Midland, Tom Green, Travis, and Bastrop Counties, as well as along the upper Coastal Prairies. **Taxonomy**: The subspecies that occurs in Texas is *T.l. sennetti* (Ridgway).

CRISSAL THRASHER *Toxostoma crissale* Henry

Uncommon to locally common resident in the Trans-Pecos. This species is rare and local in the foothills of the various mountain ranges below 5,000 feet. During the 1960s and early 1970s, Crissal Thrashers were discovered to be rare residents east to the northwestern part of the Edwards Plateau in Crockett and Irion Counties (Lockwood 2001), and they have since been found to occur as far north as Howard and Midland Counties. Seasonal movements occur in local populations, making the species hard to find after the breeding season, which spans from February to late April. **Taxonomy**: The subspecies that occurs in Texas is *T. c. crissale* Henry.

SAGE THRASHER *Oreoscoptes montanus* (Townsend)

Uncommon to rare migrant and irruptive winter resident in the Trans-Pecos and western High Plains, south through the western Edwards Plateau. The number of Sage Thrashers wintering in the state is quite variable from year to year, but they can be locally uncommon. During years when the wintering population is high in the Trans-Pecos, this species is also found with greater regularity in the Rolling Plains and Edwards Plateau and in larger numbers in the High Plains. Sage Thrasher is very rare to casual in late fall east to the Pineywoods and upper Coastal Prairies. These thrashers are also very rare to casual in the late fall and winter to the South Texas Brush Country. During the fall of 2011, very large numbers moved eastward to the eastern Edwards Plateau and Oaks and Prairies region, with some remaining through the winter. There are three July sightings from the Panhandle and another from El Paso County that are presumed to be postbreeding wanderers. **Timing of occurrence:** Fall migrants begin to arrive in late September and are recorded through early November. Spring migrants are found from late February through early April, with records of lingering birds from as late as early May. **Taxonomy:** Monotypic.

NORTHERN MOCKINGBIRD
Mimus polyglottos (Linnaeus)

Abundant to common resident throughout and aptly serves as the state bird of Texas. Northern Mockingbirds are one of the most widespread and common birds found in Texas and are conspicuous by their song. Seasonal movements have been reported in northernmost Texas, and individuals from northern migratory populations may account for apparent influxes of mockingbirds during fall and winter. Also, there are many reports of Northern Mockingbirds mimicking species that do not occur within their territories and have led some observers to suspect seasonal movements within the state. **Taxonomy:** Two subspecies occur in the state.

M. p. leucopterus (Vigors)
 Abundant to common resident in the western two-thirds of the state, intergrading with the following subspecies from Cooke County south along the eastern Edwards Plateau to the central coast. This taxon is sometimes included under *M. p. polyglottos*.

M. p. polyglottos (Linnaeus)
 Abundant to common resident in the eastern third of the state.

Family Sturnidae: Starlings

EUROPEAN STARLING *Sturnus vulgaris* Linnaeus

Common to abundant resident throughout most of the state but can be rare to uncommon in rural areas. A release of 60 individuals in New York City in 1890 is purportedly how this Old World species was introduced into the United States (M. Cooke 1925). The first Texas record was a bird found dead near Cove, Chambers County, in late December 1925 (Oberholser 1974). European Starlings are cavity nesters that compete with native species for available nest sites. For this reason, this species is the most problematic of the introduced birds found in Texas. Some observers have listed starlings as one of the factors responsible for the decline of Red-headed Woodpeckers (Ingold 1989). **Taxonomy**: The subspecies that occurs in Texas is *S. v. vulgaris* Linnaeus.

Family Motacillidae: Wagtails and Pipits

AMERICAN PIPIT *Anthus rubescens* (Tunstall)

Common to uncommon migrant and winter resident in most of the state. American Pipits are rare to locally uncommon in winter in the northern third of the Panhandle and are locally uncommon in the southernmost reaches of the state. The North American population was formerly known as Water Pipit until that complex was divided into three species. The AOU adopted the name American, but this species also occurs in eastern Asia, where the common name is Buff-bellied Pipit. **Timing of occurrence**: Fall migrants begin to arrive in the state in late September and are uncommon until mid-November, when the bulk of the winter population is present. Spring migrants begin to leave the wintering grounds in late February, but migrants are present as late as mid-May, particularly in the western half of the state. There are a few scattered records from July through mid-August that are presumed to be postbreeding dispersal from the breeding grounds in the southern Rocky Mountains. **Taxonomy**: Two subspecies occur in the state.

A. r. rubescens (Tunstall)
> As described above. This taxon includes *A. r. pacificus* Todd, which has been reported as occurring in the Trans-Pecos.

A. r. alticola Todd
> A rare to very rare migrant through the Trans-Pecos. Burleigh and Lowery (1940) noted specimens taken in October and in April, while Thompson (1953) noted a specimen taken in March. Birds resembling this form are reported farther east, but its status is clouded by individual variation within *A. r. rubescens*.

SPRAGUE'S PIPIT *Anthus spragueii* (Audubon)

Uncommon migrant, primarily through the center of the state. Sprague's Pipits are uncommon to locally common winter residents along the Coastal Prairies from Galveston

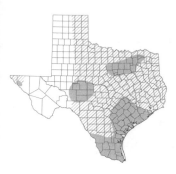

County south to the Lower Rio Grande Valley. This species is rare to locally uncommon inland to the Oaks and Prairies region from Williamson and Brazos Counties, south through much of the eastern and southern South Texas Brush Country. Wintering Sprague's Pipits are rare to locally uncommon in Hudspeth County, north-central Texas, the Concho Valley, and the northwestern Edwards Plateau. These pipits are rare migrants and casual winter visitors through the remainder of the state. **Timing of occurrence:** Fall migrants begin to arrive in the state in mid-September and are found through early November, when the winter population has arrived. Spring migrants are found from late March through late April, with records of individuals straggling into late May. **Taxonomy:** Monotypic.

Family Bombycillidae: Waxwings

TBRC Review Species

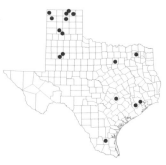

BOHEMIAN WAXWING *Bombycilla garrulus* (Linnaeus)

Very rare winter visitor to the northern Panhandle and accidental elsewhere in the state. Although there are 17 documented records, this species occurred more regularly between the late 1950s and early 1970s prior to the formation of the TBRC. During that time, numerous reports were made by experienced observers in the Panhandle and across north-central Texas. Seyffert (2001b) reports a flock of more than 300 present in Amarillo in January 1962 and states that large flocks, sometimes numbering more than 50 individuals, were present at Palo Duro Canyon SP, Randall County, in January 1967 and 1973. The species was first reported from the state in 1926, although the first documented record was not obtained until 26 December 1936 from Randall County. There are 38 reports of Bohemian Waxwings from prior to the development of the Review List in 1988 for which there is no documentation on file. **Timing of occurrence:** All the documented records are between 17 November and 1 April, which coincides with the undocumented occurrences. **Taxonomy:** The subspecies that occurs in Texas is *B. g. pallidiceps* Reichenow.

CEDAR WAXWING *Bombycilla cedrorum* Vieillot

Common to abundant migrant and winter resident east of the Pecos River but generally considered an irregular un-

common to rare winter resident in the Lower Rio Grande Valley. Cedar Waxwings are common to uncommon migrants and winter residents in the Trans-Pecos. The number of waxwings present in a given area fluctuates greatly from year to year. Juveniles have been found in the Panhandle and the Davis Mountains in August and September, which has led to speculation that they may nest (Seyffert 1991b). **Timing of occurrence:** Fall migrants have been detected as early as mid-August, but the primary migration occurs from mid-October to early December. Spring migrants are found from late March through early May, with large numbers sometimes lingering into early June and, on rare occasions, late June. There is one midsummer record of a single bird at Boerne, Kendall County, on 13 July 2007. **Taxonomy:** Monotypic.

Family Ptiliogonatidae: Silky-flycatchers

TBRC Review Species

GRAY SILKY-FLYCATCHER

Ptiliogonys cinereus Swainson

Accidental. Texas has two documented records of this species. Gray Silky-flycatchers are uncommon in northeastern Mexico as far north as southwestern Nuevo León and southern Coahuila. In northwestern Mexico they occur north to southern Chihuahua and Sonora. Northern populations are at least partially migratory and are known to wander in lowland areas during winter. **Taxonomy:** The subspecies for either record is not known. Based on distribution, it seems likely that the Lower Rio Grande Valley record would pertain to *P. c. cinereus* Swainson, and the Trans-Pecos record to *P. c. otofuscus* Moore.

31 OCT.–11 NOV. 1985, LAGUNA ATASCOSA NWR, CAMERON CO.
  (TBRC 1989-37; TPRF 363)
12 JAN.–5 MAR. 1995, EL PASO, EL PASO CO.
  (TBRC 1995-13; TPRF 1320)

PHAINOPEPLA *Phainopepla nitens* (Swainson)

Rare to locally uncommon resident throughout most of the Trans-Pecos, except in the foothills of the Chinati, Davis, Del Norte, and Guadalupe Mountains, where this species can be common. This species has a very localized distri-

bution in Texas, particularly during the breeding season. Phainopeplas routinely wander eastward into other areas of the state. They are casual visitors, in all seasons, to the western Edwards Plateau, High Plains, and western Rolling Plains north to Moore County. Vagrants have been reported from virtually all other parts of the western two-thirds of the state, including from as far east as Bosque and Tarrant Counties and as far south as Hidalgo County. Populations fluctuate greatly during drought years, although some individuals are always present. **Taxonomy:** Two subspecies occur in the state.

P. n. lepida van Tyne
    As described above, excluding the southernmost Trans-Pecos.

P. n. nitens (Swainson)
    Uncommon and local resident in southern Brewster and Presidio Counties, primarily in the foothills of the Chisos and Chinati Mountains. Winter vagrants have been collected in the southwestern Edwards Plateau and in Maverick County.

Family Peucedramidae: Olive Warbler

TBRC Review Species **OLIVE WARBLER**
Peucedramus taeniatus (Du Bus de Gisignies)

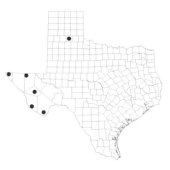

Casual. There are seven documented records from the Trans-Pecos and one from Briscoe County in the Panhandle. The first was found 3 May 1991 at Big Bend NP, Brewster County (TBRC 1991-61). The first documented with a photograph was from the Davis Mountains on 19 May 1992 (TBRC 1992-72; TPRF 1094). There are six reports of Olive Warblers from prior to the development of the Review List in 1988 for which there is no documentation on file. **Timing of occurrence:** All documented records occurred between 30 April and 30 May in the spring and from 9 August to 14 November in the fall. **Taxonomy:** The subspecies found in Texas is not confirmed. The population that occurs in northwestern Coahuila is *P. t. jaliscensis* Miller & Griscom. It is known to nest within 50 miles of the Rio Grande and is the most likely to occur in Texas.

Family Calcariidae: Longspurs and Allies

LAPLAND LONGSPUR *Calcarius lapponicus* (Linnaeus)

Uncommon to abundant migrant and winter resident in the Panhandle east to north-central Texas. Lapland Longspurs are rare to locally common migrants and winter visitors through the remainder of the eastern half of North Texas. During invasion winters, they can be locally common as far south as Bexar County and east to Harris County. This species is a rare migrant and winter visitor to the South Plains and very rare south to the Concho Valley. Lapland Longspur is a very rare to casual winter visitor in the Trans-Pecos, with most records coming from the northern portion of that region. **Timing of occurrence:** Fall migrants begin arriving in the state in late October, but the majority of the wintering population is not present until mid-November. The early date for the state is 7 October. Spring migrants begin leaving the state in mid-February, and the species has normally departed by mid-March. **Taxonomy:** Two subspecies occur in the state.

C.l. alascensis Ridgway

    Rare winter resident in the northern Panhandle. Status elsewhere in the state is uncertain.

C.l. lapponicus (Linnaeus)

    As described in main species account. This taxon includes *C.l. subcalcaratus* (Brehm).

CHESTNUT-COLLARED LONGSPUR
Calcarius ornatus (Townsend)

Common to uncommon migrant and winter resident to the High Plains, Rolling Plains, and Trans-Pecos. Chestnut-collared Longspur is also a locally common to rare migrant and winter resident east through north-central Texas and south through the Blackland Prairies and Edwards Plateau. This species is a very rare visitor during winter and early spring along the coast and south through the northern South Texas Brush Country. It is casual in spring east to the Pineywoods. **Timing of occurrence:** Fall migrants begin to arrive in early October, with an increase in abundance in late October and early November. Winter residents begin to depart in early March and have mostly departed by

mid-April, but very small numbers linger into late April.
Taxonomy: Monotypic.

SMITH'S LONGSPUR *Calcarius pictus* (Swainson)

Locally uncommon to rare migrant and winter resident to
eastern north-central and northeast Texas. Smith's Longspurs are casual winter visitors as far south as Guadalupe
and Brazoria Counties. Despite the restricted winter range,
this species has been detected in many other areas of the
state, including the central coast (Calhoun County), Edwards Plateau (Schleicher County), Pineywoods (Gregg,
Nacogdoches, Smith, and Walker Counties), South Plains
(Crosby County), and Trans-Pecos (Brewster, El Paso, and
Jeff Davis Counties). **Timing of occurrence:** Fall migrants
begin to arrive in early November. The winter population
arrives in late November and early December. Winter residents have departed by mid-March. On rare occasions
some linger as late as mid-April. **Taxonomy:** Monotypic.

MCCOWN'S LONGSPUR

Rhynchophanes mccownii (Lawrence)

Abundant to uncommon migrant and winter resident on the
High Plains and Rolling Plains, south to the northwestern
Edwards Plateau. McCown's Longspur is a locally uncom-
mon to rare migrant and winter resident through the Oaks
and Prairies region south to Bexar County, as well as in the
northern half of the Trans-Pecos. This species is casual in
winter along the coast and to the northern South Texas
Brush Country. Oberholser (1974) reported that McCown's
Longspurs were regular winter visitors to the South Texas
Brush Country prior to 1900. They nested in the western
Oklahoma Panhandle until 1914 (Sutton 1967) and may
have also nested in the northwestern Texas Panhandle. **Timing of occurrence:** Fall migrants begin to arrive in early October, and the winter population arrives in late October and
early November. Winter residents have mostly departed by
mid-March, with some lingering into early April. **Taxonomy:** Monotypic. Formerly included in the genus *Calcarius.*

TBRC Review Species   ### SNOW BUNTING *Plectrophenax nivalis* (Linnaeus)

Casual. Texas has eight documented records of this species. The first record was a bird found at Lake Livingston,

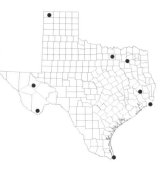

Polk County, in 1977 (Bryan, Gallucci, and Moldenhauer 1978). Snow Buntings are found south to the central Great Plains during a normal winter. When food crops are poor and sometimes after particularly strong storms, these birds may be pushed much farther south. During these invasions, Snow Buntings are sometimes found in numbers across the southeastern United States and into the southern Great Plains north of Texas. **Timing of occurrence:** Six of the records occurred in winter, as expected, between 27 November and 15 January. The final two records are unexplainable, involving alternate-plumaged males in the Chisos Basin, Brewster County, on 9 May 1988 and at Sea Rim SP, Jefferson County, on 13 June 2011. **Taxonomy:** The subspecies that occurs in Texas is *P. n. nivalis* (Linnaeus).

Family Parulidae: Wood-Warblers

OVENBIRD *Seiurus aurocapilla* (Linnaeus)

Uncommon to common migrant through the eastern half of the state, becoming increasingly less common farther west. Ovenbirds are encountered in greatest numbers along the coast. This species is a rare to uncommon winter resident in the Lower Rio Grande Valley, becoming rare northward along the coast and very rare inland in the southern half of the state east of the Edwards Plateau. Ovenbirds are casual in winter north to Lubbock County. Although they are a summer resident in Arkansas and southeastern Oklahoma very close to the Texas border, no breeding records exist for the state. **Timing of occurrence:** Spring migrants have been recorded from late March through mid-May, and a few have lingered to early June. Fall migrants pass through the state between early August and late October. Migration peaks from mid-September through early October. **Taxonomy:** Two subspecies occur in the state.

S. a. aurocapillus (Linnaeus)
   As described above.

S. a. cinereus Miller
   Rare to very rare migrant through the western half of the state.

WORM-EATING WARBLER

Helmitheros vermivorum (Gmelin)

Uncommon to locally common migrant in the eastern third of the state, although primarily encountered along the coast. Worm-eating Warblers are rare to uncommon migrants westward to the Balcones Escarpment and very rare to casual farther west. They are very rare and local summer residents in the central and southern portions of the Pineywoods and casual and very local nesters in northeast Texas. Worm-eating Warblers are casual to accidental winter visitors along the coast. **Timing of occurrence**: Spring migrants begin to arrive in mid-March. Migration peaks from early April to early May, and a few straggle into late May. Fall migrants have been recorded from late August to early November. **Taxonomy**: Monotypic.

LOUISIANA WATERTHRUSH

Parkesia motacilla (Vieillot)

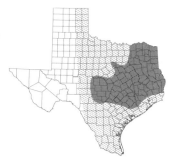

Rare to locally uncommon summer resident in the Pineywoods, easternmost Blackland Prairies, the Brazos River drainage up to Hood County, and the center of the state west to the central Edwards Plateau. Louisiana Waterthrush is a locally rare to casual summer resident through the remainder of the eastern third of the state south to the southern Edwards Plateau. This species is an uncommon migrant through the eastern two-thirds of the state, becoming rare to casual farther west. These birds are casual winter visitors to the Coastal Prairies and Lower Rio Grande Valley and accidental farther inland to Bexar, Walker, and Washington Counties. **Timing of occurrence**: Spring migrants are present from early March through late April, rarely straggling into early May. There are several records from the Coastal Prairies from February that could be early migrants or local winterers. What are presumed to be postbreeding wanderers have been found during June away from breeding areas in the Pineywoods and Post Oak Savannah. Fall migrants have been recorded from early July through late September, with some straggling as late as early November. **Taxonomy**: Monotypic.

NORTHERN WATERTHRUSH
Parkesia noveboracensis (Gmelin)

Common to uncommon migrant in the eastern half of the state, becoming uncommon to rare west through the Trans-Pecos. This species is a rare winter visitor to the Lower Rio Grande Valley, becoming very rare and more irregular northward along the coast. **Timing of occurrence:** Spring migrants are seen from early April through late May. There are records from the Coastal Prairies from the first half of March that could be early migrants or local winterers. Fall migrants have been recorded from mid-August through mid-October, with some straggling (or lingering) as late as December. **Taxonomy:** Monotypic.

GOLDEN-WINGED WARBLER
Vermivora chrysoptera (Linnaeus)

Uncommon to rare spring migrant and rare to very rare fall migrant through the eastern third of the state. This species is most frequently encountered at stopover habitats along the coast and is very rare to accidental in the western two-thirds of the state. Golden-winged Warblers are declining throughout their range, and competition with Blue-winged Warblers (*V. cyanoptera*) is thought to be one factor. Hybrids between Blue-winged Warblers and Golden-winged Warblers have been reported in Texas on numerous occasions, with most pertaining to the "Brewster's" Warbler, although the "Lawrence's" Warbler also occurs. These hybrids vary greatly in appearance, including individuals that are very similar to one of the parental types. **Timing of occurrence:** Spring migrants have been recorded as early as late March. The primary migration period is from mid-April to mid-May, and very small numbers are present into late May. Fall migrants have been found between late August and mid-October. An exceptionally late migrant was at South Padre Island, Cameron County, on 23 November 2008. **Taxonomy:** Monotypic.

BLUE-WINGED WARBLER
Vermivora cyanoptera Olson & Reveal

Common to uncommon spring migrant and uncommon to rare fall migrant in the eastern third of the state. Blue-winged Warblers become increasingly rare westward through the

Rolling Plains and Edwards Plateau and are accidental to casual in the Trans-Pecos and the Panhandle. Hybrids with Golden-winged Warblers have been reported on numerous occasions in Texas. These hybrids vary greatly in appearance, including individuals that are very similar to one of the parental types. **Timing of occurrence:** Spring migrants have been recorded from late March to mid-May and very rarely into late May. Fall migrants have been found between mid-August and mid-October, and some birds straggle into mid-December on rare occasions. **Taxonomy:** Monotypic.

BLACK-AND-WHITE WARBLER

Mniotilta varia (Linnaeus)

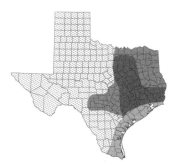

Rare to common summer resident in the Pineywoods and the Oaks and Prairies region and west through the southern Edwards Plateau, where more localized. There are a number of midsummer records from the Panhandle and South Plains without evidence of nesting. Black-and-white Warblers are uncommon to common migrants throughout most of the state, becoming rare in the western Trans-Pecos. Like many trans-Gulf migrants, they can be locally abundant along the coast during migration. This species is a rare to uncommon winter resident through the Coastal Prairies and Lower Rio Grande Valley and very rare northward through the southern Pineywoods and Oaks and Prairies region. **Timing of occurrence:** Spring migrants have been recorded from early March to late May. Territorial birds are often in place by 10 March in the southern portions of the breeding range. Fall migrants have been found between mid-July and late October, but a few have straggled into late November. **Taxonomy:** Monotypic.

PROTHONOTARY WARBLER

Protonotaria citrea (Boddaert)

Rare to locally common summer resident in the eastern half of the state, locally west to the edge of the Rolling Plains and southward to the central Coastal Prairies. Prothonotary Warblers are uncommon to common migrants through the eastern half of the state and along the coast, becoming rare to very rare farther west. During the winter, this species is casual along the coast and in the Lower Rio Grande Valley. **Timing of occurrence:** Spring migrants have been recorded from mid-March to late May. Fall migrants have been found between late July and early September, and a few have lingered until late October. **Taxonomy**: Monotypic.

SWAINSON'S WARBLER
Limnothlypis swainsonii (Audubon)

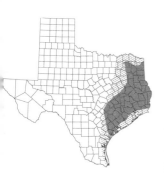

Rare to locally uncommon summer resident in the Pineywoods and rare and very local south and west to Aransas, Bastrop, and Victoria Counties. As spring migrants Swainson's Warblers are locally rare to uncommon at stopover sites along the coast from the Louisiana border south to Nueces County. Elsewhere in the eastern third of the state west to about the center of the Oaks and Prairies region, this species is a locally rare to casual spring migrant. Swainson's Warblers are very rare to accidental spring migrants in the western two-thirds of the state. As a fall migrant this species is casual to accidental at coastal stopover sites and unrecorded farther west in the state. **Timing of occurrence**: Spring migrants have been recorded from late March to mid-April. A few straggle into late April, and many of those are found west of the main migratory path in early May. Fall migrants have been found between late August and mid-October. **Taxonomy**: Monotypic.

TENNESSEE WARBLER *Oreothlypis peregrina* (Wilson)

Uncommon to common spring and uncommon to rare fall migrant in the eastern half of the state. Tennessee Warblers can be abundant at stopover sites on the upper coast as well as in the Pineywoods. This species is a rare to casual migrant in the western half of the state and is more frequent in spring than in fall. These warblers are very rare winter visitors along the coast and in the Lower Rio Grande Valley. First-winter male and adult female Tennessee Warblers can be confused with the Orange-crowned Warbler (*O. celata*), which may account for some winter reports. **Timing of occurrence**: Spring migrants have been recorded from late March to mid-May, with a few straggling into early June. Fall migrants have been found between late August and late October, with a few straggling to late November. **Taxonomy**: Monotypic.

ORANGE-CROWNED WARBLER
Oreothlypis celata (Say)

Uncommon to abundant migrant throughout the state. Orange-crowned Warblers are uncommon to common winter residents in much of the state, although they are casual above the Llano Estacado Escarpment from Lubbock County northward through the Panhandle during this season. They are especially common in winter in the southern third of the state and along the coast, where they can be locally abundant. This species is an uncommon and local summer resident in the higher elevations of the Davis and Guadalupe Mountains. **Timing of occurrence**: Fall migrants begin to arrive in the Panhandle and Trans-Pecos as early as late August, becoming more common by mid-September. Farther east in the state fall migrants do not arrive until very late September or early October. Migrants are recorded through late October, rarely straggling into early November north of the winter range. Spring migrants are found from late March through mid-May, rarely straggling into very early June. **Taxonomy**: Two subspecies have been documented in the state.

O. c. orestera (Oberholser)

Uncommon and local summer resident in the Guadalupe and Davis Mountains. Common to uncommon migrant through the western half of the state. Uncommon to rare winter resident in the same region south of the Panhandle, where it is casual. Rare to very rare winter resident in the eastern through the central third of the state south to the Lower Rio Grande Valley.

O. c. celata (Say)

Common to abundant migrant throughout the state and uncommon to common winter resident in much of the state south of the Panhandle, where it is casual.

COLIMA WARBLER
Oreothlypis crissalis (Salvin and Godman)

Uncommon and extremely localized migrant and summer resident in the upper Chisos Mountains. This population in West Texas is the only one north of Mexico, and these warblers were first discovered breeding in the Chisos Mountains in 1932 (Van Tyne 1936a). Single, apparently territorial males have been found in the upper Davis Moun-

tains at least five times since first discovered there in 1999. There is a population of presumed Colima × Virginia Warblers (*O. virginiae*) in mixed oak-juniper woodlands in the upper elevations of the Davis Mountains, which suggests that spring migrants that overshoot the Chisos Mountains are more regular than sightings would indicate. There is a sight record of a Colima Warbler from the Lower Rio Grande Valley at Santa Ana NWR, Hidalgo County, on 4 October 1978, and one photographed on the King Ranch, Kenedy County, on 9 May 2013. **Timing of occurrence:** Spring migrants arrive in Big Bend NP in mid-April, and most depart by early September, with a few occasionally found later in September. **Taxonomy:** Monotypic.

LUCY'S WARBLER *Oreothlypis luciae* (Cooper)

Rare to locally uncommon summer resident in mesquite woodlands along the Rio Grande from southern Hudspeth County to western Brewster County. Lucy's Warblers are very rare but appear to be increasing as summer visitors upriver to El Paso County. This species is accidental in winter, with two records from Presidio County and one from Brazoria County. Lucy's Warbler has also been documented east of the Pecos River on two other occasions, with single records from Kenedy and Navarro Counties. **Timing of occurrence:** Spring migrants arrive on the breeding grounds in mid-March. The breeding population is present until late August, and a few linger through September. There is an exceptional late date of three at Big Bend NP on 17 November 1991. **Taxonomy:** Monotypic.

NASHVILLE WARBLER *Oreothlypis ruficapilla* (Wilson)

Uncommon to abundant migrant throughout most of the state, generally more common in fall than in spring. These warblers are circum-Gulf migrants; in spring they are less common on the upper coast as birds head north from the central coast after rounding the Gulf. They are rare to very uncommon migrants in the western half of the Trans-Pecos. Nashville Warblers are locally uncommon to rare winter residents in the Lower Rio Grande Valley and very rare northward along the southern coast and in the South Texas Brush Country. They are casual to accidental in winter elsewhere in the state except in the Panhandle, where there are no records during this season. **Timing of occurrence: Spring migrants begin to arrive in mid-March, with the main passage from early April through early May. They rarely linger to the end of May. Fall migrants are found from mid-August to early November, but small numbers linger into early December, particularly along the Coastal Prairies. **Taxonomy:** Two subspecies occur in the state.

O. r. ruficapilla (Wilson)
As described above.

O. r. ridgwayi (van Rossem)
Status unclear, but this taxon appears to be a rare migrant through the Trans-Pecos and very rare to casual east through the Rolling Plains, Edwards Plateau, and western South Texas Brush Country.

VIRGINIA'S WARBLER *Oreothlypis virginiae* (Baird)

Uncommon to rare migrant in the Trans-Pecos and a very rare migrant through the western High Plains and western Edwards Plateau. This species is casual to accidental during migration eastward in the state, with most of the records from the upper Coastal Prairies. Virginia's Warblers are rare to uncommon summer residents in the upper elevations of the Davis and Guadalupe Mountains. The summer population in the Davis Mountains is primarily found in oak-pine woodlands at higher elevations than the population of presumed Colima × Virginia's Warblers. A Virginia's Warbler at Santa Ana NWR, Hidalgo County, from 3 January to 28 February 1991 is the only documented winter record for Texas. **Timing of occurrence:** Spring migrants begin to arrive in late March, but the main passage is from mid-April through mid-May. There are records from El Paso County from July that appear to be postbreeding wanderers. Fall migrants are found from mid-August to early October. **Taxonomy:** Monotypic.

TBRC Review Species   **CONNECTICUT WARBLER** *Oporornis agilis* (Wilson)

Casual migrant through the eastern half of the state and accidental in the Trans-Pecos. Texas has 11 accepted records, only three of which are documented with photographs. Five of the records are from the upper coast, two from north-central Texas, three from the Edwards Plateau,

and one from the Trans-Pecos. The first documented record was a bird photographed at High Island, Galveston County, on 16 September 1978 (TPRF 140). The identification difficulties associated with this species are reflected by the 21 submissions that the TBRC has deemed inadequate. There are an additional 48 reports of Connecticut Warbler from prior to the development of the Review List in 1988 for which there is no documentation on file. This species may be confused with other warblers in the genus *Geothlypis* that were formerly included in *Oporornis,* as well as Nashville Warbler. **Timing of occurrence:** The six spring records are from 30 April to 19 May, and the five from the fall are between 5 September and 10 October. **Taxonomy:** Monotypic.

TBRC Review Species

GRAY-CROWNED YELLOWTHROAT

Geothlypis poliocephala Baird

Casual visitor during all seasons to the Lower Rio Grande Valley. Prior to the twentieth century this species was a rare to uncommon spring and summer resident in the Lower Rio Grande Valley, primarily in Cameron County. The reasons for the decline of Gray-crowned Yellowthroats in Texas are unclear, but probably habitat related. Reports from that time suggest that they were present from early February to at least early July. Between May 1890 and May 1894, a minimum of 34 specimens were collected from the Brownsville area. Since 1927 there have been 11 documented records of the species, all occurring since 1988 (see map for distribution of these records). There was an unsuccessful nesting attempt in 2005 at the Sabal Palm Sanctuary, Cameron County (Lorenz, Butler, and Paz 2006). One at San Ygnacio, Zapata County, during the winter of 1995–96 was presumed to be a hybrid with Common Yellowthroat (*G. trichas*), and it is possible that other birds seen in the Lower Rio Grande Valley have involved such hybrids or back-crosses. **Timing of occurrence:** Documented records since 1988 have primarily occurred between 8 February and 12 August. The returning male found at the Sabal Palm Sanctuary was discovered in early December in 2005 and 2006, prompting speculation as to whether it actually left. **Taxonomy:** The subspecies that occurs in Texas is *G. p. ralphi* Ridgway.

MACGILLIVRAY'S WARBLER
Geothlypis tolmiei (Townsend)

Uncommon to rare migrant through the western half of the state and rare to very rare migrant eastward through the Blackland Prairies and central and lower coasts. During the winter this species is a very rare visitor to the coast, with reports from as far north as Jefferson County. The migratory pathways of MacGillivray's and Mourning Warblers (*G. philadelphia*) overlap through the eastern Rolling Plains, Edwards Plateau, and South Texas Brush Country. MacGillivray's Warbler is a rare and very local summer resident at the highest elevations of the Davis Mountains. Nesting was confirmed in the vicinity of Mount Livermore in 2002 at an elevation of 7,600 feet. **Timing of occurrence:** Spring migrants pass through the state between early April and early June. Fall migrants are present beginning in early August, with an increase in abundance between late August and mid-September. Small numbers linger through early October. **Taxonomy:** Two subspecies occur in Texas.

G. t. monticola (Phillips)
  As described above.

G. t. tolmiei (Townsend)
  Status unclear, and there is a reported specimen from Culberson County. There are other specimens from as far east as Cooke County of *G. t. intermedia* Phillips and *G. t. austinsmithi* Phillips, both of which have been subsumed into *G. t. tolmiei*.

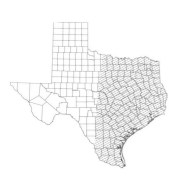

MOURNING WARBLER
Geothlypis philadelphia (Wilson)

Uncommon to rare spring and uncommon fall migrant in the eastern half of the state and very rare to casual in the western half. Mourning Warblers are circum-Gulf migrants and therefore rare at stopover sites along the immediate upper coast in spring. There is a single record in the winter, a specimen from Brownsville, Cameron County, on 21 February 1894. **Timing of occurrence:** Spring migrants are present from late April to early June. Fall migrants begin to arrive in late August, with an increase in abundance during September. Small numbers straggle through late October. **Taxonomy:** Monotypic.

KENTUCKY WARBLER *Geothlypis formosa* (Wilson)

Rare to locally uncommon summer resident in the eastern quarter of the state, locally west to Bastrop, Grayson, and McLennan Counties and southeast to northern Hardin County. Kentucky Warblers formerly were rare summer residents on the southeastern Edwards Plateau. They are uncommon to locally common migrants in the eastern third of the state, becoming increasingly less common west to the eastern Edwards Plateau and Rolling Plains. In the western half of the state, Kentucky Warblers are casual migrants only in the spring. This species is very rare in winter on the coast and in the Lower Rio Grande Valley. **Timing of occurrence**: Spring migrants begin to arrive in late March. The numbers increase from early April through mid-May, and a few straggle into the first few days of June. Fall migrants are found from mid-August through late September, with very small numbers straggling into late October. **Taxonomy**: Monotypic.

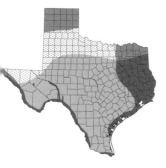

COMMON YELLOWTHROAT

Geothlypis trichas (Linnaeus)

Rare to locally common summer resident in the northeastern part of the state south to the upper coast and along the Coastal Prairies to Kleberg County on the southern coast. Common Yellowthroats are locally uncommon summer residents along the Rio Grande, up the Pecos and Devils Rivers, and at Balmorhea Lake and vicinity in Reeves County. There have been isolated nesting records from virtually all of the remaining areas of the state. They are common to uncommon migrants throughout the state. In winter, Common Yellowthroat is uncommon to locally common across much of the southern half of Texas but is generally rare to locally uncommon in the Trans-Pecos and Edwards Plateau. In the northern half of the state, these birds winter locally but are absent from most of the Panhandle. **Timing of occurrence**: Spring migrants are recorded from mid-March to mid-May. Fall migrants begin to arrive in early September and are present into early November. **Taxonomy**: Seven subspecies occur in the state.

G. t. campicola Behle & Aldrich
> Uncommon migrant in the western two-thirds of Texas east to Collin County and south to Cameron County (specimens).

Winter status is not well understood but is suspected as a rare winter resident in the western half of the state north to the South Plains and east to Kaufman County.

G. t. occidentalis Brewster
Rare to locally uncommon summer resident in the Panhandle. An uncommon migrant through the western half of the state, with specimens reported from Bexar, Cameron, and Dallas Counties. Rare to locally uncommon winter resident in the western half of the state north to the South Plains, east to Harris County, and south to Cameron County.

G. t. chryseola van Rossem
Rare to locally uncommon resident along the Rio Grande from El Paso County to Starr County, possibly in western Hidalgo County. Also found in wetlands elsewhere in the Trans-Pecos and along the Devils and Pecos Rivers. Rare migrant east and south to Victoria and Cameron Counties.

G. t. trichas (Linnaeus)
Common migrant through the eastern half of the state, west at least to the eastern Rolling Plains and eastern Edwards Plateau, and casual to Presidio County. Rare to locally common resident in the eastern third of the state and on the Coastal Prairies. Uncommon to common winter resident in the eastern half of the state north to the southern edge of the Rolling Plains and to the northern Oaks and Prairies region and northern Pineywoods.

G. t. typhicola Burleigh
Uncommon migrant and winter resident on the coastal plain and eastern half of the Lower Rio Grande Valley.

G. t. ignota Chapman
Very rare migrant and winter resident on the upper Coastal Prairies.

G. t. insperata van Tyne
Rare to locally uncommon resident in Cameron and eastern Hidalgo Counties, mostly associated with resacas lined with dense vegetation.

HOODED WARBLER *Setophaga citrina* (Boddaert)

Uncommon to locally abundant summer resident in the eastern quarter of the state. Very localized breeding populations also exist in Bastrop, Colorado, and Matagorda

Counties, the westernmost to be found in North America. They are also reported to breed in Aransas County (Jones 1992), and there is a midsummer record from Randall County in the Panhandle. This species is an uncommon to common migrant in the eastern third of the state west to the Balcones Escarpment and can be locally common along the coast. Hooded Warblers are very rare to casual migrants in the western two-thirds of the state. They are rare winter visitors in extreme southern Texas north to the central coast, with only scattered records farther north and west. **Timing of occurrence:** Spring migrants are recorded from mid-March to mid-May. Postbreeding wanderers are found away from breeding habitat as early as mid-July, and fall migrants are found from mid-August to late October, rarely lingering through November. **Taxonomy:** Monotypic.

AMERICAN REDSTART *Setophaga ruticilla* (Linnaeus)

Rare to uncommon and very local summer resident in the forested areas of East Texas southwest to Montgomery County. American Redstart is an uncommon to locally common migrant in the eastern half of the state and rare to uncommon in the western half. The breeding range has decreased significantly in recent decades, as it formerly extended south to Lavaca and Victoria Counties (Oberholser 1974). Away from the known breeding range, this species is a rare summer visitor to virtually all regions, including Dallam and Hemphill Counties in the Panhandle. American Redstart is a rare winter resident in the Lower Rio Grande Valley and along the lower coast, becoming very rare elsewhere along the Coastal Prairies inland to Grimes and Washington Counties. There are documented inland winter records from El Paso and Tarrant Counties. **Timing of occurrence:** Spring migrants have been found as early as late March. Most pass through between mid-April and mid-May, and some straggle into early June. Fall migrants are found from mid-August to the end of October, and very small numbers linger into early November. **Taxonomy:** Monotypic.

CAPE MAY WARBLER *Setophaga tigrina* (Gmelin)

Rare spring migrant and very rare fall migrant along the Coastal Prairies south to Cameron County. Even in the

Coastal Prairies this species is rarely encountered inland away from stopover sites along the coast. Cape May Warblers are casual spring and accidental fall migrants elsewhere in the eastern half of the state and accidental spring and fall migrants in the western half of the state. They are also casual winter visitors in the Lower Rio Grande Valley, north to Zapata County, and lower coast and accidental elsewhere inland to Jeff Davis County. **Timing of occurrence:** Spring migrants have been found as early as late March, with more regular occurrence between mid-April and mid-May, and a few as late as early June. Fall migrants have been found from early September to late October, with a small number of records into mid-November. **Taxonomy:** Monotypic.

CERULEAN WARBLER *Setophaga cerulea* (Wilson)

Uncommon to rare spring migrant and very rare to casual fall migrant through the eastern third of the state, but most frequent at stopover sites along the coast. There are no documented records from the Panhandle and only one from the Trans-Pecos. Cerulean Warblers were formerly summer residents in northeast Texas from Cooke and Dallas Counties eastward (Oberholser 1974). The only recent summer records involved a single bird at Caddo Lake, Marion County, on 15 July 2008 and possibly as many as four near Lake Wright Patman, Cass and Morris Counties, during the summer of 2011. There are no winter records for the state. Cerulean Warblers have declined more sharply than any other warbler, with some estimates that the population has declined as much as 75 percent in the last 40 years. **Timing of occurrence:** Spring migrants have been found as early as late March. The primary passage is between early April to early May, and a few have been seen into the latter half of May. Fall migrants have been found from late August to mid-October, with a few as late as early November. **Taxonomy:** Monotypic.

NORTHERN PARULA *Setophaga americana* (Linnaeus)

Uncommon to common summer resident in the eastern half of Texas, west to the eastern edge of the Rolling Plains and Edwards Plateau, and south to the San Antonio River on the Coastal Bend. Northern Parulas are local summer residents

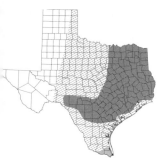

along the major rivers on the southern Edwards Plateau west to the West Nueces River and very rare west to the Devils River. This species is rare and local in summer south of the Edwards Plateau, with single nesting records from Hidalgo and Live Oak Counties. These birds are common to uncommon migrants in the eastern half of the state west to the eastern Panhandle and central Rolling Plains, becoming increasingly less common father west. They are rare winter residents in the Lower Rio Grande Valley and casual elsewhere in the southern half of the state. **Timing of occurrence**: Spring migrants begin to arrive in very small numbers in late February, and the breeding population arrives between early March and early April. A second peak of migrants passes though the state from mid-April to mid-May on the way to more northern breeding sites. Fall migrants have been found from late August to early November. **Taxonomy**: Monotypic.

TROPICAL PARULA *Setophaga pitiayumi* (Vieillot)

Rare to uncommon resident in the live oak woodlands of the Coastal Sand Plain in Kenedy and Brooks Counties and rare to very rare in the Lower Rio Grande Valley. This species is very rare and local, primarily in summer, northward along the Coastal Prairies to Nueces County and casual north to Bee, Calhoun, and Victoria Counties. Tropical Parulas are rare to locally uncommon summer residents along the Devils and Pecos Rivers in Val Verde County. They were first found to be breeding in these areas in about 2000, and there has been a coinciding increase in occurrence on the Edwards Plateau. This is particularly true in the southwestern portion of that region. Wandering Tropical Parulas have been found north to Lubbock County, west to the central Trans-Pecos, and east to the upper coast. During the summer of 2001, a pair unsuccessfully nested at Davis Mountains SP, Jeff Davis County. Individuals that are presumed hybrids between Tropical and Northern Parulas have been documented in Texas. The extent of hybridization is unknown and may be the result of secondary contact, as Tropical Parulas have colonized areas in the western Hill Country. **Timing of occurrence**: The breeding populations arrive in late March and are present through mid-September. **Taxonomy**: The subspecies that occurs in Texas is *S. p. nigrilora* (Coues).

MAGNOLIA WARBLER *Setophaga magnolia* (Wilson)

Uncommon to locally common migrant in the eastern half of the state. As in the case of many other migrant warblers, Magnolia Warblers are encountered more frequently and in larger numbers along the coast. They are uncommon migrants west through the eastern Edwards Plateau, becoming increasingly less common westward. This species is a casual to very rare winter visitor along the coast and in the Lower Rio Grande Valley. **Timing of occurrence:** Spring migrants have been found as early as mid-March, but the primary migration period is from mid-April to late May. A few linger as late as mid-June. Fall migrants have been recorded from late August to late October, with small numbers lingering into mid-November. A bird present in Tarrant County from 7 to 8 December 1986 is presumed to be a very late migrant. **Taxonomy:** Monotypic.

BAY-BREASTED WARBLER
Setophaga castanea (Wilson)

Uncommon to common spring migrant and very rare to casual fall migrant in the eastern half of the state. During the fall this species is most frequently encountered on the upper and central coasts. In the western half of the state Bay-breasted Warblers are casual migrants, becoming accidental in the Trans-Pecos. This species is accidental in winter on the Coastal Prairies. Populations of this warbler have declined in recent decades, and it is no longer as common as previously recorded. Nehrling (1882) considered Bay-breasted Warbler to be one of the most common migrants on the upper coast. **Timing of occurrence:** Spring migrants have been found in early April, but the primary period of occurrence is from late April through mid-May, with smaller numbers present into late May and very early June. Fall migrants have been recorded from early September to late October and very rarely into mid-November. One was documented at High Island, Galveston County, from 29 December 1973 through 15 January 1974. **Taxonomy:** Monotypic.

BLACKBURNIAN WARBLER
Setophaga fusca (Statius Müller)

Uncommon to locally common spring migrant in the eastern half of the state, west to the Balcones Escarpment. On

the eastern Edwards Plateau and Rolling Plains, Blackburnian Warblers are rare spring migrants, becoming very rare farther west. In fall they are casual to rare migrants throughout the state and are most frequent in the eastern third. As are many other species of warblers, Blackburnians are most commonly encountered as migrants along the immediate coast. **Timing of occurrence:** Spring migrants have been found in late March, but the primary period of occurrence is from early April through mid-May, with smaller numbers lingering as late as early June. Fall migrants occur between early September to late October and very rarely into mid-November. **Taxonomy:** Monotypic.

YELLOW WARBLER *Setophaga petechia* (Linnaeus)

Common to abundant migrant statewide, particularly in fall. The only breeding population in Texas is the very distinctive Mangrove Warbler, which is a rare to very locally uncommon resident in the mangroves surrounding South Bay in Cameron County. This distinctive subspecies was considered an accidental visitor in the state before the breeding population was discovered in 2004. Yellow Warblers were formerly rare summer residents in the state, nesting along the Rio Grande in the Trans-Pecos, the Edwards Plateau, the Panhandle, and at scattered locations in the eastern half of the state. Territorial males are still occasionally encountered, particularly in the Panhandle, but nesting has not been confirmed in the Panhandle since 1956 in Gray County. Strecker (1927) reported the species to be a "rather common summer resident" in McLennan County. Yellow Warbler is a rare to very rare winter visitor in the Lower Rio Grande Valley, becoming casual in winter north to at least McMullen County and accidental in winter elsewhere in the state. **Timing of occurrence:** Spring migrants are found beginning in late March, and migration peaks from mid-April through mid-May. Smaller numbers linger as late as early June. Fall migrants occur between late July and mid-October, with small numbers lingering into mid-November. **Taxonomy:** Six subspecies occur in the state.

S. p. rubiginosa (Pallas)

Status uncertain. Rare to very rare migrant through the western half of the state, with specimens reported east to Tarrant

(Pulich 1979) and Cameron (Griscom and Crosby 1926) Counties.

S. p. amnicola (Batchelder)

Rare to uncommon migrant through the eastern half of the state.

S. p. aestiva (Gmelin)

Common migrant in the eastern half of the state, west to at least the eastern Rolling Plains and eastern Edwards Plateau, and south to the Lower Rio Grande Valley. Casual summer visitor in summer to the Panhandle.

S. p. morcomi (Coale)

Common migrant through the western half of the state, east to Tarrant County (Pulich 1979) and the western Edwards Plateau.

S. p. sonorana (Brewster)

Formerly a rare summer resident in the western Trans-Pecos. Occasional singing males seen along or near the Rio Grande in the Trans-Pecos may pertain to this subspecies.

S. p. oraria (Parkes & Dickerman)

Part of the Mangrove Warbler group. Rare to very locally uncommon resident surrounding South Bay in Cameron County. There are two records from mangroves at the mouth of the Rio Grande. The first record for Texas was photographed at Rockport, Aransas County, on 26 May 1978, but since then there have been no records outside Cameron County

CHESTNUT-SIDED WARBLER

Setophaga pensylvanica (Linnaeus)

Uncommon to locally common spring migrant and uncommon to rare fall migrant in the eastern third of the state and rare to casual migrant westward. Chestnut-sided Warblers are most commonly encountered along the coast. This species is casual in winter, primarily in the southern third of the state, with December reports from as far north as Lubbock and Tarrant Counties. **Timing of occurrence:** Spring migrants are found from mid-April through late May, with a few into early June. Fall migrants occur as early as late July, but the primary period of occurrence is from early August to mid-October and occasionally into mid-November. **Taxonomy:** Monotypic.

BLACKPOLL WARBLER *Setophaga striata* (Forster)

Uncommon to rare spring migrant and rare fall migrant along the coast. Blackpoll Warblers are rare to very rare spring migrants inland through the eastern half of the state, becoming casual to accidental farther west through the remainder of the state. Inland they are casual in fall, with scattered records from across the state, although there are none from the Pineywoods. **Timing of occurrence:** Spring migrants are found from mid-April through late May, with a few into early June. Exceptional summer records include one in Alpine, Brewster County, on 20 June 2011 and one in Converse, Bexar County, on 24 June 2013. Fall migrants have been found from early August to late October and occasionally into mid-November. The latest reported sighting for the state was in El Paso, El Paso County, on 11 December 1993. **Taxonomy:** Monotypic.

BLACK-THROATED BLUE WARBLER
Setophaga caerulescens (Gmelin)

Rare migrant along the Coastal Prairies south to Cameron County. This species is most often encountered at stopover sites along the coast and is a rare to very rare migrant elsewhere in the state. Black-throated Blue Warblers are more frequently encountered in the fall than in the spring. They are very rare winter visitors along the Coastal Prairies and in the Lower Rio Grande Valley and casual inland, with records north to Brazos and Travis Counties. **Timing of occurrence:** Spring migrants are found from mid-April through late May. Fall migrants have been found from early September to late October, although some have lingered into mid-November. **Taxonomy:** The subspecies that occurs in Texas is *S. c. caerulescens* (Gmelin). This species is sometimes considered monotypic.

PALM WARBLER *Setophaga palmarum* (Gmelin)

Rare to locally uncommon migrant in the eastern two-thirds of the state west to the Pecos River. Palm Warblers are encountered as migrants much more frequently on the Coastal Prairies than at inland locations. In the Trans-Pecos this species is a casual migrant across the entire region. Palm Warblers are locally uncommon winter resi-

dents along the upper and central coasts, becoming rare along the lower coast and in the Lower Rio Grande Valley. Away from the Coastal Prairies they are rare to casual winter visitors inland to the Balcones Escarpment. They are accidental in winter farther inland north to Lubbock and Tarrant Counties and west to El Paso County. **Timing of occurrence:** Spring migrants are found from mid-April through mid-May, with a few as late as mid-June. Fall migrants are found between late August and late October, but small numbers linger away from wintering areas into early December. **Taxonomy:** Two subspecies occur in the state.

S. p. palmarum (Gmelin)
Status as provided above.

S. p. hypochrysea (Ridgway)
Rare to very rare migrant and winter resident on the upper coast. Accidental elsewhere in the state, with specimens from Brewster and Kenedy Counties, and reported from the Panhandle.

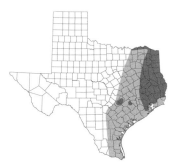

PINE WARBLER *Setophaga pinus* (Wilson)

Common resident in the pine forests of East Texas, south to isolated patches of pine habitat in the upper coastal Prairies and the "Lost Pines" area of Bastrop and northeast Caldwell Counties of Central Texas. They are also rare and local residents in Austin, Colorado, Fayette, Gonzales, and Washington Counties. Pine Warblers are uncommon to common winter residents in the eastern third of the state west to Parker and Travis Counties, becoming uncommon to rare southward to the Lower Rio Grande Valley. Pine Warblers are frequently encountered along the coast during winter and can be locally common. Wintering birds are rare to casual westward and have been noted west to El Paso County and north to Hale County. This species is a rare to very rare migrant west to the High Plains and a casual migrant in the Trans-Pecos. **Timing of occurrence:** The timing of occurrence away from breeding areas is not well defined. In general, winterers arrive in early October and are found through late March and occasionally early April. Many fall records west of the normal range are from early December. **Taxonomy:** The subspecies that occurs in Texas is *S. p. pinus* (Wilson).

YELLOW-RUMPED WARBLER
Setophaga coronata (Linnaeus)

Common to abundant migrant over the entire state. Yellow-rumped Warblers are common to abundant winter residents south of the Panhandle, where they are rare to uncommon. They are rare to locally uncommon summer residents in the higher elevations of the Davis and Guadalupe Mountains. **Timing of occurrence:** Fall migrants begin to arrive in the state in early September and become common from mid-September through mid-October when the winter population is in place. Spring migrants are found from mid-March through mid-May. Large numbers are present to late May in the western half of the state. **Taxonomy:** Two subspecies occur in the state. The Yellow-rumped Warbler was formerly considered two species, Myrtle and Audubon's Warblers.

S. c. coronata (Linnaeus)

Myrtle Warbler. Common to locally abundant migrant and winter resident in the eastern two-thirds of the state, becoming common to uncommon through the remainder. In the Panhandle this form is a common migrant and rare winter visitor.

S. c. auduboni (Townsend)

Audubon's Warbler. Uncommon to locally common migrant and winter resident in the Trans-Pecos and South Plains. In the Panhandle this subspecies is a common to locally abundant migrant in the east and central areas and a rare winter visitor throughout. These warblers are uncommon to rare migrants and winter residents through the Rolling Plains, Edwards Plateau, South Texas Brush Country, and southern Coastal Prairies. Audubon's Warblers are very rare to casual migrants and winter visitors east to the Pineywoods and along the central and upper Coastal Prairies. This subspecies is also an uncommon and local summer resident at the higher elevations of the Davis and Guadalupe Mountains. In the western half of the state, where the two subspecies overlap broadly, Audubon's Warbler is seen later into the spring and arrives earlier in the fall than Myrtle Warbler.

YELLOW-THROATED WARBLER
Setophaga dominica (Linnaeus)

Uncommon to locally common summer resident in the Pineywoods and west along the forested areas of the Red

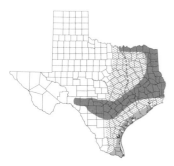

River to Cooke County. This species is also locally uncommon to common as a summer resident in the forest belt that extends from the southern Pineywoods through the southern Post Oak Savannah to the central Edwards Plateau. On the Edwards Plateau, these warblers are uncommon to rare summer residents and confined to riparian corridors, west to Edwards County. Away from the aforementioned areas Yellow-throated Warbler is a very local and rare summer resident east to Tarrant and Wise Counties and to the Devils River, Val Verde County. This species is an uncommon to locally common migrant in the eastern half of the state. In the western half of the state this species is a rare migrant, becoming casual in spring only in the Trans-Pecos and High Plains. Yellow-throated Warblers are rare to locally uncommon winter residents along the coast and in the Lower Rio Grande Valley. There are isolated winter records inland to Travis County and the southern half of the Pineywoods. **Timing of occurrence**: Spring migrants begin to arrive in late February, and the breeding population arrives between mid-March and early April. Spring migrants can be found along the coast into early May. Fall migrants have been found from late July to late September, with small numbers lingering to late October. **Taxonomy**: Two subspecies occur in the state.

S. d. albilora (Ridgway)
Status as provided above.

S. d. dominica (Linnaeus)
Status uncertain. There is a single record of this subspecies, a specimen from Aransas County on 23 March 1892.

PRAIRIE WARBLER *Setophaga discolor* (Vieillot)

Rare to locally common summer resident in the northeastern quarter of the state, south to Hardin County and west to Madison County. Prairie Warblers are uncommon migrants along the coast, becoming rare inland in the eastern third of the state. They are very rare to casual migrants in the western two-thirds of the state, where they are encountered far more frequently in the fall than in the spring. Winter sightings are most frequent along the Coastal Prairies and in the Lower Rio Grande Valley, where they are rare to very rare. They are casual to accidental in winter across the state, including the Trans-Pecos and northeast and Central Texas. **Timing of occurrence:** Spring migrants begin to arrive in mid-March, with a peak from late March through late April. Small numbers arrive through late May. Fall migrants have been found from mid-July to mid-September, with small numbers as late as early November. **Taxonomy:** The subspecies that occurs in Texas is *S. d. discolor* (Vieillot).

GRACE'S WARBLER *Setophaga graciae* (Baird)

Uncommon to very locally common summer resident in the higher elevations of the Davis and Guadalupe Mountains. Grace's Warblers are rarely seen during migration, even in the Trans-Pecos. There are very few reports east of the Pecos River and only one documented in the last 25 years, photographed near Junction, Kimble County, on 18 September 2011. **Timing of occurrence:** Spring migrants found away from breeding areas have been recorded from early April through mid-May. The breeding population arrives in mid-April, with most present by the end of the month, and they depart by mid-September. Fall migrants have been found from early August to mid-September, with very few reported into mid-October. **Taxonomy:** The subspecies that occurs in Texas is *S. g. graciae* (Baird).

BLACK-THROATED GRAY WARBLER
Setophaga nigrescens (Townsend)

Rare to uncommon migrant through the Trans-Pecos and High Plains. Black-throated Gray Warblers are casual spring and very rare to casual fall migrants east of the Pecos River, east to the Pineywoods, and south to the Lower Rio Grande Valley. They are rare to locally uncommon winter residents in the Lower Rio Grande Valley and casual winter visitors along the Coastal Prairies and inland as far north as Randall County. They are rare to very rare summer residents in Dog Canyon in the northern portion of the Guadalupe Mountains. Breeding has been suspected since the 1960s but was not documented until 2012 (Lockwood 2012). **Timing of occurrence:** Spring migrants are found from late March through late May. Fall migrants have been found from early August to early November, and small numbers have lingered into early December. **Taxonomy:** Two subspecies occur in the state.

S. *n. halseii* (Giraud)

Status as provided above.

S. *n. nigrescens* (Townsend)

Status uncertain. There is a single record of this subspecies, a specimen from Cameron County on 2 April 1910.

TOWNSEND'S WARBLER
Setophaga townsendi (Townsend)

Uncommon to common migrant in all but the easternmost Trans-Pecos, primarily at higher elevations. Townsend's Warbler is a rare to locally uncommon migrant farther east to the High Plains, becoming very rare to casual eastward to the upper coast. In winter, Townsend's Warblers are very rare to rare in riparian woodlands along the Rio Grande and its tributaries from El Paso County to Brewster County and in the Lower Rio Grande Valley. They occur casually in winter in other areas of the state, although most records are from the coast. **Timing of occurrence**: Spring migrants are found from early April through late May. Fall migrants have been found from early August to early November. **Taxonomy**: Monotypic.

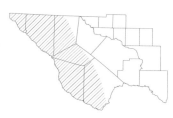

HERMIT WARBLER *Setophaga occidentalis* (Townsend)

Rare to very rare migrant in the western Trans-Pecos, primarily through the major mountain ranges. Hermit Warblers are most frequently encountered in the Chisos Mountains, although they may have a similar occurrence in the Davis Mountains. East of the Pecos River Hermit Warblers are casual to accidental migrants, with records from as far east as the upper coast and as far north as Bailey and Castro Counties on the High Plains. They are casual in winter in the Lower Rio Grande Valley, with all records from Cameron and Hidalgo Counties. There is one fully documented winter record from elsewhere in the state, a male at Killeen, Bell County, from 10 to 11 December 2011. Hybrids with Townsend's Warbler have been reported from the Trans-Pecos and Lower Rio Grande Valley. **Timing of occurrence**: Spring migrants are found from early April through late May. Fall migrants have been found from late July to late September. **Taxonomy**: Monotypic.

GOLDEN-CHEEKED WARBLER
Setophaga chrysoparia (Sclater and Salvin)

Uncommon to rare migrant and summer resident in Central Texas. The breeding range is restricted to the Balcones Canyonlands subregion of the Edwards Plateau and northward locally to Palo Pinto and Somervell Counties. Golden-cheeked Warblers temporarily occupied habitat in Dallas County in 2001 and 2002, where they were historically found. Migrants are rarely found in the state away from breeding areas but have been found in Aransas, Bastrop, Brewster, Cameron, Fayette, Galveston, Hidalgo, and Karnes Counties. This Endangered songbird is a habitat specialist and nests in diverse Ashe juniper–oak woodlands. **Timing of occurrence:** Breeding birds begin to arrive in very late February, with most males on territory by mid-March. Fall migration is early, as birds leave the breeding grounds from late June through mid-July. Very small numbers linger through mid-August. **Taxonomy:** Monotypic.

BLACK-THROATED GREEN WARBLER
Setophaga virens (Gmelin)

Common to uncommon migrant in the eastern half of the state, becoming uncommon to rare farther west and very rare in the western Trans-Pecos. Black-throated Green Warblers are locally uncommon to rare winter residents in the Lower Rio Grande Valley and along the lower coast. In winter they are very rare elsewhere on the coast and inland to Bexar County. There are a number of late June and July records from the eastern half of the state that are thought to be postbreeding wanderers from some of the southern breeding populations. **Timing of occurrence:** Spring migrants begin to arrive in mid-March, and migration peaks from mid-April until mid-May. A few linger into early June. In the fall, they have been recorded from early August through mid-November. **Taxonomy:** Monotypic.

TBRC Review Species   ### FAN-TAILED WARBLER
Basileuterus lachrymosus (Bonaparte)

Accidental. There are only 10 documented records for the United States, with eight from Arizona, one from New

Mexico, and one from Texas. The majority of these records are from late spring, in contrast with the fall occurrence of the Texas record. The Texas bird was also one of the longer-staying individuals and completed a partial molt during its stay. **Taxonomy:** The subspecies that has occurred in Texas is *B.l. tephrus* Ridgway. This species is sometimes considered monotypic.

13 AUG.–24 SEPT. 2007, PINE CANYON, BIG BEND NP, BREWSTER CO. (TBRC 2007–64; TPRF 2501)

TBRC Review Species   **RUFOUS-CAPPED WARBLER**
Basileuterus rufifrons (Swainson)

Very rare visitor. This species was first documented in Texas on 10 February 1973 at Falcon Dam, Starr County. Since then there have been 29 other records, most coming from the southwestern portion of the state. The majority of records have come from two regions: Big Bend NP, Brewster County, and the southwestern Edwards Plateau. This species was found almost annually between 1997 and 2001 along the Devils River drainage of central Val Verde County, leading to speculation that these warblers might be low-density residents, but none have been found there since. A pair apparently spent the summer of 2006 near Concan, Uvalde County, but no evidence of nesting could be found. **Timing of occurrence:** Birds have been found in every month of the year except November, with the greatest number of records from January to May. There have been a number of long-staying birds, both in summer and winter. **Taxonomy:** The subspecies that has been documented in Texas is *B. r. jouyi* Ridgway, as confirmed by the specimens from Webb County (Arnold 1980). The majority of records from the state are presumed to belong to this taxon; however, it is possible that some records from Big Bend NP could refer to the subspecies found in northwestern Mexico, *B. r. caudatus* Nelson.

TBRC Review Species   **GOLDEN-CROWNED WARBLER**
Basileuterus culicivorus (Deppe)

Very rare winter visitor to the Lower Rio Grande Valley and accidental on the lower coast. There are 21 documented records of this widespread tropical warbler for the state. The first record involved a bird collected at Brownsville, Cameron County, on 6 January 1892. Other

than a single record from Nueces County, all others are from Cameron and Hidalgo Counties. **Timing of occurrence:** The majority of records occur between 23 October and 6 April. There are two records from later in the spring: one from 30 April 1995 (Hidalgo County) and another from 25 to 29 April 2001 (Nueces County). An adult mist-netted at Laguna Atascosa NWR, Cameron County, on 12 August 2001 provided the earliest fall record. **Taxonomy:** The subspecies that occurs in Texas is *B. c. brasierii* (Giraud).

CANADA WARBLER *Cardellina canadensis* (Linnaeus)

Uncommon to rare spring migrant in the eastern half of the state and very rare to casual during this season farther west. This species is a common to uncommon fall migrant in the eastern half of the state west to the central Edwards Plateau and very rare to casual in the western half. Canada Warblers are more frequently encountered during migration along the coast, where they can be locally common. There is one documented midwinter record for the state from Cameron County, a specimen collected on 3 February 1891. **Timing of occurrence:** Spring migrants are found from mid-April to late May, with a few into early June. In the fall, they have been recorded from as early as late July. The main migration period is from mid-August to mid-October, and a few are found into early November. There is a single record of a migrant lingering into mid-December on the upper coast. **Taxonomy:** Monotypic.

WILSON'S WARBLER *Cardellina pusilla* (Wilson)

Uncommon to common migrant throughout the state. Wilson's Warblers are most common in migration in the Trans-Pecos, Panhandle, and South Plains, where they can be abundant. This species is generally less common in the spring than in the fall. Wilson's Warblers are uncommon to rare winter residents in the Lower Rio Grande Valley and northward through the southern Coastal Prairies. They are rare to very rare and more irregular in occurrence during winter elsewhere in the southern third of the state, ranging north to near the Balcones Escarpment and west along the Rio Grande to Brewster County. Exceptional was a male photographed on 30 December 2007 in Tarrant County. Two summer records of singing

Wilson's Warblers from the Panhandle probably represent late migrants, as there are no nesting records for Texas. **Timing of occurrence:** Spring migrants have been recorded away from wintering areas as early as late March, but the main period of migration is from early April through mid-May. A few linger as late as early June. Fall migrants begin to arrive as soon as early August, but the main passage is from early September through October. Small numbers linger into mid-November. **Taxonomy:** Two subspecies occur in the state. A third subspecies, *C. p. chryseola* (Ridgway), from the west coast winters south to Panama and could occur as a migrant in West Texas. Recent preliminary research has suggested that the eastern and western populations of Wilson's Warbler may be separate species (Irwin, Irwin, and Smith 2011).

C. p. pileolata (Pallas)

Common to abundant migrant in the western half of the state, with specimens from east to Hidalgo, Hunt, and Victoria Counties. Casual in winter in the Lower Rio Grande Valley based on three specimens from Cameron County, but status is uncertain.

C. p. pusilla (Wilson)

Uncommon to common migrant throughout the state. Uncommon to rare winter resident in the Lower Rio Grande Valley and Coastal Prairies.

RED-FACED WARBLER *Cardellina rubrifrons* (Giraud)

Rare fall migrant and very rare spring migrant through the western Trans-Pecos. Red-faced Warbler is an accidental migrant elsewhere in the state. The first state record was a bird collected in El Paso County on 17 August 1890. The vast majority of records are from Boot Canyon in the Chisos Mountains. There is one summer record of a female in the Guadalupe Mountains on 10 June 2003. There are seven documented records from east of the Pecos River, and all refer to single individuals found in Bastrop, Calhoun, Cameron, Comal, Denton, Nueces, and Travis Counties. **Timing of occurrence:** Spring migrants have been found primarily between 20 April and 16 May, with two records of late migrants from 29 May and 2 June. Fall migrants have been found between 23 July and 15 September, with the majority of records from 3 to 15 August. **Taxonomy**: Monotypic.

PAINTED REDSTART *Myioborus pictus* (Swainson)

Rare summer resident in the Chisos Mountains and very rare and local in summer in the Davis Mountains. Painted Redstarts are casual in summer in the Guadalupe Mountains, where there are two nesting records. The population of Painted Redstarts in Texas has fluctuated greatly over the past 40 years. Between 1976 and 1989 they were uncommon summer residents in the Chisos Mountains but were only very rare visitors there from 1990 through 2002, with no nesting records. Since then there has been a small, but stable, summer population in the Chisos Mountains, and they have been annual in the Davis Mountains. Painted Redstarts are very rare migrants elsewhere in the Trans-Pecos. They are casual migrants east of the Pecos River, with records from the High Plains, Edwards Plateau, Coastal Prairies, and Lower Rio Grande Valley. The only winter record was one present the entire season of 2002–3 in Richmond, Fort Bend County. **Timing of occurrence**: Spring migrants first arrive in late March. The breeding population is present by mid-April, and migrants are noted until late May. Fall migrants have been sighted away from breeding areas as early as late July, with most found from mid-September to early November. **Taxonomy**: The subspecies that occurs in Texas is *M. p. pictus* (Swainson).

TBRC Review Species

SLATE-THROATED REDSTART
Myioborus miniatus (Swainson)

Casual spring visitor to the Chisos Mountains and accidental elsewhere. Texas has 10 documented records, with six from the Chisos Mountains. There are two summer records from the Davis Mountains, including the first fully documented occurrence in the state (Bryan and Karges 2001). The two remaining records were from unexpected locations: the first at Corpus Christi, Nueces County, on 10 April 2002, which involved two birds; and the second at Pharr, Hidalgo County, on 12–13 March 2003. **Timing of occurrence**: The spring records from the Chisos Mountains are between 16 April and 15 May, with the majority between 25 April and 3 May. **Taxonomy**: The subspecies that has occurred in Texas is unknown. The Trans-Pecos records are likely from the population in western Mexico, *M. m. miniatus* (Swainson), but the birds found in South Texas may be from the population in eastern Mexico, *M. m. molochinus* Wetmore, which occurs as close as southern Veracruz. However, there are records of various species found primarily in western Mexico from the Lower Rio Grande Valley; thus, the two records from southern Texas of Slate-throated Redstart could also originate from that region.

YELLOW-BREASTED CHAT *Icteria virens* (Linnaeus)

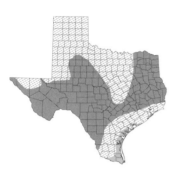

Common to uncommon summer migrant and summer resident through most of the Trans-Pecos, Edwards Plateau, and Pineywoods. Yellow-breasted Chats are uncommon and local summer residents in the western half of the South Texas Brush Country and locally in the southern Rolling Plains and southernmost Post Oak Savannah and Blackland Prairies. Although this species is considered a summer resident throughout much of the state, populations are small in many areas and breeding is highly localized. These birds are uncommon to common migrants through the eastern two-thirds of the state, becoming locally uncommon to rare farther west. Yellow-breasted Chat is a rare to very rare winter resident in the Lower Rio Grande Valley northward to the central coast. **Timing of occurrence**: Spring migrants are found from early April to late May and in fall are recorded from mid-August to late October. **Taxonomy**: Two subspecies occur in the state.

I. v. virens (Linnaeus)
   Common migrant and summer resident in the Pineywoods and uncommon and local west through portions of the Oaks and Prairies region to the Balcones Escarpment. Rare to very rare winter resident on the southern Coastal Prairies and in the Lower Rio Grande Valley.

I. v. auricollis (Deppe)
   Common migrant and summer resident in the Trans-Pecos and Edwards Plateau. Rare to locally uncommon migrant and summer resident from the northwestern Rolling Plains south through the western portion of the region. Uncommon to rare migrant through the High Plains.

Family Emberizidae: New World Sparrows

WHITE-COLLARED SEEDEATER

Sporophila torqueola (Bonaparte)

Uncommon to rare resident along the Rio Grande from northern Starr County north to southern Val Verde County. White-collared Seedeaters were virtually extirpated from the United States in the late 1950s and early 1960s. It has been suggested that the decline was closely tied to the heavy use of herbicides and pesticides, including DDT (Oberholser 1974). In the 1980s the species began a slow recovery from a population primarily restricted to a small area along the Rio Grande in Starr County. This recovery has resulted in considerable range expansion up the Rio Grande drainage. Expansion across their former range in the Lower Rio Grande Valley has been much slower, and White-collared Seedeaters are casual at best east into Hidalgo County. There are two reports from well away from the known range: five birds in Kenedy County on 30 August 1997 and one in Goliad County on 11 January 2009. **Taxonomy**: The subspecies that occurs in Texas is *S. t. sharpei* Lawrence. There are records of the subspecies found in western Mexico, *S. t. atriceps* (Lawrence), from the Lower Rio Grande Valley and El Paso County that are presumed to have escaped from captivity.

TBRC Review Species

YELLOW-FACED GRASSQUIT

Tiaris olivaceus (Linnaeus)

Accidental. There are four records for the state. Three of these are of adult males, and the second of these established a territory and constructed the shell of a nest but did not attract a female (Brush 2003). Yellow-faced Grassquits are found as far north as central Tamaulipas and Nuevo León, Mexico, and are widespread south through northern Central America. The 1990 record was the first for the United States, although there have since been several additional records from Florida of the Caribbean subspecies. **Taxonomy**: The subspecies that occurs in Texas is *T. o. pusillus* Swainson.

22–24 JAN. 1990, SANTA ANA NWR, HIDALGO CO.
  (TBRC 1990-23; TPRF 859)
11–29 JUNE 2002, BENTSEN–RIO GRANDE VALLEY SP, HIDALGO
  CO. (TBRC 2002-75; TPRF 2031)

28 SEPT. 2003, SANTA ANA NWR, HIDALGO CO. (TBRC 2003–75)
30 JAN.–20 MAR. 2011, GOOSE ISLAND SP, ARANSAS CO.
(TBRC 2011–016; TPRF 2941)

OLIVE SPARROW *Arremonops rufivirgatus* (Lawrence)

Common resident throughout the South Texas Brush Country. Olive Sparrows reach the northern limits of their range on the southern edge of the Edwards Plateau, where they have a discontinuous distribution. They are rare to locally uncommon residents from the Devils River in central Val Verde County east to southern Gonzales and western DeWitt Counties. Along the Coastal Prairies, they are found north to the western edge of Calhoun County. **Taxonomy**: The subspecies that occurs in Texas is *A. r. rufivirgatus* (Lawrence).

GREEN-TAILED TOWHEE *Pipilo chlorurus* (Audubon)

Common to uncommon migrant and winter resident in the western half of the state, reaching the eastern edge of its typical winter range in the western Panhandle south through the western Edwards Plateau into the western South Texas Brush Country. Green-tailed Towhee becomes increasingly rare and irregular as a migrant and winterer farther east but has occurred in all regions of the state during irruptive years. These towhees are rare to locally uncommon summer residents at higher elevations of the Davis and Guadalupe Mountains. **Timing of occurrence**: Fall migrants begin to arrive in early September, with an increase in occurrence from early October through early November. Spring migrants begin to leave wintering areas in early April, and migration peaks from mid-April through mid-May. A few linger to the end of May. **Taxonomy**: Monotypic.

SPOTTED TOWHEE *Pipilo maculatus* Swainson

Uncommon to locally abundant summer resident in the mountains of the Trans-Pecos. Spotted Towhees are common to uncommon migrants and winter residents throughout the western two-thirds of the state, east to the Blackland Prairies region. They become increasingly rare farther east and normally are a rare to very rare winter resident in the Pineywoods, although this species is somewhat irruptive and can be locally uncommon there. **Timing of occurrence**: Fall migrants and winterers begin to arrive in late September, and the majority of winter residents arrive by late October. Wintering birds begin to depart in late March through late April, with some lingering to the end of May. **Taxonomy:** Three subspecies occur in the state.

P. m. arcticus (Swainson)
> Abundance in winter and migration as described above, except that it becomes less abundant west of the Pecos River.

P. m. montanus Swarth
> Common to uncommon migrant and winter resident throughout the western half of the state, becoming uncommon to rare farther east to Dallas County.

P. m. gaigei van Tyne & Sutton
> Uncommon to locally abundant resident in the mountains of the Trans-Pecos. A portion of the breeding population disperses to lower elevations in the western two-thirds of the Trans-Pecos.

EASTERN TOWHEE *Pipilo erythrophthalmus* (Linnaeus)

Uncommon to rare migrant and winter resident in the eastern third of the state. Eastern Towhees are found regularly westward through the Blackland Prairies region, where they are significantly outnumbered by Spotted Towhees. They are rare west through the Rolling Plains to the eastern Panhandle and South Plains and very rare in the eastern Edwards Plateau and northeastern South Texas Brush Country. This species is casual to accidental to the western Edwards Plateau, Trans-Pecos, and remainder of the South Texas Brush Country south to the Lower Rio Grande Valley. Since the late 1990s territorial male Eastern Towhees have been found with increasing regularity during the late spring and early summer in the Pineywoods, although nesting has not been confirmed. The only documented breeding record for Texas involved a nest found in Harrison County on 31 July 1914. **Timing of occurrence:** Fall migrants begin to arrive in mid-October, and the majority of winter residents arrive by mid-November. Wintering birds depart from early March through mid-April, but some linger to mid-May. **Taxonomy:** The subspecies that occurs in Texas is *P. e. erythrophthalmus* (Linnaeus).

RUFOUS-CROWNED SPARROW
Aimophila ruficeps (Cassin)

Common to uncommon and local resident in the western two-thirds of the state. Rufous-crowned Sparrow ranges eastward through the Rolling Plains to Tarrant and Johnson Counties in north-central Texas and to the eastern and southern edges of the Edwards Plateau. These sparrows are rare and local residents on the South Plains and portions of the Panhandle. They are rare residents in the western half of the South Texas Brush Country as far south as Starr County and very locally east to Live Oak County. Rufous-crowned Sparrows are casual to accidental to the east, with scattered records from Dallas, Harris, Fayette, and Wise Counties. **Taxonomy:** Two subspecies occur in the state.

A. r. scottii (Sennett)
> Common to uncommon resident in the western Trans-Pecos east to Reeves County and south to northern Brewster and northern Presidio Counties.

A. r. eremoeca (Brown)
> Common to uncommon throughout remainder of range as described above.

CANYON TOWHEE *Melozone fusca* (Swainson)

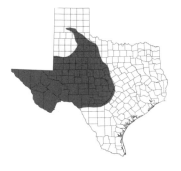

Common to uncommon resident in the Trans-Pecos east to the canyonlands of the southern Panhandle, the South Plains, western Rolling Plains, and western Edwards Plateau. Canyon Towhees become uncommon to rare and local on the eastern Rolling Plains and eastern Edwards Plateau. There have been isolated reports, primarily during the winter, eastward to Tarrant County and in the northern South Texas Brush Country. **Taxonomy:** Two subspecies occur in the state. A third subspecies, *M. f. mesata* (Oberholser), may occur in the northwestern Panhandle as a very local resident. This taxon nests within a few miles of the Texas border in northeastern New Mexico.

M. f. mesoleuca (Baird)
> Common resident from El Paso County east through Culberson County. Intergrades to the east and south with the following subspecies.

M. f. texana (van Rossem)
> Common to uncommon resident from the southern Trans-

Pecos east through the Edwards Plateau and northward locally to the southeastern Panhandle.

BOTTERI'S SPARROW *Peucaea botterii* (Sclater)

Uncommon to locally common summer resident on the Coastal Prairies from southern Kleberg County southward. There have been isolated breeding records up the coast to San Patricio County and inland to Duval County. The Texas population retreats into northern Mexico in winter and is not believed to occur in Texas during that season. An unexpected find was a nesting pair of Botteri's Sparrows in Presidio County in June 1997 (Adams and Bryan 1999). **Timing of occurrence:** Spring birds return to the breeding grounds in mid-April and are present through mid-September, although a few linger into late September. **Taxonomy:** Two subspecies occur in the state.

P. b. arizonae Ridgway
> Accidental. The single record from Presidio County probably refers to this subspecies, although *P. b. mexicana* (Lawrence) cannot be entirely ruled out.

P. b. texana (Phillips)
> The range and status as described in the main text above.

CASSIN'S SPARROW *Peucaea cassinii* (Woodhouse)

Common to abundant summer resident in the Panhandle and western South Plains, becoming locally common to rare in the remainder of the western two-thirds of the state. The eastern edge of the breeding range extends from western Tarrant County south to Matagorda Bay. This species is an uncommon to rare winter resident within the breeding range in the Trans-Pecos, Edwards Plateau, and south through the South Texas Brush Country and southern Coastal Prairies. During mild winters Cassin's Sparrows are rare to very rare farther north through the South Plains and have irrupted east of the expected breeding range in the spring and summer on rare occasions. This occurred most recently in 2006 and 2012, when birds were found at numerous locations in the Oaks and Prairies region east to Delta, Rains, Van Zandt, and Waller Counties and exceptionally east to Harrison and San Augustine Counties. **Timing of occurrence:** Spring migrants return to the breeding grounds from mid-March to mid-April. They are present through early September, when fall migrants begin to disperse, leaving the smaller wintering population in many areas. **Taxonomy:** Monotypic.

BACHMAN'S SPARROW
Peucaea aestivalis (Lichtenstein)

Very rare to locally uncommon resident in the Pineywoods. Bachman's Sparrows are found west to about Leon and Henderson Counties and south to Harris County. They occurred as far west as Cooke, Lee, and Navarro Counties until the late 1880s (Oberholser 1974). Bachman's Sparrow is a habitat specialist found primarily in open pine or oak forests where the understory is composed of tall grasses, sometimes with scattered patches of shrubs and pine regeneration. Management with prescribed fire maintains this habitat type (Conner et al. 2005). Some portion of the migratory population in eastern Oklahoma and western Arkansas presumably must winter in Texas, but evidence of an increase in abundance during that season has not been found. **Taxonomy:** The subspecies that occurs in Texas is *P. a. illinoensis* Ridgway.

AMERICAN TREE SPARROW
Spizella arborea (Wilson)

Locally uncommon winter resident in the Panhandle. The winter range of American Tree Sparrow has shifted northward over the past 20 years, and this species is now a very rare to casual winter visitor to the South Plains and eastward across the northern edge of the state to the Arkansas border. In the 1950s American Tree Sparrows occurred at least irregularly as far south as the northern Edwards Plateau. This species is now casual to accidental in the southern three-quarters of the state, and there are well-documented records into the South Texas Brush Country and southern Coastal Prairies. These records include a specimen from South Padre Island, Cameron County, on 28 October 1908 and one photographed on North Padre Island, Nueces County, on 19 December 2004. **Timing of occurrence:** Fall migrants begin to arrive on the wintering grounds in late October, but the normal arrival time is early November. Wintering birds begin to move northward in early March, with very few present by early April. A single bird recorded in Randall County on 3 May 1973 provided an exceptionally late record. **Taxonomy**: Two subspecies occur in the state.

S. a. arborea (Wilson)

Very rare to casual winter visitor to northeast Texas west to Denton and Dallas Counties (Pulich 1988). There is a specimen from Hardin County from 1929 that had been banded in Massachusetts.

S. a. ochracea Brewster

As described above in the Panhandle and eastward to Dallas and Denton Counties, where both subspecies have been documented.

CHIPPING SPARROW *Spizella passerina* (Bechstein)

Common to abundant migrant and winter resident in nearly all parts of the state. This species is locally uncommon as a winter resident on the South Plains and is rare during that season in the Panhandle. Chipping Sparrows have a rather disjunct breeding distribution in the state. They are common residents on the Edwards Plateau and locally abundant in the Guadalupe and Davis Mountains. In the Pineywoods, Chipping Sparrows are uncommon residents. **Timing of occurrence**: Fall migrants begin to arrive as early as mid-August, with an increase in migrants beginning in early September. Migration peaks from mid-September to mid-October. Spring migrants are recorded from mid-March through mid-May, with a few straggling into late May. **Taxonomy**: Two subspecies occur in the state.

S. p. passerina (Bechstein)

Common to abundant migrant and winter resident in the eastern three-quarters of the state, extending west to the Pecos River but uncommon to rare in the Panhandle. Uncommon resident in the Pineywoods and a common resident on the Edwards Plateau.

S. p. arizonae Coues

Common to abundant migrant through the western half of the state. Common winter resident in the Trans-Pecos and probably in the western Edwards Plateau and western South Texas Brush Country. Locally abundant summer resident in the Guadalupe and Davis Mountains.

CLAY-COLORED SPARROW

Spizella pallida (Swainson)

Common to uncommon migrant through the center of the state. Clay-colored Sparrows are an uncommon to rare migrant in the western half of the Trans-Pecos and rare to very rare in the eastern third of the state. This species is an uncommon to rare winter resident in the western South Texas Brush Country. These sparrows are also rare and irregular winter visitors north through the western Edwards Plateau to the southern Rolling Plains and west to the Davis Mountains. Clay-colored Sparrows are rare to very rare winter visitors along the coast and in the western half of the Trans-Pecos and accidental north to the Panhandle. **Timing of occurrence:** Fall migrants begin to arrive as early as early September, but the main passage is from mid-September to early November. Spring migrants are recorded from mid-March through mid-May, with a small number lingering into late May. **Taxonomy:** Monotypic.

BREWER'S SPARROW *Spizella breweri* Cassin

Common to uncommon migrant in the western third of the state. This species is a rare to locally abundant winter resident in the Trans-Pecos and a rare and irregular winter visitor from the southern Panhandle south to the Lower Rio Grande Valley. Brewer's Sparrows have been found farther east, both as winter visitors and as migrants, with records from the upper coast and Brazos County. Most surprising is the lone Pineywoods record from San Augustine County on 21 October 2000. Brewer's Sparrows were found breeding in the Panhandle in the 1870s, but there are no recent nesting records (Seyffert 1985a, 2001b). **Timing of occurrence:** Fall migrants begin to arrive in late August, and migration peaks from late September through early November. Spring migrants are recorded from early April through mid-May, with a small number lingering into late May. **Taxonomy:** Two subspecies occur in the state.

S. b. taverneri Swarth & Brooks

Rare migrant through the Trans-Pecos and possibly the western Edwards Plateau. There is a specimen from Irion County and a photographed record from Live Oak County on 12 December 2011. The winter range is unknown but may include parts of the Trans-Pecos and the western South Texas Brush Country.

S. b. breweri Cassin

As described in main species account.

FIELD SPARROW *Spizella pusilla* (Wilson)

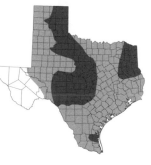

Common to uncommon migrant and winter resident throughout most of the state, becoming rare in the Lower Rio Grande Valley. In the central and western Trans-Pecos, Field Sparrow is a rare to very rare migrant and winter resident. The breeding distribution is disjunct, with isolated populations in various areas across the state. Field Sparrows are uncommon residents in the northern half of the Pineywoods, the eastern Panhandle westward along the Canadian River Valley to Oldham County, the Edwards Plateau north into the eastern Rolling Plains, and also in Brooks and Kenedy Counties in the South Texas Brush Country. Field Sparrows might also breed on the Stockton Plateau in the eastern Trans-Pecos, although Thornton (1951) did not report them. They are suspected of breeding on the Coastal Prairies from Victoria to Nueces Counties. **Timing of occurrence**: Fall migrants are recorded from mid-September through mid-November, by which time the winter residents have arrived. Spring migrants are found from early March through early May. **Taxonomy**: Two subspecies occur in the state.

S. p. pusilla (Wilson)

Common to uncommon migrant and winter resident throughout the state west to the eastern Trans-Pecos. Summer distribution as described above.

S. p. arenacea Chadbourne

Rare to uncommon migrant and winter resident in the Panhandle southward through the western Edwards Plateau and rare to very rare west through the Trans-Pecos. This subspecies appears to be rare in the eastern Rolling Plains and eastern Edwards Plateau and casual east to Aransas, Tarrant, and Wood Counties.

BLACK-CHINNED SPARROW
Spizella atrogularis (Cabanis)

Uncommon summer resident in the mountains of the central Trans-Pecos. The population in Texas exhibits an altitudinal migration, as the higher-elevation breeding population moves to lower elevations in winter. Black-chinned

Sparrows are generally rare to locally uncommon in winter, suggesting that some of the population moves south out of the state. In the Franklin Mountains, El Paso County, Black-chinned Sparrows are rare summer residents but are uncommon in winter as migrants from more northerly populations move into the region. They are casual visitors east of the Pecos River, including Garza, Howard (TPRF 197), Lubbock, Midland, Randall, and San Patricio Counties. **Timing of occurrence:** Migrants are virtually never reported, but birds moving to lower elevations for winter are found from late August to late March. Vagrants from east of the Pecos River have been found from early October to May. **Taxonomy:** The subspecies that occurs in Texas is *S. a. evura* Coues.

VESPER SPARROW *Pooecetes gramineus* (Gmelin)

Uncommon to common migrant and winter resident throughout most of the state. Vesper Sparrows are rare winter residents in the southern Panhandle and generally absent farther north. This species is a very rare summer visitor to the Panhandle and South Plains and is casual in the northern Trans-Pecos. Lloyd (1887) reported collecting two egg sets from Tom Green County in 1885. **Timing of occurrence:** Fall migrants have been recorded as early as mid-August, but the typical migration period starts in mid-September and extends through early November. Spring migrants are found from early March through early May, and small numbers linger into late May. **Taxonomy:** Two subspecies occur in the state.

P. g. gramineus (Gmelin)
   Uncommon to common migrant and winter resident in the eastern half of the state, west through the eastern Rolling Plains and eastern Edwards Plateau, and south through the Coastal Prairies to the Rio Grande.

P. g. confinis Baird
   Uncommon to common migrant and winter resident in the western half of the state, east through the Rolling Plains and Edwards Plateau, and south through the South Texas Brush Country.

LARK SPARROW *Chondestes grammacus* (Say)

Common to uncommon migrant and summer resident throughout most of the state. Lark Sparrows are uncom-

mon and local breeders in southern Hudspeth County but are generally absent from the most arid habitats in the southern and western Trans-Pecos. They are rare to locally uncommon during summer in open habitats in the Piney-woods. This species is a rare to locally uncommon winter resident in the southern Oaks and Prairies region, Coastal Prairies, and South Texas Brush Country. Lark Sparrows are rare to casual winter residents in the remainder of the state. **Timing of occurrence**: Spring migrants are recorded from late March through early May. Fall migrants are found from mid-August through early October. **Taxonomy**: Two subspecies occur in the state.

C. g. grammacus (Say)

Rare to locally uncommon summer resident in the Piney-woods. Uncommon migrant through the eastern two-thirds of the state and rare to locally uncommon winter resident in the southern Oaks and Prairies region, Coastal Prairies, and South Texas Brush Country.

C. g. strigatus Swainson

Common to uncommon migrant and summer resident in the western two-thirds of the state, east through the Oaks and Prairies region. Rare to locally uncommon winter resident in the South Texas Brush Country.

BLACK-THROATED SPARROW
Amphispiza bilineata (Cassin)

Common to abundant resident in the Trans-Pecos east through the western Edwards Plateau and south through the South Texas Brush Country. This species is uncommon and more local on the eastern Edwards Plateau north through the Rolling Plains to Bosque and Erath Counties and rare farther north to Shackelford and Palo Pinto Counties. Black-throated Sparrows were formerly a rare summer resident in the northwestern Rolling Plains in the canyon-lands of the southeastern Panhandle, and there have been recent breeding records from the western Rolling Plains north to Crosby and Garza Counties. They are rare to very rare in winter and spring north of the breeding range to the southern Panhandle and southeast along the Coastal Prairies to the upper coast. There are scattered records from east of the breeding range into the Oaks and Prairies region. **Timing of occurrence**: Records away from the breeding range are primarily from early November through mid-April. **Taxonomy:** Three subspecies occur in the state.

A. b. bilineata (Cassin)
> Common to uncommon resident as described above east of the Pecos River.

A. b. opuntia Burleigh & Lowery
> Common to abundant resident in the Trans-Pecos. Includes *A. b. dapiola* Oberholser.

A. b. deserticola Ridgway
> Status uncertain. Oberholser (1974) lists specimens from El Paso and Culberson Counties, saying this subspecies is "fairly common." Modern confirmation is needed.

SAGEBRUSH SPARROW
Artemisiospiza nevadensis (Ridgway)

Uncommon migrant and local winter resident in the western and central Trans-Pecos. Sagebrush Sparrow is a very rare to casual migrant and winter visitor to the western Panhandle south through the western South Plains to Andrews County. There are documented records east to Midland and exceptionally Wilbarger Counties. Sagebrush Sparrows have also been reported from the western edge of the Edwards Plateau but without supporting documentation (Lockwood 2001). **Timing of occurrence:** Fall migrants arrive in mid-October, and winter residents are present until mid-March, rarely lingering into early April. **Taxonomy:** Monotypic.

LARK BUNTING *Calamospiza melanocorys* Stejneger

Locally uncommon to rare summer resident in the Panhandle, primarily in the northern half of the region. Lark Buntings are rare and irregular summer residents south through the South Plains to the northwestern portion of the Edwards Plateau. This species is an abundant to uncommon migrant and winter resident in the western half of the state, east through the western Rolling Plains and western Edwards Plateau, and south to the central coast and Lower Rio Grande Valley. Lark Buntings are rare migrants and winter visitors to the upper Coastal Prairies and Oaks and Prairies region and are accidental spring and fall visitors to the Pineywoods. **Timing of occurrence:** Spring migrants are recorded from late March through early May, but small numbers are found as late as early June. Fall migrants begin to arrive in the state as early as mid-July, including alternate-plumaged males. The primary migration period is from mid-August through late October, when the wintering population reaches the southern portions of the state. **Taxonomy:** Monotypic.

SAVANNAH SPARROW

Passerculus sandwichensis (Gmelin)

Abundant to uncommon migrant throughout the state. Savannah Sparrows are abundant to uncommon winter residents from the southern Panhandle southward through the remainder of the state. In the northern Panhandle they are very rare in occurrence in winter. Savannah Sparrows are casual summer visitors to the western Panhandle and accidental elsewhere, with records south to Brazoria and Kleberg Counties. **Timing of occurrence:** Fall migrants have been recorded as early as mid-August, but the typical migration period starts in mid-September and extends through early November. Spring migrants are found from mid-March through early May, with small numbers lingering into late May. **Taxonomy:** Six subspecies occur in the state. Some authors have included all of these under one subspecies, *P. s. sandwichensis* (Jaramillo et al. 2011).

P. s. oblitus Peters & Griscom

Uncommon migrant and winter resident from the southern Panhandle south to the South Texas Brush Country and east to the Pineywoods.

P. s. mediogriseus Aldrich

Common to abundant migrant and winter resident in the eastern half of the state, west to the central Edwards Plateau, and south to the eastern Lower Rio Grande Valley. This subspecies is sometimes included under *P. s. savanna* (Wilson).

P. s. labradorius Howe

Rare migrant through the eastern third of the state. There are specimen records west to Calhoun and Dallas Counties.

P. s. athinus Bonaparte

Rare migrant and winter resident in the western half of the state, east through the Panhandle, and south through the western South Texas Brush Country.

P. s. nevadensis Grinnell

> Common to abundant migrant and winter resident in the western half of the state south of the Panhandle and east to at least the Blackland Prairies and central Coastal Prairies.

P. s. rufofuscus Camras

> Rare migrant and winter resident in the Trans-Pecos and western Edwards Plateau.

GRASSHOPPER SPARROW
Ammodramus savannarum (Gmelin)

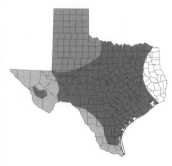

Rare to locally common summer resident in most of the state east of the Pecos River, although generally absent from the southern and western South Texas Brush Country. Breeding populations fluctuate greatly depending on environmental conditions. In areas where conditions are favorable, breeding activities commence almost immediately, but when conditions are less favorable, evidence of breeding can be difficult to find. In the Trans-Pecos, Grasshopper Sparrows are rare to locally uncommon residents in the mid-elevation grasslands in northern Brewster, Jeff Davis, and northern Presidio Counties. In the Pineywoods, they are very rare to rare and local breeders in the northern half of the region but are generally absent farther south. Grasshopper Sparrows are uncommon to common migrants throughout the state west of the Pineywoods, where they are rare. This species is a rare to locally uncommon winter resident west of the Pineywoods and south of the Panhandle but can be locally common in the southern South Texas Brush Country. **Timing of occurrence:** Spring migrants are found from early April through mid-May, but a few linger into early June. Fall migrants have been recorded from mid-August through early November. The breeding population in the northern portions of the state depart in early to mid-September. **Taxonomy:** Two subspecies occur in the state.

A. s. perpallidus (Coues)

> Rare to common summer resident in the western two-thirds of the state east through the Rolling Plains and south through the northeastern South Texas Brush Country. Common to uncommon migrant and uncommon to rare winter resident as described above.

A. s. pratensis (Vieillot)

> Rare to very rare summer resident in the northeastern corner of the state south through the northern Pineywoods. Uncommon migrant through the Oaks and Prairies region and an uncommon to rare winter resident in that region south to the coast and very rare during winter in the Pineywoods.

BAIRD'S SPARROW *Ammodramus bairdii* (Audubon)

Rare to very rare winter resident in diverse mid-elevation grasslands in northern Brewster, southern Jeff Davis, and northern Presidio Counties. Baird's Sparrow may also be a very rare winter resident northward into the western South Plains. The only documented winter records from the High Plains are three specimens collected out of 14 individuals observed between November 1976 and January 1977 at Muleshoe NWR, Bailey County (Grzybowski 1982). Observers' lack of access to high-quality habitat, along with the very elusive and secretive behavior of this species, may cloud the true winter status of this species, which may be more common than is currently known. These sparrows are rare migrants through the western High Plains south through Val Verde County. They are very rare farther east through the remainder of the Panhandle south through the western Edwards Plateau. Baird's Sparrow is a casual spring migrant and accidental fall migrant farther east, with records of migrants east to the Oaks and Prairies region from Parker and Tarrant Counties and the eastern Edwards Plateau from Bexar and Burnet Counties. The species was removed from the Review List in 2008, with 61 documented records. **Timing of occurrence:** Fall migrants have been recorded from late August through late October. Spring migrants have been found from late March through late May, with an apparent peak from 20 April through 10 May. **Taxonomy:** Monotypic.

HENSLOW'S SPARROW
Ammodramus henslowii (Audubon)

Rare to very rare migrant and winter resident in the eastern and southern Pineywoods. Henslow's Sparrows are locally rare to casual during the winter along the Coastal Prairies south to San Patricio County. They are casual farther west to Cooke, Tarrant, and Williamson Counties, and there is

one record from the Panhandle involving a single bird found in Oldham County on 6 May 2000. This species was formerly a rare and very local resident in Harris County. This population was discovered in 1973 and was last detected in 1981. It was described as a separate subspecies, *A. h. houstonensis* Arnold (Arnold 1983); however, it is most often considered as part of *A. h. henslowii*. The species was removed from the Review List in 1994 after surveys showed it to be a regular winter resident. **Timing of occurrence:** Fall migrants have been recorded from early October through late November. Spring migrants have been found from early March through mid-April. **Taxonomy:** The subspecies that occurs in Texas is *A. h. henslowii* (Audubon).

LE CONTE'S SPARROW
Ammodramus leconteii (Audubon)

Uncommon to common winter resident in the eastern third of the state, including the Coastal Prairies south to Kleberg County. Le Conte's Sparrows are uncommon to rare in winter westward to the eastern Edwards Plateau and the Concho Valley and southward to Cameron County. This species is a very rare and irregular winter resident west to the central Trans-Pecos and southern Panhandle, becoming casual to the north. A singing male found in Ochiltree County on 25 June 1994 provided the only summer record for the state. **Timing of occurrence:** Fall migrants have been found as early as mid-September, but mid-October is the normal period for the first migrants. The main passage of fall migrants occurs from late October through late November. Spring migrants have been found from mid-February to mid-April, although small numbers have lingered into early May. **Taxonomy:** Monotypic.

NELSON'S SPARROW *Ammodramus nelsoni* Allen

Uncommon to locally common winter resident along the coast south to Nueces County, then rare and local south to the Rio Grande. Despite generally being an uncommon winter resident on the coast, this species is rarely detected inland and is considered a very rare to locally rare migrant through the Pineywoods and eastern Oaks and Prairies region. There is one documented record from as far west as Reeves County (TPRF 221). Specimens of the Saltmarsh Sparrow (*A. caudacutus*) had been previously reported from Texas (TOS 1995); however, upon examination they were identified as *A. nelsoni*. Formerly known as Nelson's Sharp-tailed Sparrow. **Timing of occurrence:** Inland fall migrants have been found from late September through early November. Inland spring migrants have been recorded between late April and mid-May. **Taxonomy:** Two subspecies occur in the state.

A. n. nelsoni Allen
As described above.

A. n. alter (Todd)
Status uncertain. Specimen records suggest that this taxon is at least an accidental to casual winter visitor along the upper and central coasts.

SEASIDE SPARROW *Ammodramus maritimus* (Wilson)

Uncommon to locally common resident in coastal marshes south to Calhoun County, becoming rare and local south to the Rio Grande. The only documented inland record of Seaside Sparrow for Texas was a single bird mist-netted and photographed on 14 December 1974 near Waco, McLennan County (TPRF 84). The bird was present until at least mid-March 1975. **Taxonomy:** Two subspecies occur in the state.

A. m. fisheri Chapman
Uncommon to locally common resident in the coastal marshes from Jefferson County to about Matagorda County.

A. m. sennetti Allen
This endemic subspecies is a rare and local resident in the coastal marshes from about San Antonio Bay, Calhoun County, to the mouth of the Rio Grande, becoming very local near the southern limit.

FOX SPARROW *Passerella iliaca* (Merrem)

Uncommon to common migrant and winter resident from the eastern Panhandle and central Edwards Plateau eastward and on the Coastal Prairies south to San Patricio County. This species is a rare to locally uncommon migrant and winter resident on the western High Plains, western Edwards Plateau, and the Trans-Pecos. Fox Sparrows are generally absent from the South Texas Brush Country, with only a few scattered winter records. Recent work has suggested that the subspecies groups of Fox Sparrow may represent three or four species (Zink 1994). **Timing of occurrence:** Fall migrants have been found from mid-October through late November. Spring migrants have been recorded between late February and late March and very rarely into mid-April. **Taxonomy:** Three subspecies occur in the state, representing two subspecies groups (Zink 1994).

P. i. schistacea Baird

Part of the Slate-colored Fox Sparrow group. Accidental. There is one documented record from El Paso, El Paso County, on 20–27 September 2012 (TBRC 2012–59; TPRF 2974). This taxon has been reported from the Trans-Pecos and the western High Plains, but those instances with documentation have referred to dark-plumaged *P. i. zaboria*. The TBRC requests documentation for any reports of this subspecies.

P. i. zaboria Oberholser

Part of the Red Fox Sparrow group. Locally common to uncommon migrant and winter resident in the eastern two-thirds of the state, west to the eastern Panhandle and central Edwards Plateau, and south on the Coastal Prairies to San Patricio County. Rare to very rare migrant and winter resident in the western third of the state. There are specimens from the South Texas Brush Country but only as far south as San Patricio County.

P. i. iliaca (Merrem)

Part of the Red Fox Sparrow group. Rare migrant and winter resident in the eastern third of Texas, casually west to the eastern edge of the Edwards Plateau and south to Calhoun County.

SONG SPARROW *Melospiza melodia* (Wilson)

Common to uncommon migrant and winter resident throughout most of the state. Song Sparrow is rare to very rare in winter in the southern Coastal Prairies and South Texas Brush Country, especially in the Lower Rio Grande Valley. **Timing of occurrence:** Fall migrants begin to arrive in late September, but the primary migration period is from early October through mid-November. Spring migrants have been recorded between early March and late April and very rarely as late as early May. **Taxonomy:** Two subspecies occur in the state. This treatment follows the revision of Song Sparrow subspecies by Patten and Pruett (2009).

M. m. montana Henshaw
> Uncommon to rare migrant and winter resident in the western half of the state east through the western Oaks and Prairies region.

M. m. melodia (Wilson)
> Common to uncommon migrant and winter resident through the eastern two-thirds of the state, becoming rare in the Trans-Pecos.

LINCOLN'S SPARROW *Melospiza lincolnii* (Audubon)

Common to uncommon migrant throughout the state. Lincoln's Sparrow is an uncommon to locally rare winter resident in much of the Panhandle and eastward across north-central Texas to the Pineywoods. This species is common to locally abundant as a winter resident throughout the remainder of the state. **Timing of occurrence:** Fall migrants begin to arrive in late August, but the primary migration period is from late September through early November. Spring migrants have been recorded between mid-March and mid-May, with small numbers lingering into late May. **Taxonomy:** Two subspecies occur in the state.

M.l. lincolnii (Audubon)
> As described above.

M.l. alticola (Miller & McCabe)
> Uncommon to rare migrant and winter resident through the western half of the state, with specimens from east to the western Oaks and Prairies region.

SWAMP SPARROW *Melospiza georgiana* (Lathum)

Common migrant through the eastern half of the state, becoming uncommon and increasingly local westward through the Trans-Pecos. Swamp Sparrows are uncommon winter residents in most areas of the state. They occur most commonly in the Pineywoods and along the upper and central coasts and are rare in the northern Panhandle and northwestern Trans-Pecos. **Timing of occurrence:** Fall migrants begin to arrive in late September and are present through early November. Spring migrants have been recorded between early March and mid-April, although small numbers have lingered into early May and rarely into mid-May. **Taxonomy:** Two subspecies occur in the state.

M. g. ericrypta Oberholser
As described above.

M. g. georgiana (Lathum)
Rare to locally uncommon migrant and winter resident in the eastern half of the state, with specimens from as far west as Val Verde County and south into the northeastern South Texas Brush Country.

WHITE-THROATED SPARROW
Zonotrichia albicollis (Gmelin)

Abundant to common migrant and winter resident in the eastern half of the state. White-throated Sparrows are locally uncommon to rare migrants and winter visitors from the central Rolling Plains and Edwards Plateau, west through the High Plains to the Pecos River, and south through the South Texas Brush Country to the Rio Grande. In the eastern Trans-Pecos they are rare and local winter visitors restricted primarily to riparian areas, becoming very rare in the western half of the region. There are five summer records for the state from the northeastern portion of the state, three of which involved birds found in late June and a fourth that remained through the summer in Dallas in 2011. The fifth summer record is from Delta County on 7 August 1998, which was probably an extraordinarily early migrant. **Timing of occurrence:** Fall migrants begin to arrive in late September and are rare until mid-October. The main passage of migrants is from mid-October through mid-November. Spring migrants have been recorded between mid-March and early May and rarely into mid-May. **Taxonomy:** Monotypic.

HARRIS'S SPARROW *Zonotrichia querula* (Nuttall)

Uncommon to locally common migrant and winter resident in the central portion of the state. Harris's Sparrows are found from the central Panhandle south to the eastern Edwards Plateau and east through the Oaks and Prairies region. They are rare winter visitors on the Coastal Prairies south to San Patricio County and are casual to Cameron County. This species is a rare to uncommon migrant and winter resident from northeast Texas south through the

Pineywoods. West of the primary wintering range, this species is rare in the western Edwards Plateau and northern South Texas Brush Country and very rare in the Trans-Pecos. Harris's Sparrows have declined in the past 20 years in the western portion of the winter range, particularly on the South Plains south through the central Edwards Plateau. One in Collin County from 14 July through early August 2011 is the only summer record for the state. **Timing of occurrence**: Fall migrants begin to arrive in the last few days of October, but the primary migration period is from early November to early December. Spring migrants have been recorded between early March and mid-April. Small numbers have lingered into late April and rarely into mid-May. **Taxonomy**: Monotypic.

WHITE-CROWNED SPARROW

Zonotrichia leucophrys (Forster)

Abundant to uncommon migrant and winter resident throughout the northern two-thirds of the state, becoming uncommon southward to the Rio Grande. White-crowned Sparrows are somewhat local in distribution within the Pineywoods during winter. **Timing of occurrence**: Fall migrants begin to arrive in mid-September but are generally rare before the primary period from early October to mid-November. Spring migrants have been recorded between very late March and mid-May, with small numbers into late May. Particularly in the western third of the state, they are found into early June routinely but rarely into mid-June. **Taxonomy**: Three subspecies occur in the state.

Z.l. leucophrys (Forster)

> Abundant to uncommon migrant and winter resident throughout the eastern half of the state west to the eastern Rolling Plains and eastern Edwards Plateau and uncommon to rare farther west. The identification of dark-lored birds in the Panhandle south through the western Edwards Plateau may refer to this or the following subspecies, and more research is needed.

Z.l. oriantha Oberholser

> Common to uncommon migrant and winter resident in the Trans-Pecos. Dark-lored, dusky-billed birds reported uncommonly east to the western Oaks and Prairies region and south through the eastern Edwards Plateau may refer to this subspecies.

Z.l. gambelii (Nuttall)

Abundant to common migrant and winter resident throughout the western half of the state, becoming uncommon to rare in the Oaks and Prairies region south through the Coastal Prairies and South Texas Brush Country. Intergrades with *Z. i. leucophrys* may occur on the eastern part of the range.

TBRC Review Species

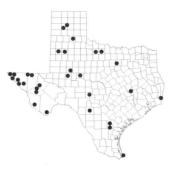

GOLDEN-CROWNED SPARROW
Zonotrichia atricapilla (Gmelin)

Very rare winter and spring visitor to the western half of the state, becoming casual farther east. There are 37 documented records from the state; almost half are from the Trans-Pecos. Ten records are scattered from the Panhandle east to the western Oaks and Prairies region and south to the Edwards Plateau. Those remaining include three from the South Texas Brush Country, one from Brazos County, one from Archer County, and three, including two specimens, from Orange County. There are 16 reports of Golden-crowned Sparrows from prior to the development of the Review List in 1988 for which there is no documentation on file. **Timing of occurrence:** The documented records for the state have all occurred between 23 October and 24 May. **Taxonomy:** Monotypic.

DARK-EYED JUNCO *Junco hyemalis* (Linnaeus)

Uncommon to abundant migrant and winter resident throughout the northern two-thirds of the state. This species is rare to locally uncommon in winter on the Texas coast and rare in the South Texas Brush Country. Dark-eyed Juncos are uncommon summer residents in the upper elevations of the Guadalupe Mountains. There are scattered summer records from the High Plains and Trans-Pecos pertaining to presumed postbreeding wanderers from the Rocky Mountains breeding populations. **Timing of occurrence:** Fall migrants are found from early October through mid-November. Spring migrants have been recorded between early March and mid-April, with small numbers lingering into mid-May, primarily in the High Plains and Trans-Pecos. **Taxonomy:** Seven subspecies occur in the state. The subspecies were formerly considered four separate species.

J. h. hyemalis (Linnaeus)

Slate-colored Junco group. Common to abundant migrant and winter visitor in the eastern two-thirds of the state south through the Edwards Plateau, becoming increasingly less common westward. Slate-colored Junco is rare to locally uncommon in the Trans-Pecos and along the coast and rare to very rare in the South Texas Brush Country.

J. h. cismontanus Dwight

Slate-colored Junco group. Uncommon to rare migrant and winter resident from the eastern Trans-Pecos east to the western Oaks and Prairies region.

J. h. montanus Ridgway

Oregon Junco group. Uncommon to common migrant and winter resident in the western half of the state east through the Rolling Plains and Edwards Plateau. Rare to very rare east through the Oaks and Prairies region and south through the South Texas Brush Country.

J. h. mearnsi Ridgway

Pink-sided Junco, part of the larger Oregon Junco group. Common to locally uncommon migrant and winter resident in the Trans-Pecos and High Plains and rare east through the Rolling Plains and Edwards Plateau. Casual to accidental east to the Pineywoods.

J. h. aikeni Ridgway

White-winged Junco. Casual to very rare visitor to the Panhandle. There are 10 reports, including four documented records, from the Panhandle. There is a single report from the South Plains and three records from the Trans-Pecos. The first confirmed occurrence in the state was a specimen collected in Briscoe County on 19 December 1968 (Weske 1974). The TBRC requests documentation for any report of this subspecies in Texas.

J. h. caniceps (Woodhouse)

Gray-headed Junco. Common to rare migrant and winter resident in the Trans-Pecos and Panhandle south to the Concho Valley. Accidental to casual visitor east to the northern half of the Pineywoods.

J. h. dorsalis Henry

Red-backed Junco, part of the larger Gray-headed Junco group. Uncommon resident in the higher elevations of the

Guadalupe Mountains and rare winter visitor to the Davis Mountains.

TBRC Review Species

YELLOW-EYED JUNCO *Junco phaeonotus* Wagler

Casual visitor to the Trans-Pecos. There are seven documented records for the state: four from Guadalupe Mountains NP, two from Big Bend NP, and one from El Paso County. There are records from all seasons, creating no discernible pattern of occurrence. **Taxonomy:** The subspecies that occurs in Texas is *J. p. palliatus* Ridgway.

Family Cardinalidae: Cardinals and Allies

HEPATIC TANAGER *Piranga flava* (Vieillot)

Uncommon to locally common summer resident in the Chisos, Davis, and Guadalupe Mountains. Hepatic Tanagers are rare to very rare migrants through the western Trans-Pecos, with most records coming from El Paso County. To the east, they are accidental during migration, with records scattered through the western half of the state, from Bailey and Castro Counties to the north, from Bexar and Gonzales Counties to the east, and south to the Lower Rio Grande Valley. Hepatic Tanagers have been casual as postbreeding wanderers in the western Edwards Plateau during the mid- to late summer. They are also casual in winter to the southern third of the state inland to Bastrop County and up the Coastal Prairies to Brazoria County. **Timing of occurrence:** The breeding population arrives in mid-April and is present until mid-October. However, the majority of spring migrants elsewhere in the Trans-Pecos have been recorded in May. Fall migrants have been found from mid-August through late October. **Taxonomy:** The subspecies that occurs in Texas is *P. f. dextra* Bangs.

SUMMER TANAGER *Piranga rubra* (Linnaeus)

Rare to locally common summer resident in the eastern half of the state west to the eastern edge of the Rolling Plains and south through the entire Edwards Plateau to the South Texas Brush Country, excluding the southernmost portion. Summer Tanagers are also locally uncommon summer residents through most of the Trans-Pecos,

although absent from the highest elevations. Summer Tanager is casual in summer in the Panhandle, and there are a few isolated breeding records from the Rolling Plains, west to southwestern Crosby County. Summer Tanagers are uncommon to common migrants in all regions. They are rare winter residents in the Lower Rio Grande Valley and very rare winter visitors to the Coastal Prairies. There are many scattered inland records during the winter; most come from the southern Pineywoods east to the Balcones Escarpment and then farther south. **Timing of occurrence:** Spring migrants are found from early April through late May, with a few into early June. Fall migrants are recorded from mid-August to late October, but small numbers linger into mid-November. **Taxonomy:** Two subspecies occur in the state.

P. r. cooperi Ridgway

Locally uncommon summer resident in the Trans-Pecos region. Uncommon migrant through the Trans-Pecos and western Edwards Plateau south through the northwestern South Texas Brush Country.

P. r. rubra (Linnaeus)

Rare to common summer resident east of the Pecos River as described above. Common migrant east of the Pecos River and a rare winter resident in the Lower Rio Grande Valley. A very rare winter visitor north through the Coastal Prairies and at other scattered locations in the east.

SCARLET TANAGER *Piranga olivacea* (Gmelin)

Rare to locally common spring and rare fall migrant in the eastern third of the state and a very rare to casual migrant father west. In the fall, Scarlet Tanagers are rarely found away from the coast. They are casual summer visitors to the eastern third of the state. There have also been mid- to late June records from the Panhandle and Trans-Pecos that may have been exceptionally late migrants. They are accidental in winter, with documented records from San Antonio, Bexar County from 31 December 2004 to 2 January 2005 and from Lake Jackson, Brazoria County, from 18 to 23 January 2012. **Timing of occurrence:** Spring migrants are found from early April through mid-May, and a few linger as late as mid-June. Fall migrants have been recorded as early as late July, with most found from late August through late October. Small numbers remain into early November and exceptionally to early December. **Taxonomy**: Monotypic.

WESTERN TANAGER *Piranga ludoviciana* (Wilson)

Uncommon to locally common summer resident in the Guadalupe and Davis Mountains, typically above 6,000 feet in elevation. A nesting pair with four young was found in Bailey County on 8 July 2001, providing the only such record for the Panhandle and South Plains. Western Tanagers are uncommon to common migrants in the Trans-Pecos and High Plains, becoming increasingly less common eastward. In the eastern half of the state, migrant Western Tanagers are rare in the spring and very rare to casual in the fall. This species is rare to very rare in the winter along the Coastal Prairies and in the Lower Rio Grande Valley. These birds are regular in occurrence inland during winter, where they are very rare to casual. Records are from virtually all other areas of the state, including as far north as Randall County in the Panhandle. **Timing of occurrence**: Spring migrants are found from mid-April through late May. Fall migrants, or postbreeding wanderers, have been found as early as early July, particularly in the Trans-Pecos. The main fall migration period is from late August until early October, with small numbers found as late as mid-November. **Taxonomy**: Monotypic.

TBRC Review Species

FLAME-COLORED TANAGER
Piranga bidentata Swainson

Casual spring and summer visitor and accidental in fall and winter. There are 11 documented records in Texas, the first an adult male in Pine Canyon in the Chisos Mountains, Brewster County, from 14 to 19 April 1996. There are seven other records from the Davis and Chisos Mountains, including three additional spring records, three from the summer, and one from the fall. The most recent record involved a male present from 6 May into July 2013 that appeared to be paired with a female Western Tanager. Surprisingly, there are three records from the Lower Rio Grande Valley. A bird found dead at McAllen, Hidalgo County, provides the only specimen record for the United States (Arnold, Marks, and Gustafson 2011). **Timing of occurrence**: There are four spring records between 11 and

22 April. There are late winter and early spring records from the Lower Rio Grande Valley from 28 February and 3 March. The summer records from the Trans-Pecos are from 12 June and 28 July, while the sole fall record is from early October. **Taxonomy:** Two subspecies have occurred in the state.

P. b. bidentata Swainson

Casual visitor to the Davis and Chisos Mountains, mostly in spring. It is possible that some of the Trans-Pecos records are of the following subspecies.

P. b. sanguinolenta Lafresnaye

Accidental late winter and spring visitor to the Lower Rio Grande Valley.

TBRC Review Species

CRIMSON-COLLARED GROSBEAK

Rhodothraupis celaeno (Deppe)

Very rare but increasingly regular winter visitor to the Lower Rio Grande Valley. Crimson-collared Grosbeaks are endemic to northeastern Mexico. There are 34 documented records, with most found from Cameron County upriver to Webb County. There are four records from north of this area of primary occurrence, all from coastal counties. The northernmost is a female-type bird from Galveston County. The first record for Texas and the United States was at Bentsen–Rio Grande Valley sp, Hidalgo County, from 28 June to 1 July 1974 (TPRF 224). All subsequent records prior to the fall of 2004 occurred between 1985–86 and 1987–88. There was a significant northward movement from the fall of 2004 through the spring of 2005, when there were 10 documented occurrences involving at least 18 individuals. They have been annual since 2008–9, and a small winter incursion took place in the late fall and winters of 2010–11 and 2011–12. **Timing of occurrence:** The primary period is from early November through late April. There are three records of birds present in May. The only summer record was the very first one, as mentioned above. **Taxonomy:** Monotypic.

NORTHERN CARDINAL

Cardinalis cardinalis (Linnaeus)

Common to abundant resident throughout most of the state. In the Trans-Pecos, Northern Cardinals are uncom-

mon to locally common residents in the southern half of the region north to southern Reeves County and along the Rio Grande into southern Hudspeth County. They are locally uncommon to rare residents in the remainder of Reeves County and rare visitors in all seasons to Culberson, the remainder of Hudspeth, and El Paso Counties. Despite the irregular occurrence of this species in the northwestern portion of the Trans-Pecos, there are scattered nesting records. In the Panhandle, they are rare and local residents in the far western counties. **Taxonomy:** Two subspecies occur in the state.

C. c. magnirostris Bangs
Common to abundant resident in the eastern third of the state west through the Oaks and Prairies region.

C. c. canicaudus Chapman
Uncommon to common resident in the remainder of the state west of the described ranges of the preceding subspecies. Status in the Trans-Pecos as described in the species account.

PYRRHULOXIA *Cardinalis sinuatus* Bonaparte

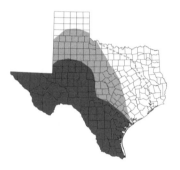

Common to uncommon resident in the southwestern half of the state. Resident populations occur throughout the Trans-Pecos, as far north as the southern South Plains, east to the central Edwards Plateau and Guadalupe River delta, and south to the Lower Rio Grande Valley. They are well known for wandering north after the breeding season and are uncommon winter visitors to the entire South Plains. They are rare and irregular in winter north to the southern Panhandle. During migration and winter this species is a rare, irruptive wanderer eastward into north-central Texas and to the upper coast. Pyrrhuloxias have significantly expanded their range in Texas over the past century, which Oberholser (1974) postulated was the result of the northward invasion of honey mesquite (*Prosopis glandulosa*). **Taxonomy:** The subspecies that occurs in Texas is *C. s. sinuatus* Bonaparte.

ROSE-BREASTED GROSBEAK
Pheucticus ludovicianus (Linnaeus)

Common to uncommon migrant in the eastern half of the state, becoming increasingly rare farther west until rare in spring and casual in fall in the Trans-Pecos. Rose-breasted

Grosbeaks are rare winter visitors along the Coastal Prairies and in the Lower Rio Grande Valley. A few isolated reports of wintering individuals have come from inland locations, including as far north as Amarillo, Randall County. This species is also a casual summer visitor to the High Plains and east across the northern edge of the state and accidental farther south in this season. **Timing of occurrence:** Spring migrants are found from early April through late May. Fall migrants begin to arrive in late August, but the primary migration period is from mid-September through early November. **Taxonomy:** Monotypic.

BLACK-HEADED GROSBEAK

Pheucticus melanocephalus (Swainson)

Common migrant and summer resident in the mountains of the Trans-Pecos and a casual summer visitor to the Panhandle and South Plains. Black-headed Grosbeaks are uncommon to rare migrants throughout the Trans-Pecos east to the High Plains, western Rolling Plains, and western Edwards Plateau south through the northwestern South Texas Brush Country, becoming increasingly rare farther east. Black-headed Grosbeaks are rare during migration east to north-central Texas and the central coast and are casual to very rare in the Pineywoods. This species is a very rare and irregular winter visitor along the Coastal Prairies and casual to accidental inland, including the Pineywoods, and as far north as Randall County. **Timing of occurrence:** Spring migrants are found from early April through late May. Fall migrants begin to arrive in early August, with most recorded from mid-September through late October and small numbers into mid-November. **Taxonomy:** The subspecies that occurs in Texas is *P. m. melanocephalus* (Swainson).

TBRC Review Species **BLUE BUNTING** *Cyanocompsa parellina* (Bonaparte)

Very rare and irregular winter visitor to the Lower Rio Grande Valley. There are 45 documented records for the state, and this species has been a more frequent visitor to Texas since 1995. However, this species is not annual in occurrence even though there have been periods when Blue Buntings have occurred each winter for three or four years in a row. There are 30 records from Hidalgo County, with most of those found in remnant patches of thorn-scrub. There are two records from higher up the Rio Grande in

Webb and Zapata Counties and three records from the Coastal Prairies: two from Brazoria County and one from Nueces County. **Timing of occurrence:** The majority of records occur from mid-November to mid-March, although there is one record from mid-October, and individuals have lingered as late as 1 May. **Taxonomy:** The subspecies that occurs in Texas is *C. p. beneplacita* Bangs.

BLUE GROSBEAK *Passerina caerulea* (Linnaeus)

Locally common to uncommon summer resident throughout most of the state. Blue Grosbeaks are uncommon to rare in the Oaks and Prairies region, as well as the Pineywoods and the Coastal Prairies north of Calhoun County. They are common to uncommon migrants throughout the state. Blue Grosbeaks occasionally linger into early winter, particularly along the coast and in extreme southern Texas. In the Trans-Pecos they are very rare winter residents near the Rio Grande in southern Brewster and Presidio Counties. **Timing of occurrence:** Spring migrants are found from early April through mid-May, with some as late as early June. The migration period is later in the Trans-Pecos, where these birds are uncommon until early May, when the majority of the migrants are seen and the breeding population arrives. Fall migrants are recorded from mid-August through mid-October, with small numbers into mid-November. **Taxonomy:** Three subspecies occur in the state.

P. c. caerulea (Linnaeus)
  Locally common to rare summer resident in the eastern half of the state, west to the eastern edge of the Rolling Plains and Edwards Plateau, and south through the South Texas Brush Country.

P. c. eurhyncha (Coues)
  Status uncertain. Oberholser (1974) lists a specimen from Brewster County for 2 July 1901. This subspecies is found in northeastern Mexico fairly close to the Texas border.

P. c. interfusa (Dwight & Griscom)
  Locally common to uncommon summer resident in the western half of the state, east through the Rolling Plains and Edwards Plateau, and into the northwestern South Texas Brush Country.

LAZULI BUNTING *Passerina amoena* (Say)

Uncommon migrant through the Trans-Pecos, Panhandle, and South Plains, becoming increasingly less common eastward. Lazuli Buntings are rare, but annual, east through the western Oaks and Prairies region and casual to very rare in the Pineywoods. They are much more common migrants in fall than spring. In summer this species is a rare visitor in the Panhandle and very rare in the Trans-Pecos. There is a nesting record from Kerr County, where Lacey (1911) reportedly collected egg sets from two nests in 1903. A male Lazuli Bunting paired with a female Indigo Bunting at Fort Hood, Bell County, in 2003 but did not fledge young (Kostecke et al. 2004). They are very rare winter residents in the Lower Rio Grande Valley and along the coast north to Aransas County. This species is casual in winter elsewhere along the coast and inland south of the Balcones Escarpment. There is a winter report from Randall County in the Panhandle on 3 December 1984. **Timing of occurrence:** Spring migrants are found from early April through late May. Fall migrants are recorded from early August through early October, and small numbers linger into mid-November. **Taxonomy:** Monotypic.

INDIGO BUNTING *Passerina cyanea* (Linnaeus)

Common to locally abundant summer resident in the eastern half of the state south through the Edwards Plateau and southernmost Oaks and Prairies region. Breeding populations of Indigo Buntings are found west to the central Edwards Plateau and along the Pecos River drainage, the eastern Rolling Plains, and the eastern Panhandle. They are also present in the western Panhandle along the Canadian River drainage to Oldham County. This species occasionally breeds in riparian habitats in the Trans-Pecos, with nesting records from El Paso, Hudspeth, and Jeff Davis Counties. Indigo Buntings are abundant migrants through the eastern half of Texas, becoming increasingly less common westward until rare in the Trans-Pecos. This species is rare to locally uncommon in winter in the Lower Rio Grande Valley and, to a lesser extent, along the coast. There are also isolated records of overwintering birds inland to the southern Edwards Plateau and east to the northern Pineywoods. **Timing of occurrence:** Spring migrants are found from late March through late May, but migration peaks from mid-April through mid-May. Fall migrants are recorded from mid-August through mid-October, with some seen through November. **Taxonomy**: Monotypic.

VARIED BUNTING *Passerina versicolor* (Bonaparte)

Locally uncommon to rare summer resident in the southern Trans-Pecos and eastward to the southwestern Edwards Plateau. This species is a rare and local summer resident in the western half of the South Texas Brush Country and the Lower Rio Grande Valley. There are numerous recent records from eastern El Paso County, suggesting that the species may be a rare and local breeder. There are isolated summer records away from the breeding range in the eastern Edwards Plateau and Tom Green County. Varied Bunting is a very rare winter visitor in southern Brewster and Presidio Counties, primarily during wet cycles. These birds are very rare to rare spring migrants along the lower and central coasts. Spring migrants have been noted well beyond the normal breeding areas, including to the upper coast, Oaks and Prairies region, South Plains, and Panhandle. **Timing of occurrence**: Spring migrants are found from mid-April through mid-May. Fall migrants are recorded from early August through mid-September, and some linger through mid-October. **Taxonomy**: The subspecies that occurs in Texas is *P. v. versicolor* (Bonaparte).

PAINTED BUNTING *Passerina ciris* (Linnaeus)

Uncommon to common summer resident and migrant throughout most of the state. Painted Buntings are generally absent as breeding birds in the arid habitats of the northwestern Trans-Pecos and on the High Plains away from more suitable habitat found in the canyonlands of the western Rolling Plains or the Canadian River drainage. They are very rare to casual winter visitors in the Lower Rio Grande Valley and along the coast, becoming casual inland with records north to Lubbock County and west to Presidio County. **Timing of occurrence**: Spring migrants begin to arrive in the eastern half of the state in very late March, with an increase in occurrence from early April through mid-May. Spring migrants are not found until mid-April in the western portions of the state and are present through late May. Fall migrants begin to depart from mid-July through mid-August, and most adult males are gone by the end of August. Some of the population, mostly hatch-year birds, are still present through mid-September, and small numbers linger as late as early November. **Taxonomy**: Two subspecies occur in the state.

P. c. ciris (Linnaeus)

Uncommon to common summer resident in the eastern third of the state, west to about Aransas, Brazos, and Tarrant Counties. Migrants have been reported west into the eastern Rolling Plains and eastern Edwards Plateau.

P. c. pallidior Mearns

Locally common to uncommon summer resident as described above west of the preceding subspecies. Migrants have been reported east through the Oaks and Prairies region and Coastal Prairies.

DICKCISSEL *Spiza americana* (Gmelin)

Uncommon to locally abundant summer resident and migrant east of the Pecos River. Breeding populations fluctuate greatly depending on environmental conditions. If conditions are favorable when migrants arrive, breeding activities commence almost immediately. During drought years when conditions are less favorable, these birds continue their migration, and evidence of breeding can be difficult to find. They are uncommon to rare fall migrants and accidental spring migrants in the Trans-Pecos and are very rare and local summer residents in the eastern half of the region. A few well-documented winter records have come from the Coastal Prairies and inland to north-central Texas and west to the South Plains. One of the more interesting occurrences in the state was an unprecedented influx of birds along the upper and central coasts during late January and February 1999, when more than 30 individuals were documented. **Timing of occurrence**: Spring migrants are found from early April to late May. Postbreeding wanderers are found well away from breeding areas beginning in mid-July, and fall migrants are present from late August to late October. Small numbers linger until late December. **Taxonomy**: Monotypic.

Family Icteridae: Blackbirds, Meadowlarks, and Orioles

BOBOLINK *Dolichonyx oryzivorus* (Linnaeus)

Uncommon to rare spring and casual fall migrant in the eastern half of the state west to the southern Oaks and Prairies region and Coastal Prairies. Bobolink is a casual spring migrant west through the Rolling Plains and eastern Edwards Plateau to the Panhandle and South Plains. It is accidental in the western Edwards Plateau and Trans-Pecos. A Bobolink photographed at Brazos Bend SP, Fort Bend County, on 27 December 1995 is the only documented winter occurrence in the state. **Timing of occurrence**: Spring migrants begin to arrive in early April, but the peak in migration is from mid-April through mid-May, with small numbers into late May. Fall migrants have been found between late August and mid-October. **Taxonomy**: Monotypic.

RED-WINGED BLACKBIRD
Agelaius phoeniceus (Linnaeus)

Abundant to locally uncommon migrant and resident throughout the state. Red-winged Blackbirds are more localized during the breeding season because of the limited availability of nesting habitat, particularly in the western half of the state. They are more widespread and often found in large flocks during the remainder of the year, as migrants join the resident population. **Timing of occurrence**: Fall migrants are found from late August through mid-November. Spring migrants have been recorded between late February and early May. **Taxonomy**: Five subspecies occur in the state.

A. p. arctolegus Oberholser
Common to locally abundant migrant and winter resident in the eastern half of the state, west through the Oaks and Prairies region.

A. p. fortis Ridgway
Common to uncommon migrant and resident in the Panhandle, south through the South Plains and western Edwards Plateau, and east through the western Rolling Plains. Locally uncommon to rare migrant and resident in the Trans-Pecos.

Common winter resident east through the Oaks and Prairies region and south through the South Texas Brush Country. This form is sometimes included under the following subspecies.

A. p. phoeniceus (Linnaeus)

Common to locally abundant resident in the eastern half of the state north of the Coastal Prairies and west through the eastern Rolling Plains and eastern Edwards Plateau. Common to uncommon migrant and winter resident in the Coastal Prairies south to Cameron County.

A. p. littoralis Howell & van Rossem

Common resident on the Coastal Prairies west to Fort Bend and Waller Counties.

A. p. megapotamus Oberholser

Uncommon to locally abundant resident in the southern Edwards Plateau and southernmost Oaks and Prairies region, south through the Coastal Prairies from Wharton and Matagorda Counties to the South Texas Brush Country and Lower Rio Grande Valley.

EASTERN MEADOWLARK *Sturnella magna* (Linnaeus)

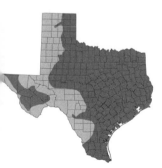

Locally common to rare resident through the eastern half of the state west to the eastern Panhandle and south along the Coastal Prairies to the Rio Grande. Eastern Meadowlark is a rare to very rare and local summer resident in the eastern half of the High Plains south through the eastern Edwards Plateau to the eastern South Texas Brush Country, including all of the Lower Rio Grande Valley. Eastern Meadowlarks are uncommon residents in the mid- and upper-elevation grasslands of the central Trans-Pecos west to northeastern El Paso County. They are locally common to rare migrants throughout the state. During winter, they are uncommon to rare and more widespread through the western half of the state, including the western South Texas Brush Country. The overall population of Eastern Meadowlark has declined almost 72 percent over the past 40 years and even more sharply in some areas of Texas. **Timing of occurrence:** Fall migrants are found from early September through late October. Spring migrants have been recorded between late February and early April. **Taxonomy:** Four subspecies occur in the state.

S. m. magna (Linnaeus)

Common to locally uncommon summer resident from the eastern Panhandle south to the northwestern Edwards Plateau and east through the Rolling Plains to the western Oaks and Prairies region. Uncommon to rare as a migrant and in winter but is more widespread, including the western Panhandle, South Plains, and remainder of the Edwards Plateau.

S. m. argutula Bangs

Common to locally uncommon resident in the eastern third of the state west through the eastern Oaks and Prairies region and south through the Coastal Prairies to Nueces County. Uncommon to rare migrant through resident range and south through the Coastal Prairies to Cameron County. Intergrades with *S. m. magna* and *S. m. hoopesi*.

S. m. hoopesi Stone

Common resident in the southern Coastal Prairies and eastern South Texas Brush Country. Uncommon migrant and winter resident in the western South Texas Brush Country and southwestern Edwards Plateau.

S. m. lilianae Oberholser

Common resident in the central Trans-Pecos west to northeastern El Paso County. This taxon has been documented in Kinney and Val Verde Counties in the winter. Sometimes considered a separate species, Lilian's Meadowlark (Barker, Vandergon, and Lanyon 2008).

WESTERN MEADOWLARK
Sturnella neglecta Audubon

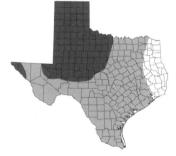

Common to uncommon resident in portions of the western half of the state, east through the Rolling Plains and western Edwards Plateau. There is an isolated nesting record from Brooks County. The summer distribution in the Trans-Pecos is limited primarily to agricultural areas of Hudspeth and El Paso Counties and similar habitats in the northeastern portion of the region. There is only minimal overlap in nesting habitat with Eastern Meadowlark in Hudspeth County. Western Meadowlark is common to abundant during migration and in the winter through the western half of the state, east to the Oaks and Prairies region, and south through the South Texas Brush Country. They are very rare but regular in winter through the Pineywoods. Western Meadowlarks appear to use agricultural fields more extensively during winter than Eastern Meadowlarks do. **Timing of occurrence:** Fall migrants are found from late September through mid-November. Spring migrants have been recorded between mid-February and early April. **Taxonomy:** Two subspecies occur in the state.

S. n. confiuenta Rathbun
> Status uncertain. Oberholser (1974) lists three specimens from San Patricio County in 1955 and 1956, but given the range of this subspecies, modern confirmation is desirable. Sometimes considered part of the following subspecies.

S. n. neglecta Audubon
> As described in species account.

YELLOW-HEADED BLACKBIRD
Xanthocephalus xanthocephalus (Bonaparte)

Common to uncommon migrant in the western half of the state, becoming increasingly less common eastward to the Pineywoods and Upper Texas Coast, where it is rare. Yellow-headed Blackbirds are also locally common summer residents in the southwestern Panhandle, with established breeding colonies primarily in Castro County (Seyffert 2001b). There have been additional nesting sites documented in Parmer, Swisher, and Bailey Counties and north to Hansford County. These colonies are at playa lakes with permanent water because of agricultural operations in the vicinity. They are rare and irregular in winter in most areas of the state, except in the western Trans-Pecos, where they are locally abundant in El Paso and Hudspeth Counties along and near the Rio Grande in flocks that can sometimes exceed 20,000 individuals. **Timing of occurrence:** Fall migrants begin to arrive in mid-July, with the peak in abundance during September. They are still present through October and rarely into early November. Spring migrants have been recorded between mid-March and mid-May, with a peak from mid-April through early May. **Taxonomy:** Monotypic.

RUSTY BLACKBIRD *Euphagus carolinus* (Statius Müller)

Rare to locally uncommon migrant and winter resident in the eastern third of the state. The majority of wintering birds are found from north-central Texas eastward and southward through the Pineywoods. The population of

this species has been in sharp decline for the past 40 years. This decrease in abundance has been reflected in the number of individuals wintering in Texas. Pulich (1961a) reported winter roosts of up to 5,000 individuals in the mid-1950s in Tarrant County, but concentrations of 100 to 250 birds are now noteworthy. Rusty Blackbird is a rare winter visitor west across the Rolling Plains to the southern Panhandle and South Plains. Elsewhere in the western half of the state these blackbirds are very rare to casual winter visitors, with records as far west as El Paso County. They are accidental in extreme southern Texas, with only one report from the Lower Rio Grande Valley. **Timing of occurrence:** Fall migrants are present from mid-November through mid-December, when the wintering population has arrived. Spring migrants have been recorded from early February to mid-March, with a small number into late March. **Taxonomy:** The subspecies that occurs in Texas is *E. c. carolinus* (Statius Müller).

BREWER'S BLACKBIRD

Euphagus cyanocephalus (Wagler)

Common to locally abundant migrant and winter resident throughout most of Texas. Brewer's Blackbirds are most common in the western half of the state, becoming less common farther east. They are locally uncommon to rare in the Pineywoods and the southern South Texas Brush Country, including the Lower Rio Grande Valley. The only two documented nesting records for the state are a nest with eggs collected in Wilbarger County in 1928 and a nest discovered in Jeff Davis County in 2000 (Bryan and Karges 2001). Additional reported breeding attempts have come from Crosby, Jeff Davis, and Lubbock Counties. **Timing of occurrence:** Fall migrants are present from mid-September through mid-November. Reports from the Panhandle of small flocks present in late July and August are either the result of postbreeding dispersal or very early migrants. Spring migrants have been recorded from late March through mid-May. **Taxonomy:** Monotypic.

COMMON GRACKLE *Quiscalus quiscula* (Linnaeus)

Common to uncommon summer resident in the eastern two-thirds of the state, west through the High Plains and central Edwards Plateau, and south to the Guadalupe River

drainage in the Coastal Prairies. The breeding range has expanded westward since the 1950s, and Common Grackles are increasingly common in urban areas on the High Plains and locally on the western Edwards Plateau. This species is a locally common to abundant migrant and winter resident in the eastern third of the state, becoming increasingly less common westward through the eastern Rolling Plains and Edwards Plateau. Common Grackles are rare and local in winter throughout the South Texas Brush Country and the eastern half of the Trans-Pecos; however, they are largely absent from the Panhandle. Beginning in the mid-1990s, small numbers of Common Grackles have wintered irregularly in El Paso County. **Timing of occurrence**: Fall migrants are present from mid-September through early November. Spring migrants have been recorded from mid-February to late May. **Taxonomy**: Two subspecies occur in the state.

Q. q. versicolor Vieillot
   As described above.

Q. q. stonei Chapman
   Accidental, with one specimen collected at Sour Lake, Hardin County, on 15 July 1902.

BOAT-TAILED GRACKLE *Quiscalus major* Vieillot

Uncommon to locally abundant resident along the Coastal Prairies from Jefferson to Aransas Counties. Boat-tailed Grackles are most common on the upper coast, becoming increasingly less so south to the central coast. This species is casual to accidental south to Willacy and Cameron Counties. Boat-tailed Grackles are found almost exclusively within 30 miles of the Gulf of Mexico. **Taxonomy**: The subspecies that occurs in Texas is *Q. m. major* Vieillot.

GREAT-TAILED GRACKLE

Quiscalus mexicanus (Gmelin)

Abundant resident throughout most of the southern half of the state, excluding the Pineywoods, where this species is uncommon to rare and found mostly in urban settings. Great-tailed Grackles are also abundant, though largely limited to urban areas, northward through the state to the Oklahoma border, including the Panhandle. In the Trans-Pecos they are common to abundant in urban areas and

rare to absent elsewhere. They have greatly expanded their range over the past century. Found primarily in the South Texas Brush Country and western Trans-Pecos prior to 1910, by 1960 they had colonized northward through the Coastal Prairies and much of the central portion of the state, with breeding documented as far north as north-central Texas and the southern Panhandle. With the exception of the Pineywoods, they had colonized the remainder of the state by the mid-1980s. **Taxonomy:** Three subspecies occur in the state.

Q. m. mexicanus (Gmelin)

Status uncertain. Oberholser (1974) lists one specimen from near Brownsville, Cameron County, on 4 January 1927. This subspecies is found in Nuevo León and southern Tamaulipas, Mexico, and could be a visitor to South Texas.

Q. m. monsoni (Phillips)

Common to abundant resident throughout the state east of the Trans-Pecos.

Q. m. prosopidicola (Lowery)

Common to locally abundant resident in the Trans-Pecos.

TBRC Review Species

SHINY COWBIRD *Molothrus bonariensis* (Gmelin)

Casual. Texas has 12 accepted records of this species. The first two were males captured in cowbird traps on Fort Hood in Bell and Coryell Counties, but there have been no records since May 2004. Half of the records have come from the upper and central coasts, with others inland to Kendall County and south to Cameron County. Shiny Cowbirds underwent a range expansion through the Caribbean into southern Florida. Many speculated that this species would soon colonize the southeastern United States; however, that has not happened, and Shiny Cowbirds are still rare even in southern Florida. All Texas records are of the distinctive males. **Timing of occurrence:** Almost all of the records are from 3 February to 12 June, with one fall record. **Taxonomy:** The subspecies that occurs in Texas is *M. b. minimus* Dalmas.

BRONZED COWBIRD *Molothrus aeneus* (Wagler)

Common to uncommon summer resident in the southwestern part of the state. Bronzed Cowbirds are found in sum-

mer through the Trans-Pecos and north to the South Plains and western Rolling Plains and east up the coast to Harris County. The population in Texas is expanding northward, and Bronzed Cowbirds are now uncommon to rare migrants and summer residents through the South Plains and are very rare farther north into the southern Panhandle. They withdraw from the northern portions of their range, including the Trans-Pecos and Edwards Plateau, during the winter. Bronzed Cowbirds are locally common to uncommon winter residents throughout the South Texas Brush Country and up the Coastal Prairies to Harris County. They are casual in the Oaks and Prairies region in the spring and summer and accidental in the central Pineywoods during the winter. **Timing of occurrence:** Spring migrants are present from mid-March through mid-May. Fall migrants have been recorded mid-July to early September. **Taxonomy:** Two subspecies occur in the state.

M. a. loyei Parkes & Blake
Locally common resident in the western half of the Trans-Pecos, then uncommon to rare east to Terrell County.

M. a. aeneus (Wagler)
As described above from the eastern Trans-Pecos eastward.

BROWN-HEADED COWBIRD
Molothrus ater (Boddaert)

Common to locally abundant summer resident throughout the state, except in the southern South Texas Brush Country, where it is uncommon. During the winter, the majority of Brown-headed Cowbirds withdraw from the Trans-Pecos, from upland habitats of the Edwards Plateau, and the Pineywoods. Small numbers remain in these regions throughout the season and may even be locally abundant at roost sites such as feedlots. In the eastern half of the state, the number of wintering birds increases greatly as migrants bolster the resident population. In the Panhandle, these migrants congregate in large flocks during the fall and remain into early winter before dispersing (Seyffert 2001b). Brown-headed Cowbirds form large winter roosts, especially in urban areas, with numbers at some sites estimated as high as 2 million individuals. **Timing of occurrence:** In areas where the majority of the population departs in the winter, spring migrants are reported from mid-February through mid-April. Fall migrants have been recorded from late August to early October. **Taxonomy:** Three subspecies occur in the state.

M. a. artemisiae Grinnell

Common migrant and locally uncommon winter resident in the western half of the state, east through the eastern Oaks and Prairies region, and south through the South Texas Brush Country.

M. a. obscurus (Gmelin)

Uncommon to common resident in the southern half of the state.

M. a. ater (Boddaert)

Common to locally abundant resident in the northern half of the state, south to the northern edge of the Edwards Plateau, and east to the Pineywoods. Present throughout the state in winter.

TBRC Review Species

BLACK-VENTED ORIOLE *Icterus wagleri* Sclater

Casual. Texas has nine documented records, three of which involve the same returning summering individual and two of which involve the same returning wintering individual. The first record was an adult discovered at Rio Grande Village, Big Bend NP, on 27 September 1968. What is believed to be the same individual returned for the summer of 1969, during which time it was captured and banded. This bird once again returned to the same location during the summer of 1970 (Wauer 1970). The remaining records include three from the Lower Rio Grande Valley, one from Kleberg County, and another from Big Bend NP. **Timing of occurrence:** Six of these records were of birds discovered in the spring or summer, with several of them remaining through the summer. There is one October record, and the remaining two records were discovered in the fall or early winter. Each remained for an extended period. **Taxonomy:** The subspecies that occurs in Texas is *I. w. wagleri* Sclater.

ORCHARD ORIOLE *Icterus spurius* (Linnaeus)

Uncommon to locally common summer resident in the eastern two-thirds of the state and locally uncommon to rare farther west. In the South Texas Brush Country, Orchard Orioles are generally absent as breeding birds away

114. Studies of the Curve-billed Thrasher (*Toxostoma curvirostre*) have suggested that the populations in Texas and eastern New Mexico may be a separate species from those in Arizona. This adult was at Davis Mountains SP, Jeff Davis County, on 17 December 2005. *Photograph by Mark W. Lockwood.*

115. Sage Thrasher (*Oreoscoptes montanus*) is a winter resident in the Trans-Pecos and eastward though the western portions of the Panhandle and South Plains. The number of birds present in a given year fluctuates greatly. On rare occasions there are invasions farther east in the state, as happened during the winter of 2011–12. This one was at Balmorhea Lake, Reeves County, on 30 January 2009. *Photograph by Mark W. Lockwood.*

116. Sprague's Pipit (*Anthus spragueii*) is a much-sought-after migrant and winter resident in Texas. Though it is not rare in some areas of the state, familiarity with the preferred winter habitat is required to locate this cryptic species. This one was near La Joya, Hidalgo County, on 22 January 2010. *Photograph by Robert Epstein.*

117. Bohemian Waxwing (*Bombycilla garrulus*) appears to have been a much more regular visitor to the Panhandle from the mid-1950s to the mid-1970s. These waxwings are now casual visitors, although there was an influx of birds in November 2004, including this one in Dalhart, Hartley County, on 21 November 2004. *Photograph by Brian Gibbons.*

118. Lapland Longspurs (*Calcarius lapponicus*) are regular winter residents in the Panhandle and northern South Plains but can be found as far south as the Coastal Prairies during invasion years. As it does in other species of longspurs, their restless behavior makes getting satisfactory looks at them difficult. This winter-plumaged adult was near Texline, Dallam County, on 10 January 2009. *Photograph by Greg W. Lasley.*

119. Smith's Longspurs (*Calcarius pictus*) have a rather restricted winter range so are much sought after in Texas. Specific habitat requirements give this cryptic species a patchy distribution, although Smith's Longspurs can be locally uncommon. Despite the local core winter range, the species has been found in many other parts of the state, as is evidenced by this one in Houston, Harris County, on 14 January 2012. *Photograph by Martin Reid.*

120. Snow Buntings (*Plectrophenax nivalis*) are casual visitors to Texas, and one might expect that they would be found solely in the northernmost regions of the state in midwinter. However, there is a record from South Padre Island, and perhaps more astounding was this alternate-plumaged male at Sea Rim SP, Jefferson County, on 13 June 2011. *Photograph by Terry Ferguson.*

121. The Swainson's Warbler (*Limnothlypis swainsonii*) is more easily heard than seen in its preferred riparian habitats. The Texas population breeds early, and these warblers are often hard to locate after mid-July. This territorial male was in the Big Thicket National Preserve, Hardin County, on 9 May 2009. *Photograph by Greg W. Lasley.*

122. In the United States, the Colima Warbler (*Oreothlypis crissalis*) has long been associated only with the Chisos Mountains of Big Bend. However, it is now known to occasionally overshoot that range and end up in the Davis Mountains. This adult Colima Warbler was photographed in the Chisos Mountains, Brewster County, on 1 May 2010. *Photograph by Mark W. Lockwood.*

123. The Tropical Parula (*Setophaga pitiayumi*) has expanded its range into the Devils and Pecos River drainages in the southwestern portion of the state since about 2000. This area, along with the live oak mottes of the Coastal Sand Plain in Kenedy County, holds the largest populations in Texas of this tropical warbler. This male was photographed near Sarita, Kenedy County, on 23 April 2009. *Photograph by Greg W. Lasley.*

124. The distinctive Mangrove Warbler is one of three groups of Yellow Warblers (*Setophaga petechia*) that are sometimes considered separate species. In 2002 a population of Mangrove Warblers was discovered around the southern tip of the Laguna Madre in Cameron County. Prior to this discovery they were considered accidental in the United States. This male was photographed there on 3 May 2007. *Photograph by Greg W. Lasley.*

125. Blackpoll Warblers (*Setophaga striata*) winter in northern South America and migrate primarily through the Caribbean, making Texas the western edge of that route. This male was at South Padre Island, Cameron County, on 25 April 2009. *Photograph by Greg W. Lasley.*

126. Prairie Warblers (*Setophaga discolor*) rely on disturbance to produce the early successional vegetation that they use for nesting habitat. As a result, they are often found in areas that have been previously clear-cut for timber harvest. Similar actions are taken on conservation lands, such as where this male was found in the Angelina National Forest, Angelina County, on 8 May 2009. *Photograph by Greg W. Lasley*

127. The Grace's Warblers (*Setophaga graciae*) nest in the ponderosa and white pine woodlands of the Guadalupe and Davis Mountains. Even though they also nest in similar habitats throughout New Mexico, they are rarely found during migration. The male was on the TNC's Davis Mountains Preserve, Jeff Davis County, on 15 April 2008. *Photograph by Mark W. Lockwood.*

128. Hermit Warbler (*Setophaga occidentalis*) is a rare migrant through the western half of the Trans-Pecos. There are a number of records of migrants farther east in the state as well as a few winter records from the Lower Rio Grande Valley. This male at Killeen, Bell County, from 10 to 11 December 2011 was well outside the normal pattern of occurrence. *Photograph by Gil Eckrich.*

129. The Golden-cheeked Warbler (*Setophaga chrysoparia*) is one of the earliest of the breeding warblers to return in the spring, with the first few arriving in early March. It is often referred to as the quintessential Texas specialty since its entire breeding range is con-fined within the borders of Texas. This male was at Garner SP, Uvalde County, on 11 April 2007. *Photograph by Mark W. Lockwood.*

130. Texas has one record of Fan-tailed Warbler (*Basileuterus lachrymosus*), a rather long-staying bird that was in Pine Canyon of the Chisos Mountains, Brewster County, from 13 August through 24 September 2007. This image was taken on 23 August. Note that it was molting a new tail. *Photograph by Mark W. Lockwood.*

131. Rufous-capped Warblers (*Basileuterus rufifrons*) are enigmatic visitors to the state, with records from all seasons, which has even led to speculation of nesting. This adult was near Pearsall, Frio County, on 5 January 2005. *Photograph by Martin Reid.*

132. The first record of Golden-crowned Warbler (*Basileuterus culicivorus*) involved a specimen that was dismissed as being mislabeled for almost 90 years. There are now more than 20 records, including this bird at the Frontera Audubon Sanctuary in Weslaco, Hidalgo County, on 19 January 2012. *Photograph by Erik Breden, The Otter Side.*

133. White-collared Seedeaters (*Sporophila torqueola*) were virtually extirpated from the United States in the 1950s but are making a comeback. They are now found in many locations from Starr to Webb Counties and have recently been discovered much farther north in Val Verde County. This female was in Del Rio, Val Verde County, on 20 February 2010. *Photograph by Karen Gleason.*

134. The first three records of Yellow-faced Grassquit (*Tiaris olivaceus*) for the state are from the Lower Rio Grande Valley. However, this male, the most recent record, was found much farther north at Goose Island SP, Aransas County, from 30 January through 20 March 2011. *Photograph by Robert Epstein.*

135. Bachman's Sparrow (*Peucaea aestivalis*) is a habitat specialist found in open pine woodlands of the East Texas Pineywoods. Territorial males sing from an exposed perch, dropping to the grassy understory when disturbed. This adult was at the Angelina National Forest, Jasper County, on 13 June 2008. *Photograph by Greg W. Lasley.*

136. Black-chinned Sparrows (*Spizella atrogularis*) nest in fairly open habitats above 5,000 feet in the various mountain ranges of the Trans-Pecos. They wander to lower elevations in the winter, but they are rarely found far from breeding areas. This male was at Dog Canyon, Guadalupe Mountains NP, Culberson County, on 26 May 2012. *Photograph by Mark W. Lockwood.*

137. Le Conte's Sparrow (*Ammodramus leconteii*) is a true grassland species and is often associated with little bluestem (*Schizachyrium scoparium*) prairies or meadows. These winter residents are most common in the eastern half of the state, but occasionally they can be found at isolated locations much farther west. This individual was at the far western end of the expanded winter range at Balmorhea Lake, Reeves County, on 16 December 2012. *Photograph by Mark W. Lockwood.*

138. Although Nelson's Sparrows (*Ammodramus nelsoni*) nest in the northern Great Plains and winter along the Gulf Coast, they are rarely detected during migration. This adult was near Sabine Pass, Jefferson County, on 31 October 2009. *Photograph by Greg W. Lasley.*

139. The winter range of Harris's Sparrow (*Zonotrichia querula*) extends from South Dakota southward through the Great Plains to the Oaks and Prairies region of Texas. During some years, many more of these birds move south into Texas, which can lead to a wider distribution. This one was on the far western edge of that expanded winter range in Alpine, Brewster County, on 26 November 2011. *Photograph by Mark W. Lockwood.*

140. Hepatic Tanager (*Piranga flava*) is an uncommon resident in woodland habitats above 6,000 feet in the major mountain ranges of the Trans-Pecos. This female was at Dog Canyon, Guadalupe Mountains NP, Culberson County, on 26 May 2012. *Photograph by Mark W. Lockwood.*

141. The Crimson-collared Grosbeak (*Rhodothraupis celaeno*) is an endemic bird of northeastern Mexico but has occurred in Texas almost annually since 2005. There was a veritable invasion during the winter of 2004–5, when 17 individuals were documented. This female-type bird was at Pharr, Hidalgo County, from 15 November 2010 to 11 April 2011. *Photograph by Robert Epstein.*

142. Blue Buntings (*Cyanocompsa parellina*) were first documented in Texas in the early 1980s. This species has become a near-annual winter visitor to the Lower Rio Grande Valley since 1995, although in most years only one or two individuals are found. This female was at Rio Grande City, Starr County, on 27 January 2011. *Photograph by Alan Wormington.*

143. The Varied Bunting (*Passerina versicolor*) is closely related to the Painted Bunting and has a very similar song. Another characteristic both species share is that males do not attain their bright body plumage until their second summer. First-summer males look like adult females but sing and defend territories. This adult male was in Big Bend NP, Brewster County, on 6 May 2010. *Photograph by Greg W. Lasley.*

144. Eastern Meadowlark (*Sturnella magna*) is a widespread species in Texas, including the mid-elevation grasslands of the Trans-Pecos. Some research into the resident populations in the desert Southwest have suggested that they may be a separate species referred to as Lilian's Meadowlark. This male was at Balmorhea SP, Reeves County, on 29 May 2009. *Photograph by Mark W. Lockwood.*

145. The overall population of Rusty Blackbirds (*Euphagus carolinus*) has declined sharply in recent decades, causing considerable concern. This female was in Austin, Travis County, on 27 December 2011. *Photograph by Greg W. Lasley.*

146. There are nine records for Black-vented Oriole (*Icterus wagleri*) in Texas. Interestingly, most of these records involve birds that were present for an extended period. Such was the case with this adult near Bentsen–Rio Grande Valley SP, Hidalgo County, from 13 December 2010 to 11 March 2011. *Photograph by Robert Epstein.*

147. Cassin's Finches (*Haemorhous cassinii*) are annual visitors to Texas only in the Davis and Guadalupe Mountains. The number of birds wandering into Texas appears to be declining and may reflect an overall northward shift in their winter range. This male was at Davis Mountains SP, Jeff Davis County, on 21 March 2008. *Photograph by Mark W. Lockwood.*

148. Red Crossbills (*Loxia curvirostra*) are rare residents in the Davis and Guadalupe Mountains and irregular wanderers into the state during winter invasions. The winter of 2012–13 was such a year, and many, if not all, of the birds found in the Trans-Pecos and High Plains were the Type 2 birds that specialize on ponderosa pines. This Type 2 male was in the Davis Mountains, Jeff Davis County, on 12 March 2011. *Photograph by Maryann Eastman.*

149. Any record of Common Redpoll (*Acanthis flammea*) is a noteworthy occurrence in the state. There are only 12 records, three of which were documented in January and February 2012. This male was near Decatur, Wise County, from 22 January to 1 March 2007. *Photograph by Kenneth Hunt.*

150. Evening Grosbeaks (*Coccothraustes vespertinus*) were once a more regular visitor to the state. This species made incursion in the state for more than 10 winters between 1968 and 1987. However, there are only six documented records since the species was added to the Review List in 2008. This male was in the Chisos Mountains, Brewster County, on 12 December 2009. *Photograph by Mark W. Lockwood.*

from riparian corridors. In the Trans-Pecos, they are locally uncommon summer residents in riparian habitats in the southern half of the region, including along the Rio Grande floodplain to extreme southeastern El Paso County. They are common to uncommon migrants in the state west to the central Trans-Pecos, becoming rare in the remainder of that region. This species is casual to accidental in winter, with isolated reports primarily along the Coastal Prairies. Many of these winter reports are probably late migrants, but there are confirmed records of birds overwintering. They are declining as a breeding bird in the Lower Rio Grande Valley, likely due to habitat alterations and parasitism by Bronzed Cowbirds. **Timing of occurrence:** Spring migrants are found from very late March through mid-May, with some into early June. Fall migrants have been recorded from early July to late September, with small numbers into mid-November. **Taxonomy:** Two subspecies occur in the state.

I. s. spurius (Linnaeus)
   As described above.

I. s. fuertesi Chapman
   There are three documented records from Cameron County. This distinctive subspecies is known as Fuertes's or Ochre Oriole and is sometimes considered a separate species (Baker et al. 2003). The first record for Texas and the United States is a specimen from Brownsville, Cameron County, on 3 April 1894. The only other record involves a bird that summered in Arroyo City, Cameron County, in 1998 and returned the next year. The TBRC requests documentation for all sightings of this subspecies.

HOODED ORIOLE *Icterus cucullatus* Swainson

Uncommon and local migrant and summer resident on the southwestern Edwards Plateau, south through the western and southern South Texas Brush Country to the Lower Rio Grande Valley, and in the southern Coastal Prairies north to Nueces County. Hooded Orioles are rare and local summer residents in El Paso County and along the Rio Grande from southern Presidio County eastward to Terrell County. The highest breeding density occurs in the oak woodlands of Kenedy and Brooks Counties. Hooded Orioles are very rare migrants and summer visitors to the Davis and Guadalupe Mountains. They are rare to locally uncommon winter residents in the Lower Rio Grande Valley and casual during that season in El Paso County. Hooded Orioles wander up the Coastal Prairies, with most records from spring and summer, but there are also several winter records. They are accidental north to Bailey, Crosby, Dallas, Randall, and Tarrant Counties. **Timing of occurrence**: Spring migrants are present from late March through mid-May. Fall migrants have been recorded from mid-August to early October. **Taxonomy**: Three subspecies occur in the state.

I. c. sennetti Ridgway

Uncommon to locally common summer resident in the western and southern South Texas Brush Country, north to Maverick and Zavala Counties, and in the southern Coastal Prairies north to Nueces County. Rare to locally uncommon in winter in the Lower Rio Grande Valley and very rare northward in any season farther north along the Coastal Prairies to Harris County.

I. c. cucullatus Swainson

Uncommon summer resident on the southwestern Edwards Plateau and west to southern Presidio County.

I. c. nelsoni Ridgway

Rare and local summer resident in El Paso County and possibly east to Brewster County. Casual winter visitor in El Paso County, with first occurrence from the winter of 2008–9.

TBRC Review Species

STREAK-BACKED ORIOLE *Icterus pustulatus* (Wagler)

Accidental. There are two documented records for the state. The first record was very unexpectedly found in Fort Bend County and provided, by far, the easternmost record for this species. The closest this species regularly occurs to Texas is northern Sonora, Mexico, and these orioles are less than annual visitors to southern Arizona, where they have occurred in all seasons. Away from the established pattern of occurrence in the United States there are records from northern Arizona, California, and exceptionally Colorado, in addition to those from Texas. **Taxonomy**: The subspecies that has occurred in Texas is not known, but the northernmost is *I. p. microstictus* Griscom, which is the most likely for at least the El Paso record.

12 DEC. 2004–8 APR. 2005, BRAZOS BEND SP, FORT BEND CO.
(TBRC 2005–32; TPRF 2276)
16 SEPT. 2005, EL PASO, EL PASO CO. (TBRC 2005–118)

BULLOCK'S ORIOLE *Icterus bullockii* (Swainson)

Common to uncommon summer resident in the western half of the state. The breeding range extends east to the central Rolling Plains, central Edwards Plateau, and South Texas Brush Country. Bullock's Orioles are rare to locally uncommon summer residents in the higher elevations of the central Trans-Pecos and the Lower Rio Grande Valley. They are very rare to casual breeders through the eastern half of the Rolling Plains, eastern Edwards Plateau, and south to the central coast. Bullock's Orioles are common to uncommon migrants through the western half of the state, becoming locally uncommon to rare east to the edge of the Pineywoods, where they are rare to very rare. This species is a very rare winter visitor in virtually all areas of the state. **Timing of occurrence:** Spring migrants begin to arrive in late March, but the main passage is from early April through mid-May. Fall migrants have been recorded from late July to late September, with a few lingering to early October. **Taxonomy:** The subspecies that occurs in Texas is *I. b. bullockii* (Swainson). Formerly considered conspecific with Baltimore Oriole under the name Northern Oriole (*I. galbula*).

ALTAMIRA ORIOLE *Icterus gularis* (Wagler)

Uncommon resident in the Lower Rio Grande Valley and northward along the Rio Grande to northern Zapata County. Altamira Orioles have been reported as far north as Kleberg (TPRF 619) and La Salle Counties, but records north of the Lower Rio Grande Valley are scant and require documentation. Altamira Oriole was first documented in Texas in 1938 (Burleigh 1939) and was seen infrequently through the 1940s. The first documented nesting record did not come until 1951. This species underwent an impressive population growth beginning in the early 1960s. Some observers suspect, however, that Altamira Oriole numbers in Texas are now declining. **Taxonomy:** The subspecies that occurs in Texas is *I. g. tamaulipensis* Ridgway.

AUDUBON'S ORIOLE *Icterus graduacauda* Lesson

Locally uncommon to rare resident in the South Texas Brush Country north to Goliad and Uvalde Counties, although generally absent from Cameron and eastern Hidalgo Counties. Audubon's Oriole has expanded its range northward in the western South Texas Brush Country during the last 10 years and is found north to the Balcones Escarpment. These orioles are now rare and local on the southern Edwards Plateau north to Bandera and Real Counties. This expansion may be continuing, with records from Val Verde County to the west and to Kendall, Kerr, and Tom Green Counties to the north and east. Vagrants have been found in Bastrop, Guadalupe, and Midland (TPRF 66) Counties. The highest concentrations of Audubon's Orioles in Texas are found in the open oak woodlands of Kenedy and Brooks Counties. **Taxonomy:** The subspecies that occurs in Texas is *I. g. audubonii* Giraud.

BALTIMORE ORIOLE *Icterus galbula* (Linnaeus)

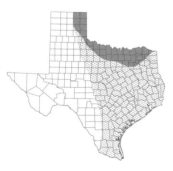

Locally uncommon summer resident in the eastern third of the Panhandle. This species is a rare to uncommon and local summer resident eastward across the northern Rolling Plains to north-central Texas, becoming very rare and local south through the Pineywoods. Baltimore Orioles are common to uncommon migrants in the eastern half of the state, becoming increasingly rare west to the Pecos River and casual farther west. This species is a rare winter visitor along the coast, and there are several winter records from inland locations. **Timing of occurrence:** Spring migrants are found from early April though late May and rarely into early June. Fall migrants have been recorded from late July to late October, and a few have lingered to mid-November. **Taxonomy:** Monotypic. Formerly considered conspecific with Bullock's Oriole under the name Northern Oriole.

SCOTT'S ORIOLE *Icterus parisorum* Bonaparte

Uncommon to common migrant and summer resident in the Trans-Pecos eastward across the southern Edwards Plateau. There are numerous winter records within the breeding range that primarily refer to lingering fall migrants or very early spring migrants; however, midwinter records also exist. Unexpected winter records include birds

in Collingsworth, Tarrant, and Walker Counties. This species is casual to accidental as a migrant well away from the breeding range, including all regions of the state except the Pineywoods. **Timing of occurrence:** Spring migrants have been found as soon as early February, with the primary migration period from mid-March through mid-May. Fall migrants have been recorded from late August to early October, and small numbers have lingered as late as mid-December. **Taxonomy:** Monotypic.

Family Fringillidae: Finches and Allies

TBRC Review Species   **PINE GROSBEAK** *Pinicola enucleator* (Linnaeus)

Accidental. Texas has six documented records, the first pertaining to a specimen taken in Dallas County. There are 10 additional sightings from northwest and West Texas that were made prior to the development of the Review List in 1988. Among these is a report of a female found alive in Pampa, Gray County, during December 1933 that subsequently died. **Timing of occurrence:** The majority of sightings, including reports and documented records, have occurred between 10 October and 15 March. There are three sightings between 24 April and 5 May, with one from late September. **Taxonomy:** The subspecies that has occurred in Texas is *P. e. montana* Ridgway.

HOUSE FINCH *Haemorhous mexicanus* (Statius Müller)

Uncommon to locally common resident throughout most of the state. During the 1990s the population of House Finches introduced to the eastern United States expanded westward, and the native western population continued to expand eastward. The two populations met in Central Texas. This species has now been reported from nearly every county in the state. House Finches are still rare to locally uncommon along the contact zone, which includes the area just east of the Balcones Escarpment and south to the central coast. They are very rare in the South Texas Brush Country south and east of Laredo, Webb County. **Taxonomy:** Two subspecies occur in the state.

H. m. frontalis (Say)
    As described above.

H. m. potosinus (Griscom)

> Uncommon to rare resident along the Rio Grande from the southeastern Trans-Pecos in Crane County south through the western South Texas Brush Country to Starr County and possibly to Hidalgo County.

PURPLE FINCH *Haemorhous purpureus* (Gmelin)

Uncommon to rare and irregular migrant and winter visitor to the northeastern portion of the state south through the central Pineywoods. Purple Finches were a more regular and widespread winter resident in Texas prior to the 1980s. Formerly they regularly occurred west to the eastern Rolling Plains and Edwards Plateau and south to the upper Coastal Prairies, but such occurrences are now associated with invasions, such as in the winter of 2004–5. They are rare but regular migrants and winter visitors in the eastern Panhandle and South Plains and very rare to casual elsewhere in the western half of the state, the central Coastal Prairies, and the extreme northern South Texas Brush Country. **Timing of occurrence:** Fall migrants begin to arrive in early November, with numbers increasing by mid-December. Spring migrants have been recorded from mid-February through mid-March, and a few birds have lingered into late April and exceptionally into early May. **Taxonomy:** The subspecies that occurs in Texas is *H. p. purpureus* (Gmelin).

CASSIN'S FINCH *Haemorhous cassinii* (Baird)

Rare to very rare winter visitor in the Trans-Pecos. Cassin's Finches exhibit an irruptive pattern of occurrence throughout most of that region but do occur annually in small numbers in the Davis and Guadalupe Mountains. This species formerly occurred in greater numbers in the state. There have been exceptional winters with large incursions but not since the winter of 1996–97. During past incursions a few birds have even remained through the summer. Cassin's Finch is a very rare to casual winter visitor to the High Plains and western Edwards Plateau and is accidental farther east. There is one record of a single bird on South Padre Island, Cameron County, from 8 to 10 November 2000 (TPRF 1766). **Timing of occurrence:** Wintering birds begin to arrive in early November and are present through mid-March, but a few birds linger into late April and exceptionally to early May. **Taxonomy:** Monotypic.

RED CROSSBILL *Loxia curvirostra* Linnaeus

Rare resident in the Davis and Guadalupe Mountains and irregular winter visitor in the northern two-thirds of the state. There have been several small invasions that have pushed into the northern two-thirds of the state. A major invasion during the winter of 1996–97 produced many first local records, including in the Pineywoods (Schaefer 1998), where Red Crossbills were found in 10 counties. Following that winter incursion, Red Crossbills remained through the summer in many locations and were strongly suspected of breeding in the Pineywoods and Taylor County. Another major invasion occurred during the winter of 2011–12, with birds found in various areas of the High Plains and Trans-Pecos and flocks located as far east as Hood, Kaufman, and Wilbarger Counties. **Timing of occurrence:** Due to their irruptive nature the timing of occurrence varies greatly. Fall migrants have arrived from early August to mid-October, with later arrival dates being more typical. Wintering birds depart from mid-February to early May. **Taxonomy:** Groth (1993) identified six types of Red Crossbill in North America based on call notes and morphology. That number has swelled to at least 10. How these types will be treated taxonomically is unresolved. Based on audio recordings, the population that resides in the Davis Mountains is Type 2. Type 2 Red Crossbills were also recorded during the fall and winter of 2012–13 in Brewster, El Paso, Jeff Davis, and Lubbock Counties. There are specimens of three subspecies under the traditional classification that have been collected in Texas, but the validity of these taxa is open to question.

L. c. minor (Brehm)

Specimens have been reported from El Paso and McLennan Counties. This subspecies is often associated with Type 3 birds, but that call type may also be associated with other named subspecies.

L. c. bendirei Ridgway

Rare resident of the higher elevations of the Guadalupe and Davis Mountains. This subspecies is associated with both Type 5 and Type 2 birds and thus is causing uncertainty in taxonomic placement.

L. c. benti Griscom

Wauer (1985) reported a specimen from Brewster County. This subspecies is associated with both Type 2 and Type 5 birds and thus is causing uncertainty in taxonomic placement.

WHITE-WINGED CROSSBILL *Loxia leucoptera* Gmelin

Casual. There are nine documented records for the state. White-winged Crossbills are rare residents in the southern Rocky Mountains to northern New Mexico, although the birds occurring in Texas probably originate from more northern populations. A substantial southward movement of White-winged Crossbills occurred during the winter of 2001–2. Many individuals reached the southern Great Plains, including the Texas Panhandle, where single individuals were documented from four separate locations. The first record for the state was a male in Lubbock, Lubbock County, from 28 December 1975 to 8 March 1976 (TPRF 94). There are three records away from the Panhandle and South Plains, with single birds documented in Bastrop, Bandera, and Parker Counties. **Timing of occurrence:** The documented records are between 1 December and 16 March. **Taxonomy:** The subspecies that occurs in Texas is *L.l. leucoptera* Gmelin.

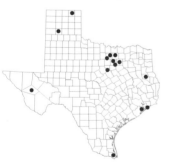

COMMON REDPOLL *Acanthis flammea* (Linnaeus)

Casual. Texas has 14 documented records of this species. The first state record involved six birds at Buffalo Lake NWR, Randall County, from 25 November 1965 to 16 January 1966 (TBRC 1978-2). Ten of the records have occurred during the winter and are scattered across the northern half of the state, with the southernmost occurrences in Jeff Davis and Nacogdoches Counties. Amazingly, there are three records in late May to early June from the Coastal Prairies, including an adult at Laguna Vista, Cameron County, from 28 to 30 May 2002 (TBRC 2002-66; TPRF 2023). Five reports were made prior to the development of the Review List in 1988. **Timing of occurrence:** The winter records are from 25 November to 1 March, and the spring/ summer records are from 28 May to 19 June. **Taxonomy:** The subspecies that occurs in Texas is *A. f. flammea* (Linnaeus).

PINE SISKIN *Spinus pinus* (Wilson)

Common to abundant migrant and winter visitor through-out the northern two-thirds of the state, becoming uncom-mon to rare farther south. The occurrence of Pine Siskins in any given area is unpredictable. They can be abundant one winter yet virtually absent the next. In general, they are more regular in occurrence in the northern third of the state, the Pineywoods, and the Trans-Pecos. There are iso-lated breeding records from the Trans-Pecos and Pan-handle and one from the South Plains. In most cases these nesting records are from May and early June. **Timing of occurrence:** Fall migrants begin to arrive in early October, but the primary migration occurs between late October and early December. Spring migrants have been recorded from late March through mid-May, with some lingering into early June. **Taxonomy:** The subspecies that occurs in Texas is *S. p. pinus* (Wilson).

LESSER GOLDFINCH *Spinus psaltria* (Say)

Uncommon to locally abundant migrant and summer resi-dent in the Trans-Pecos and Edwards Plateau south to the Lower Rio Grande Valley. Lesser Goldfinches are locally uncommon to rare migrants and summer residents north-ward through the Rolling Plains to the southeastern Pan-handle. They are rare to casual migrants and summer visi-tors eastward to the northern Pineywoods and to the upper coast. Most individuals retreat from the northern half of the breeding range during winter, but a few individuals may remain. Lesser Goldfinches are locally uncommon to rare winter residents in the southern Trans-Pecos and east-ward through the South Texas Brush Country and south-ern Coastal Prairies. **Timing of occurrence:** Spring migrants start to arrive in the northern portions of the range in mid-March, and the breeding population has arrived by mid-May. Fall migrants have been recorded from mid-September through mid-November. **Taxonomy:** Two subspecies occur in the state.

S. p. hesperophila (Oberholser)
> Status uncertain. Fully adult males with green backs have been reported from El Paso County. More research is needed to determine if those birds belong to this taxon.

S. p. psaltria (Say)
    As described in main species account.

TBRC Review Species

LAWRENCE'S GOLDFINCH *Spinus lawrencei* Cassin

Very rare fall and winter visitor to the Trans-Pecos. Texas has 19 documented records of this western finch. The occurrence of Lawrence's Goldfinch in Texas has been primarily linked to irruptive events. One of these incursions took place in the 1950s; unfortunately, there is no supporting documentation. The first documented record for the state was a male photographed at Hueco Tanks State Historical Park, El Paso County, on 7 December 1984 (TBRC 1988-101; TPRF 533). There are 11 records from the Trans-Pecos that were part of an invasion during the winter of 1996–97, and some of these records involved multiple birds, including a flock of up to 23 in El Paso County during January 1997. There is one very unexpected summer record of an adult male at Guadalupe Mountains NP, Culberson County, from 5 to 7 June 2002 (TBRC 2002-73; TPRF 2023). Equally unexpected was one at Rockport, Aransas County, from 21 to 22 February 2005 (TBRC 2005–52; TPRF 2300). Eleven reports were made prior to the development of the Review List in 1988. **Timing of occurrence:** The winter records have all been between 12 October and 27 March. **Taxonomy:** Monotypic.

AMERICAN GOLDFINCH *Spinus tristis* (Linnaeus)

Uncommon to abundant migrant and winter resident throughout the state. The occurrence of American Goldfinch varies based on food availability; consequently, there can be considerable fluctuation in the number of birds present in a given area from one year to the next. This species is a very rare and local summer resident in portions of northeast Texas and in Hemphill County in the northeastern Panhandle. **Timing of occurrence:** Fall migrants begin to arrive in early September but are not common until late October or early November. Spring migrants have been recorded from early March through mid-May, and some have lingered as late as early June. **Taxonomy:** Two subspecies occur in the state.

S. t. pallidus (Mearns)

Uncommon to abundant migrant and winter resident in the western half of the state east to the western Oaks and Prairies region.

S. t. tristis (Linnaeus)

Uncommon to abundant migrant and winter resident to all regions. Very rare and local summer resident as described in the species account.

TBRC Review Species

EVENING GROSBEAK

Coccothraustes vespertinus (Cooper)

Casual to very rare winter visitor to the state. Evening Grosbeak was formerly a rare and irruptive winter visitor to the northern half of the state and the Trans-Pecos. There were five major and two minor invasions into the Pineywoods between 1968 and 1987. Since 1968 there have also been several minor incursions into the Panhandle and Trans-Pecos, the last during the winter of 1996–97. Evening Grosbeak was added to the Review List in 2008, and there have been only six records since that time (see map). **Timing of occurrence**: Historically they have appeared in Texas as early as late August and lingered as late as mid-June. There are two July reports, including two from the Guadalupe Mountains on 10 July 1988 and one at Amarillo, Randall County, from 27 to 28 July 2000. **Taxonomy**: Both of the subspecies found across the northern United States and Canada have been reported from the state.

C. v. vespertinus (Cooper)

Specimens associated with the influxes into the eastern half of the state have been assigned to this subspecies. Pulich (1971) details an invasion in 1968–69 in which five specimens taken in Gregg County have been assigned to this subspecies. Two specimens from Kerr County have also been identified as this subspecies.

C. v. brooksi Grinnell

Associated with influxes to the western half of the state, with specimens reported from the Guadalupe Mountains (Burleigh and Lowery 1940) and from Kerr County (Buechner 1946–47). A female at San Saba, San Saba County, from 25 February to 8 March 2010 (TBRC 2010–21; TPRF 2801) cannot be assigned to either subspecies.

Family Passeridae: Old World Sparrows

HOUSE SPARROW *Passer domesticus* (Linnaeus)

Common to locally abundant resident in urban areas across the state. House Sparrows are common to locally rare in rural areas and are generally found near human habitations. They are native to Europe and were introduced to the United States in several releases during the early 1850s in Brooklyn, New York. House Sparrows were released in Galveston, Galveston County, beginning in 1867 and had established an expanding population by 1880 (Casto 2004). By 1905 this species was reported from isolated locations throughout the state. These populations contributed to the rapid expansion across North America. House Sparrows are one of the most successful introductions to the continent. **Taxonomy**: The original releases were of birds from England that belong to the subspecies *P. d. domesticus* (Linnaeus).

APPENDIX A:
PRESUMPTIVE SPECIES LIST

For a species to be accepted on the Texas list, at least one record must be supported by specimen, photograph, or tape-recording. However, the Texas Bird Records Committee may accept sight records for species not currently on the state list that are supported by written details only. Such species are added to this Presumptive Species List, so named because it can be safely presumed that they have, in all probability, occurred in Texas. If subsequent records are documented by specimen, photograph, or tape-recording, the species is then moved to the state list.

murre species (*Uria* sp.)
There is a single sight record of a murre for Texas. An apparent adult murre was observed at Lake O' the Pines, Marion County, from 19 to 20 March 1994 (TBRC 1994-70). The bird was initially identified as a Common Murre (*U. aalge*). The committee accepted the sight record only as murre species because Thick-billed Murre (*U. lomvia*) could not be ruled out. Detailed descriptions of the bird did eliminate Long-billed Murrelet (*Brachyramphus perdix*) as an alternative identification, however.

Razorbill *Alca torda* Linnaeus
There is one sight record of a Razorbill from Texas. A winter-plumaged adult was observed at length at the West Flower Garden National Marine Sanctuary approximately 115 miles south of Sabine Pass, Jefferson County, on 16 March 2013 (TBRC 2013-26). There was a major movement of Razorbills into the eastern Gulf of Mexico during the winter of 2012–13, with individuals well documented in Alabama and Louisiana, as well as this sighting in Texas waters.

White-crowned Pigeon *Patagioenas leucocephala* Linnaeus
There is one accepted sight record of White-crowned Pigeon for Texas. A single adult was observed on Green Island, Cameron County, on 24 June and again on 2 July 1989 (TBRC 1989-186). The closest populations of this spe-

cies to Texas are found in southern Florida and along the eastern coast of the Yucatán Peninsula, Mexico.

Black Swift *Cypseloides niger* (Gmelin)
There is one accepted sight record of Black Swift in Texas. A small group of Black Swifts was observed in the Franklin Mountains, El Paso County, on 22 August 1985 (TBRC 1994-172). Black Swift is probably a very rare, but regular, fall migrant through the Trans-Pecos. This species has been reported on several other occasions. The inherent difficulty in photographing swifts makes obtaining documentation needed to add this species to the main list even more difficult. Individuals fitted with geolocators reportedly passed through the state on their migration from Colorado to South America (Beason et al. 2012). The accuracy of those data is unknown at this time, as is whether there is any additional documentation of this potential occurrence.

Crescent-chested Warbler *Oreothlypis superciliosa* (Hartlaub)
There is a single accepted sight record for this species for Texas. A singing male was discovered in the Chisos Mountains of Big Bend NP, Brewster County, on 2 June 1993 (TBRC 1993-90). Crescent-chested Warbler is found as far north as central Nuevo León and western Chihuahua in Mexico.

APPENDIX B:
NON-ACCEPTED SPECIES

There are published reports of a number of species for Texas that have not, for a variety of reasons, been accepted by the Texas Bird Records Committee. The following list includes species that have been frequently attributed to Texas.

Sharp-tailed Grouse *Tympanuchus phasianellus* (Linnaeus)
Oberholser (1974) listed the Sharp-tailed Grouse as hypothetical based on reports of the species from the northwestern Panhandle from the nineteenth century. This species had apparently disappeared from Texas by 1906. Sharp-tailed Grouse was formerly a resident in the extreme western Panhandle of Oklahoma (Sutton 1967), and Ligon (1961) reported this species from northeastern New Mexico in the vicinity of Raton, Colfax County. Today, none of these populations are extant. Apparently, no specimens collected in Texas of Sharp-tailed Grouse were ever preserved.

Wilson's Storm-Petrel *Oceanites oceanicus* (Kuhl)
Formerly included on the official state list in good standing (Lockwood and Freeman 2004). The photographs that were the basis of the single previously accepted record were reexamined in 2010, and Band-rumped Storm-Petrel could not be eliminated as an alternative identification (Lockwood 2010). There are eight additional reports of this species without supporting documentation, some of which are of multiple birds and date as far back as 1912.

White-tailed Tropicbird *Phaethon lepturus* Daudin
Prior to the first documented record of Red-billed Tropicbird, this species was thought to be the tropicbird most likely to occur in the state.

Scarlet Ibis *Eudocimus ruber* (Linnaeus)
Oberholser (1974) mentions two specimens collected from the state, although neither was apparently saved. A Scarlet Ibis was photographed at Green Island, Cameron County, in June 1972. Further investigation determined that a Scar-

let Ibis had escaped earlier that spring from the Gladys Porter Zoo in Brownsville.

Limpkin *Aramus guarauna* (Linnaeus)
There are three reports of Limpkin from Texas (Oberholser 1974). One of these refers to a specimen reportedly collected at Brownsville, Cameron County, now housed at the American Museum of Natural History (*AMNH 79,775). Colonel Field and E. C. Greenwood supposedly collected this specimen on 23 May 1889. Recent investigations have determined that Greenwood was on a collecting trip to the marshes near Tampico, Mexico, during the spring of 1889. While on this expedition, Greenwood contracted a fever and returned to Brownsville, where he died later that summer (S. Casto, pers. comm.). It seems clear that Field and Greenwood were not together during the spring of 1889. They were business partners selling bird and mammal specimens, which may account for the label data listing them both as the collectors of the Limpkin specimen. There is a distinct possibility that the bird was collected in Tampico. For this reason, the TBRC has not included Limpkin on the official state list.

Yellow-footed Gull *Larus livens* Dwight
Formerly included on the official state list in good standing (Lockwood and Freeman 2004). A second-year bird found near Surfside, Brazoria County, in 1998 was thought to pertain to this species (Weeks and Patten 2000). However, the publication of analysis of plumage variation of Herring × Kelp Gulls on the Chandeleur Islands off the coast of Louisiana (Dittmann and Cardiff 2005) caused the record to be reexamined.

Vaux's Swift *Chaetura vauxi* (Townsend)
Quite likely to occur as a very rare fall or winter vagrant (e.g., specimens from Louisiana), but identification difficulties continue to cloud all reports. There are more than 25 sight reports to date, many with written notes. The difficulty in obtaining documentation hampers acceptance.

Antillean Crested Hummingbird
Orthorhynchus cristatus (Linnaeus)
The origin of the single specimen from Galveston County, housed at the American Museum of Natural History, is dubious (Pulich 1968; Oberholser 1974).

Rufous-tailed Hummingbird *Amazilia tzacatl* (De la Llave)
Oberholser (1974) reported that J. C. Merrill and Robert Ridgway identified two separate Rufous-tailed Hummingbirds during June and July 1876 at Fort Brown, Cameron County. Neither bird was preserved, nor is there compelling evidence that would eliminate Buff-bellied Hummingbird.

Saltmarsh Sparrow *Ammodramus caudacutus* (Gmelin)
Oberholser (1974) reported two specimens of this species from Texas, but subsequent examination of both specimens showed them to be Nelson's Sparrows.

APPENDIX C: EXOTICS AND BIRDS OF UNCERTAIN PROVENANCE

This short list includes species that are not native to Texas and for which no established populations exist. This group can be divided into two categories. The first category consists of species that originate from individuals intentionally released into the wild and includes Mute Swan, African Collared-Dove, Red-vented Bulbul, Orange Bishop, and Nutmeg Mannikin. The second group includes several psittacids that have been encountered in the Lower Rio Grande Valley. There are occasional nests of these parrots found, but there does not appear to be sustainable reproduction for the population at this time, and only a small number of individuals are present. There is a remote possibility that some of these species could occur as vagrants from northeastern Mexico.

Mute Swan *Cygnus olor* (Gmelin)
Feral Mute Swans can be found in urban and other areas throughout the state. Some of these individuals are breeding and free-flying. They are not considered to be established exotics in Texas and therefore are not included on the official state list. Any swan discovered in Texas, however, should be carefully examined to eliminate the possibility of Mute Swan. The largest concentrations of these birds are in Travis County and the eastern Hill Country.

African Collared-Dove *Streptopelia roseogrisea (Linnaeus)*
African Collared-Doves, formerly known as Ringed Turtle Dove (*S. risoria*), have been replaced by the expansion of Eurasian Collared-Doves into urban habitats within Texas. In areas where Eurasian Collared-Doves are common, they outcompete smaller doves and are known to displace Mourning Doves in urban habitats. The last populations of African Collared-Doves in Texas were found on the Upper Texas Coast. All recent records appear to pertain to local escapees.

Lilac-crowned Parrot *Amazona finschi* (Sclater)
There is a small population in El Paso, El Paso County, consisting of approximately 40 individuals. They are sus-

pected of nesting, but there has been no confirmation. Lilac-crowned Parrot also occurs in small numbers in the Lower Rio Grande Valley, although some reports may refer to female Red-crowned Parrots. This species is normally found in the thorn-forests of western Mexico.

Red-lored Parrot *Amazona autumnalis* (Linnaeus)
Red-lored Parrots are usually found in flocks with other parrots in the Lower Rio Grande Valley. Like some of the other exotic psittacids, they are generally found singly, although as many as 15 have been reported from a single winter flock. This species occurs in northeastern Mexico but is more narrowly confined to heavy tropical forests than is the Red-crowned Parrot.

Yellow-headed Parrot *Amazona oratrix* Ridgway
Yellow-headed Parrots occur in northeastern Mexico within 150 miles of the Rio Grande. This species is regularly encountered and occasionally seen in small flocks in the Lower Rio Grande Valley.

Red-vented Bulbul *Pycnonotus cafer* (Linnaeus)
Red-vented Bulbul was first reported in Texas in 1998, and the first nest was discovered in 1999. Since then a small population has persisted in Houston, Harris County. The only established population found in the United States of this Asian species is on Oahu, Hawaii. Whether the population is large enough to be self-sustaining is unknown, but the persistent occurrence within urban habitats in Houston suggests that these birds are reproducing and that the population may be sufficient to become established.

Orange Bishop *Euplectes franciscanus* (Isert)
The alternate-plumaged male Orange Bishop is readily identifiable and attracts attention with its bright plumage. Basic-plumaged males and females are harder to identify and have not been included in North American field guides until recently. The first reports of these birds in Texas began in 2002, and since then a small population has been found in Harris County. They have been suspected of breeding since 2005. Orange Bishops are found in sub-Saharan Africa and have been introduced to Hawaii. There has been a population near Los Angeles, California, since the 1980s.

Nutmeg Mannikin *Lonchura punctulata* (Linnaeus)

The first reports that could not be easily dismissed as a local escapee came from Harris County in 2004. Since then a sizable population has been found in southern and western Harris County, with reports from Waller and Orange Counties as well. There have been reports of suspected breeding in several locations since 2005. This species seems to have a high potential to become established. Nutmeg Mannikins are native to Southeast Asia. They are established in Hawaii and are also present in large numbers in coastal southern California.

APPENDIX D:
LIST OF REVIEW SPECIES

REVIEW LIST A

The list below includes species that have occurred four or fewer times per year in Texas over a 10-year average. The TBRC requests documentation for review for these species, as well as any bird not yet accepted on the Texas State List.

Brant
Trumpeter Swan
Eurasian Wigeon
American Black Duck
White-cheeked Pintail
Garganey
King Eider
Common Eider
Harlequin Duck
Barrow's Goldeneye
Masked Duck
Yellow-billed Loon
Red-necked Grebe
American Flamingo
Yellow-nosed Albatross
Black-capped Petrel
Stejneger's Petrel
White-chinned Petrel
Great Shearwater
Sooty Shearwater
Manx Shearwater
Leach's Storm-Petrel
Red-billed Tropicbird
Jabiru
Blue-footed Booby
Brown Booby
Red-footed Booby
Bare-throated Tiger-Heron
Snail Kite
Double-toothed Kite
Northern Goshawk
Crane Hawk
Roadside Hawk
Short-tailed Hawk

Paint-billed Crake
Spotted Rail
Double-striped Thick-knee
Pacific Golden-Plover
Collared Plover
Northern Jacana
Wandering Tattler
Spotted Redshank
Eskimo Curlew
Black-tailed Godwit
Surfbird
Sharp-tailed Sandpiper
Red-necked Stint
Curlew Sandpiper
Purple Sandpiper
Ruff
Red Phalarope
South Polar Skua
Long-tailed Jaeger
Black-legged Kittiwake
Black-headed Gull
Black-tailed Gull
Heermann's Gull
Mew Gull
Western Gull
Yellow-legged Gull
Iceland Gull
Slaty-backed Gull
Glaucous-winged Gull
Great Black-backed Gull
Kelp Gull
Brown Noddy
Black Noddy
Roseate Tern

Arctic Tern
Elegant Tern
Ruddy Ground-Dove
Ruddy Quail-Dove
Dark-billed Cuckoo
Mangrove Cuckoo
Snowy Owl
Northern Pygmy-Owl
Mottled Owl
Stygian Owl
Northern Saw-whet Owl
White-collared Swift
Green Violetear
Green-breasted Mango
Costa's Hummingbird
Berylline Hummingbird
Violet-crowned Hummingbird
White-eared Hummingbird
Elegant Trogon
Amazon Kingfisher
Red-breasted Sapsucker
Ivory-billed Woodpecker
Collared Forest-Falcon
Gyrfalcon
Greenish Elaenia
White-crested Elaenia
Tufted Flycatcher
Greater Pewee
Buff-breasted Flycatcher
Nutting's Flycatcher
Social Flycatcher
Sulphur-bellied Flycatcher
Piratic Flycatcher
Thick-billed Kingbird
Gray Kingbird
Fork-tailed Flycatcher
Masked Tityra
Rose-throated Becard
Black-whiskered Vireo
Yucatan Vireo
Brown Jay
Pinyon Jay

Clark's Nutcracker
Black-billed Magpie
Tamaulipas Crow
Gray-breasted Martin
Black-capped Chickadee
American Dipper
Northern Wheatear
Orange-billed Nightingale-Thrush
Black-headed Nightingale-Thrush
White-throated Thrush
Rufous-backed Robin
Varied Thrush
Aztec Thrush
Blue Mockingbird
Black Catbird
Bohemian Waxwing
Gray Silky-flycatcher
Olive Warbler
Snow Bunting
Connecticut Warbler
Gray-crowned Yellowthroat
Fan-tailed Warbler
Rufous-capped Warbler
Golden-crowned Warbler
Slate-throated Redstart
Yellow-faced Grassquit
Golden-crowned Sparrow
Yellow-eyed Junco
Flame-colored Tanager
Crimson-collared Grosbeak
Blue Bunting
Shiny Cowbird
Black-vented Oriole
Streak-backed Oriole
Pine Grosbeak
White-winged Crossbill
Common Redpoll
Lawrence's Goldfinch
Evening Grosbeak

REVIEW LIST B

The following subspecies are under special study by the TBRC, which will formally review the records.

Green-winged (Eurasian) Teal

(Vega) Herring Gull

(Lawrence's) Dusky-capped
  Flycatcher

(Slate-colored) Fox Sparrow

Dark-eyed (White-winged)
  Junco

Orchard (Fuertes's) Oriole

SELECTED REFERENCES

Adams, M. T., and K. B. Bryan. 1999. Botteri's Sparrow in Trans-Pecos, Texas. *Tex. Birds* 1:6–13.

Albers, R. P., and F. R. Gehlbach. 1990. Choices of feeding habitat by relict Montezuma Quail in Central Texas. *Wilson Bull.* 102:300–308.

American Birding Association. 2011. *ABA checklist: Birds of the continental United States and Canada.* 7th ed. Colorado Springs: American Birding Association.

American Ornithologists' Union. 1957. *A.O.U. check-list of North American birds.* 5th ed. Baltimore: American Ornithologists' Union.

———. 1983. *A.O.U. check-list of North American birds.* 6th ed. Lawrence, Kans.: Allen Press.

———. 1998. *A.O.U. check-list of North American birds.* 7th ed. Washington, D.C.: American Ornithologists' Union.

Arnold, K. A. 1968. Olivaceous Flycatcher in the Davis Mountains of Texas. *Bull. Tex. Ornith. Soc.* 2:28.

———. 1972. Crested titmice in Cottle and Foard Counties. *Bull. Tex. Ornith. Soc.* 5:23.

———. 1975. First record of the Greater Shearwater from the Gulf of Mexico. *Auk* 92:394–95.

———. 1978a. First United States record of Paint-billed Crake (*Neocrex erythrops*). *Auk* 95:745–46.

———. 1978b. A Jabiru (*Jabiru mycteria*) specimen from Texas. *Auk* 95:611–12.

———. 1980. Rufous-capped Warbler and White-collared Seedeater from Webb County, Texas. *Bull. Tex. Ornith. Soc.* 13:27.

———. 1983. New subspecies of Henslow's Sparrow (*Ammodramus henslowii*). *Auk* 100:504–5.

———. 1984. *Check-list of the birds of Texas.* Waco: Texas Ornithological Society.

———. 1994. First specimen of Clark's Grebe for Texas: An environmental casualty. *Bull. Tex. Ornith. Soc.* 27:26–28.

Arnold, K. A., and N. C. Garza Jr. 1998. Populations and habitat requirements of breeding Henslow's Sparrow in Harris County, Texas. *Bull. Tex. Ornith. Soc.* 31:42–49.

Arnold, K. A., and J. C. Henderson. 1973. First specimen of Arctic Loon from Texas. *Auk* 90:420–21.

Arnold, K. A., and E. A. Kutac, eds. 1974. *Check-list of the birds of Texas.* Waco: Texas Ornithological Society.

Arnold, K. A., and B. D. Marks. 2009. Recent Texas specimens of Red-footed and Brown Boobies. *Bull. Tex. Ornith. Soc.* 42:95–96.

Arnold, K. A., B. D. Marks, and M. Gustafson. 2011. First specimen of Flame-colored Tanager (*Piranga bidentata*) for the United States. *Bull. Tex. Ornith. Soc.* 44:97–99.

Arvin, J., C. Cottam, and G. Unland. 1975. Mexican Crow invades South Texas. *Auk* 92:387–90.

Arvin, J. C. 1980. The Golden-crowned Warbler: An 88-year-old "new" species for the avifauna of the United States. *Birding* 12:10–11.

———. 2005. *Birds of the South Texas brushlands: A field checklist.* Austin: Texas Parks and Wildlife Department.

Arvin, J. C., and M. W. Lockwood. 2006. First photographically documented record of Social Flycatcher (*Myiozetetes similis*) for the United States. *N. Am. Birds* 60:180–81.

Attwater, H. P. 1887. Nesting habits of Texas birds. *Ornithologist and Oologist* 12:103–5, 123–25.

———. 1892. List of birds observed in the vicinity of San Antonio, Bexar County, Texas. *Auk* 9:229–38, 337–45.

Bailey, V. 1905. *Biological survey of Texas.* Washington, D.C.: United States Department of Agriculture.

Baker, J. M., E. López-Medrano, A. G. Navarro-Sigüenza, O. R. Rojas-Soto, and K. E. Omland. 2003. Recent speciation in the Orchard Oriole group: Divergence of *Icterus spurius spurius* and *Icterus spurius fuertesi. Auk* 120:848–59.

Banks, R. C. 1988. Geographic variation in the Yellow-billed Cuckoo. *Condor* 90:473–77.

———. 2011. Taxonomy of Greater White-fronted Geese (Aves: Anatidae). *Proc. Biol. Soc. Wash.* 124: 226–33.

Banks, R. C., C. Cicero, J. L. Dunn, A. W. Kratter, P. C. Rasmussen, J. V. Remsen Jr., J. D. Rising, and D. F. Stotz. 2002. Forty-third supplement to the American Ornithologists' Union check-list of North American birds. *Auk* 119:897–906.

Banks, R. C., and R. Hole Jr. 1991. Taxonomic review of the Mangrove Cuckoo, *Coccyzus minor* (Gmelin). *Caribbean J. Sci.* 27:54–62.

Barker, F. K., A. J. Vandergon, and S. M. Lanyon. 2008. Assessment of species limits among Yellow-breasted Meadowlarks (*Sturnella* spp.) using mitochondrial and sex-linked markers. *Auk* 125:869–79.

Beason, J. P., C. Gunn, K. M. Potter, R. A. Sparks, and J. W. Fox. 2012. The Northern Black Swift: Migration path and wintering area revealed. *Wilson J. Ornith.* 124:1–8.

Beavers, R. A. 1977. First specimen of Allen's Hummingbird, *Selasphorus sasin* (Trochilidae) from Texas. *Southwest. Nat.* 21:285.

Bent, A. C. 1940. *Life histories of North American birds, cuckoos, goatsuckers, hummingbirds and their allies.* United States National Museum Bulletin 176. Washington, D.C.: US Government Printing Office.

Biaggi, V., Jr. 1960. The birds of Culberson County, Texas, with notes on ecological aspects. Part II. Annotated list of the birds (continued). Newsletter, *Tex. Ornith. Soc.* 8 (10): 1–20.

Bielefeld, R. R., M. G. Brasher, T. E. Moorman, and P. N. Gray. 2010. Mottled Duck (*Anas fulvigula*). In *The birds of North America online*, edited by A. Poole. Ithaca, N.Y.: Cornell Lab of Ornithology. Available at http://bna.birds .cornell.edu/bna/species/081.

Blacklock, G. W., and J. Peabody. 1983. Two specimen records of Leach's Storm-Petrel for Texas. *Bull. Tex. Ornith. Soc.* 16:34.

Blankenship, T. L., and J. T. Anderson. 1993. A large concentration of Masked Duck (*Oxyura dominica*) on the Welder Wildlife Refuge, San Patricio County, Texas. *Bull. Tex. Ornith. Soc.* 26:19–21.

Bleitz, D. 1962. Photographing the Eskimo Curlew. *Western Bird Bander* 37 (3): 42–45.

Bolen, E. G. 1987. A specimen record of the Barnacle Goose in Texas. *Southwest. Nat.* 32:506–7.

Bolton, M. 2007. Playback experiments indicate absence of vocal recognition among temporally and geographically separated populations of Madeiran Storm-Petrels *Oceanodroma castro. Ibis* 149:255–63.

Braun, M. J., and V. Emanuel. 1982. Records of Crimson-collared Grosbeak (*Rhodothraupis celaeno*) in Texas. *Auk* 99:787.

Brewer, D. 2001. *Wrens, dippers and thrashers.* New Haven, Conn.: Yale University Press.

Brooks, A. 1933. Some notes on the birds of Brownsville, Texas. *Auk* 52:59–63.

Brown, N. C. 1882a. Description of a new race of *Peucaea ruficeps* from Texas. *Bull. Nutt. Ornith. Club* 7:26.

———. 1882b. A reconnaissance in southwestern Texas. *Bull. Nutt. Ornith. Club* 7:33–42.

———. 1884. A second season in Texas. *Auk* 1:120–24.

Brush, T. 1998. Recent nesting and current status of Red-billed Pigeon along the Lower Rio Grande in southern Texas. *Bull. Tex. Ornith. Soc.* 31:22–26.

———. 1999a. Current status of Northern Beardless-Tyrannulet and Tropical Parula in Bentsen–Rio Grande Valley State

Park and Santa Ana National Wildlife Refuge, southern Texas. *Bull. Tex. Ornith. Soc.* 32:2–12.

———. 1999b. The Hook-billed Kite: A reclusive, snail eating raptor of the Lower Rio Grande Valley. *Tex. Birds* 1 (2): 26–32.

———. 2000. First nesting record of Blue Jay (*Cyanocitta cristata*) in Hidalgo County. *Bull. Tex. Ornith. Soc.* 33:35–36.

———. 2003. First nesting attempt of Yellow-faced Grassquit (*Tiaris olivacea*) in the United States. *Bull. Tex. Ornith. Soc.* 46:237–38.

———. 2005. *Nesting birds of a tropical frontier: The Lower Rio Grande Valley of Texas.* College Station: Texas A&M University Press.

———. 2008. Additions to the breeding avifauna of the Lower Rio Grande Valley of Texas. *Stud. Avian Biol.* 37:11–19.

Brush, T., and M. H. Conway. 2012. Range expansion of Clay-colored Thrush (*Turdus grayi*) in Texas. *N. Am. Birds* 65:700–703.

Brush, T., and J. C. Eitniear. 2002. Status and recent nesting of Muscovy Duck (*Cairina moschata*) in the Rio Grande Valley, Texas. *Bull. Tex. Ornith. Soc.* 35:12–14.

Bryan, K., T. Gallucci, G. Lasley, M. Lockwood, and D. H. Riskind. 2003. *A checklist of Texas birds.* Austin: Natural Resources Program, Texas Parks and Wildlife Department.

Bryan, K., T. Gallucci, and R. Moldenhauer. 1978. First record of the Snow Bunting for Texas. *Am. Birds* 32:1070.

Bryan, K. B. 1999. *Birds of Big Bend Ranch and vicinity: A field checklist.* Austin: Natural Resources Program, Texas Parks and Wildlife Department.

———. 2002. *Birds of the Trans-Pecos: A field checklist.* Austin: Natural Resources Program, Texas Parks and Wildlife Department.

Bryan, K. B., and J. Karges. 2001. Recent changes to the Davis Mountains avifauna. *Tex. Birds* 3 (1): 41–53.

Bryan, K. B., and M. W. Lockwood. 2000. Gray Vireo in Texas. *Tex. Birds* 2 (2): 18–24.

Buechner, H. K. 1946–47. Birds of Kerr County, Texas. *Trans. Kans. Acad. Sci.* 49:357–62.

Burleigh, T. D. 1939. Alta Mira Oriole in Texas: An addition to the A. O. U. "check-list." *Auk* 56:87–88.

Burleigh, T. D., and G. H. Lowery Jr. 1940. Birds of the Guadalupe Mountain region of western Texas. *Occas. Papers Mus. Zool., La. State Univ.,* No. 8.

Burt, D. B., D. Burt, T. C. Maxwell, and D. G. Tarter. 1987. Clapper Rail (*Rallus longirostris*) in west-central Texas. *Tex. J. Sci.* 39:378.

Cahn, A. R. 1921. Summer birds in the vicinity of Lake Caddo, Harrison County, Texas. *Wilson Bull.* 33:165–76.

———. 1922. Notes on the summer avifauna of Bird Island, Texas. *Condor* 24:169–80.

Carpenter, E. 2013. Texas Bird Records Committee report for 2011. *Bull. Tex. Ornith. Soc.* 45:33–39.

Carroll, J. J. 1900. Notes on the birds of Refugio County, Texas. *Auk* 17:337–48.

Casto, S., and H. W. Garner. 1969. Photographic evidence for the occurrence of the Dipper in West Texas. *Bull. Tex. Ornith. Soc.* 3:29.

Casto, S. D. 2001. Additional records of the Passenger Pigeon in Texas. *Bull. Tex. Ornith. Soc.* 34:5–16.

———. 2002. The early history of ornithology in Texas. *Occas. Publ. Tex. Ornith. Soc.*, No. 4.

———. 2004. The House Sparrow in Texas, 1867–1905. *Bull. Tex. Ornith. Soc.* 37:3–8.

Chapman, B. P., P. A. Buckley, and F. G. Buckley. 1979. First photographic record of Greater Flamingo in Texas. *Bull. Tex. Ornith. Soc.* 12:20–21.

Chesser, R. T., R. C. Banks, F. K. Baker, C. Cicero, J. L. Dunn, A. W. Kratter, I. J. Lovette, P. C. Rasmussen, J. V. Remsen Jr., J. D. Rising, D. F. Stotz, and K. Winkler. 2010. Fifty-first supplement to the American Ornithologists' Union *Check-list of North American birds. Auk* 127:726–44.

———. 2011. Fifty-second supplement to the American Ornithologists' Union *Check-list of North American birds. Auk* 128:600–613.

———. 2012. Fifty-third supplement to the American Ornithologists' Union *Check-list of North American birds. Auk* 129:573–88.

Childs, J. L. 1909. The Ruff in Texas. *Warbler* 5:31.

Clark, C. 1982. Jabiru in the United States. *Birding* 14:8–9.

Clark, W. S. 2004. First dark-morph Hook-billed Kite fledged in the United States. *N. Am. Birds* 58:170.

Clements, J. F., T. S. Schulenberg, M. J. Iliff, B. L. Sullivan, C. L. Wood, and D. Roberson. 2012. *The Clements checklist of birds of the world: Version 6.7.* Available at http://www.birds.cornell.edu/clementschecklist/down loadable-clements-checklist.

Clum, N. J., and T. J. Cade. 1994. Gyrfalcon. In *Birds of North America*, No. 114, edited by A. Poole and F. Gill. Washington, D.C.: Academy of Natural Sciences, Philadelphia, and American Ornithologists' Union.

Collar, N. 2005. Family Turdidae (Thrushes): In *Handbook of the birds of the world*, vol. 10. Barcelona: Lynx Edicions.

Conner, R. N., D. Saenz, and D. C. Rudolph. 2006. Population trends in Red-cockaded Woodpeckers in Texas. *Bull. Tex. Ornith. Soc.* 39:42–48.

Conner, R. N., C. E. Shackelford, R. R. Schaefer, and D. Saenz. 2005. The effects of fire suppression on Bachman's Sparrows in upland pine forests of eastern Texas. *Bull. Tex. Ornith. Soc.* 38:6–11.

Cooke, M. T. 1925. *Spread of the European Starling in North America.* U.S. Dept. Agr. Circ. 336.

Cooke, W. W. 1885. A re-discovery for Texas. *Ornithologist and Oologist* 10:172–73.

Cooksey, M. 1998. A pre-1996 North American record of Stygian Owl. *Field Notes* 52:265–66.

Cottam, C., E. Bolen, and R. Zink. 1975. Sabine's Gull on South Texas coast. *Southwest. Nat.* 20:134–35.

Dalquest, W. W., and L. D. Lewis. 1955. Whistling Swan and Snowy Owl in Texas. *Condor* 57:243.

D'Anna, W., D. DiTommaso, B. Potter, J. Skelly, and S. Skelly. 1999. First Texas record of Black-tailed Gull. *Tex. Birds* 1(2):20–24.

Davis, L. I. 1945a. Brasher's Warbler in Texas. *Auk* 62:146.

———. 1945b. Rose-throated Becard nesting in Cameron County, Texas. *Auk* 62:316–17.

———. 1945c. Yellow-green Vireo nesting in Cameron County, Texas. *Auk* 62:146.

Davis, W. B. 1940. Birds of Brazos County, Texas. *Condor* 42:81–85.

———. 1961. Woodcock nesting in Brazos County, Texas. *Auk* 78:272–73.

DeBenedictis, P. A., J. L. Dunn, K. Kaufman, G. Lasley, J. V. Remsen, S. Tingley, and T. Tobish. 1994. ABA checklist report, 1992. *Birding* 26:92–102.

Delacour, J. 1954. *The waterfowl of the world.* Vol. 1. London: Country Life.

Delnicki, D. 1978. Second occurrence and first successful nesting record of the Hook-billed Kite in the United States. *Auk* 95:427.

Dennis, J. V. 1979. The Ivory-billed Woodpecker *Campephilus principalis. Avic. Mag.* 85:75–84.

Devine, A. B., D. Gendron, and D. G. Smith. 1978. Occurrence of the Coppery-tailed Trogon in Hidalgo County, Texas. *Bull. Tex. Ornith. Soc.* 11:52.

Dickerman, R. W. 1964. A specimen of Fuertes' Oriole, *Icterus fuertesi,* from Texas. *Auk* 81:433.

———. 2004a. Distribution of subspecies of Great Horned Owls in Texas. *Bull. Tex. Ornith. Soc.* 37:1-4.

———. 2004b. A review of North American subspecies of Great

Blue Heron (*Ardea herodias*). *Proc. Biol. Soc. Wash.* 117:242–50.

Dickerman, R. W., and A. B. Johnson. 2013. Notes on the Elf Owls of Trans-Pecos Texas and adjacent Coahuila and New Mexico. *Bull. Tex. Ornith. Soc.* 45:1–5.

Dittmann, D. L., and S. W. Cardiff. 2005. The "Chandeleur" Gull: Origins and identification of Kelp × Herring Gull hybrids. *Birding* 37:266–76.

Dittmann, D. L., and G. W. Lasley. 1992. How to document rare birds. *Birding* 24:145–59.

Dixon, K. L. 1990. Constancy of margins of the hybrid zone in Titmice of the *Parus bicolor* complex in coastal Texas. *Auk* 107:184–88.

Draheim, H. M., M. P. Miller, P. Baird, and S. M. Haig. 2010. Subspecific status and genetic structure of Least Tern (*Sternula antillarum*) inferred by mitochondrial DNA control-region sequences and microsatellite DNA. *Auk* 127:807–19.

Dresser, H. E. 1865. Notes on birds of southern Texas. *Ibis* 1:312–30.

Dunn, J. L., and K. L. Garrett. 1997. *A field guide to warblers of North America*. Boston: Houghton Mifflin.

Dunn, J. L., K. L. Garrett, and J. K. Alderfer. 1995. White-crowned Sparrow subspecies: Identification and distribution. *Birding* 27:182–200.

Duvall, A. J. 1943. Breeding Savannah Sparrows of the southwestern United States. *Condor* 45:237–38.

Easterla, D. A. 1972. Specimens of Black-throated Blue Warbler and Yellow-green Vireo from West Texas. *Condor* 74:489.

Eisenmann, E., J. I. Richardson, and G. I. Child. 1968. Yellow-green Vireo collected in Texas. *Wilson Bull.* 80:235.

Eitniear, J., and D. Schaezler. 2012. Hutton's Vireo nesting in Guadalupe County, Texas. *Bull. Tex. Ornith. Soc.* 45:40–42.

Eitniear, J. C., and T. Rueckle. 1996. Noteworthy avian breeding records from Zapata County, Texas. *Bull. Tex. Ornith. Soc.* 29:43–44.

Emanuel, V. L. 1961. Another probable record of an Eskimo Curlew on Galveston Island, Texas. *Auk* 78:259–60.

———. 1962. Texans rediscover the nearly extinct Eskimo Curlew. *Audubon Mag.* 64:162–65.

Engelmoer, M., and C. S. Roselaar. 1998. *Geographical variation in waders*. Dordrecht, Netherlands: Kluwer Academic Publishers.

Eubanks, T. L., Jr. 1977. Black-legged Kittiwake sightings on the Upper Texas Coast in winter of 1976–1977. *Bull. Tex. Ornith. Soc.* 10:42–43.

———. 1994. Status and distribution of the Piping Plover in Texas. *Bull. Tex. Ornith. Soc.* 27:19–25.

Eubanks, T. L., Jr., R. A. Behrstock, and R. J. Weeks. 2006. *Birdlife of Houston, Galveston, and the Upper Texas Coast.* College Station: Texas A&M University Press.

Eubanks, T. L., Jr., and J. Morgan. 1989. First photographic documentation of a live White-collared Swift from the United States. *Am. Birds* 43:258–59.

Fall, B. A. 1973. Noteworthy bird records from South Texas (Kenedy Co.). *Southwest. Nat.* 18:244–47.

Farmer, M. 1990. A Herring Gull nest in Texas. *Bull. Tex. Ornith. Soc.* 23:27–28.

Fischer, D. L. 1979. Black-billed Cuckoo (*Coccyzus erythrophthalmus*) breeding in South Texas. *Bull. Tex. Ornith. Soc.* 12 (1): 25.

Fisher, A. K. 1895. The Masked Duck in the Lower Rio Grande Valley, Texas. *Auk* 12:297.

Fleetwood, R. J. 1973. Jacana breeding in Brazoria County, Texas. *Auk* 90:422–23.

Fleetwood, R. J., and J. L. Hamilton. 1967. Occurrence and nesting of the Hook-billed Kite (*Chondrohierax uncinatus*) in Texas. *Auk* 84:598–601.

Fleetwood, R. J., and C. E. Hudson Jr. 1965. Probable Green Violet-ear Hummingbird in Cameron County, Texas. *Southwest. Nat.* 10:312.

Ford, E. R. 1938. Notes from the Lower Rio Grande Valley. *Auk* 55:132–34.

Freeman, B. 1996. *Birds of Bastrop, Buescher, and Lake Bastrop State Parks: A field checklist.* Austin: Natural Resources Program, Texas Parks and Wildlife Department.

———. 2003. *Birds of the Oaks and Prairies and Osage Plains of Texas: A field checklist.* Austin: Texas Parks and Wildlife Department.

Friedmann, H. 1925. Notes on the birds observed in the Lower Rio Grande Valley of Texas during May, 1924. *Auk* 42:537–54.

Galley, J. E. 1951. Clark's Nutcracker in the Chisos Mountains, Texas. *Wilson Bull.* 63:115.

Gallucci, T. 1979a. County records for bird specimens in the collection of the Museum of Arid Land Biology (Univ. of Texas at El Paso) and two other West Texas collections. *Bull. Tex. Ornith. Soc.* 12:26–27.

———. 1979b. Successful breeding of Lucy's Warbler in Texas. *Bull. Tex. Ornith. Soc.* 12:37–41.

Gallucci, T., and J. G. Morgan. 1987. First documented record of the Mangrove Cuckoo for Texas. *Bull. Tex. Ornith. Soc.* 20:2–6.

Gee, J. P., and C. E. Edwards. 2000. Interesting gull records from north-east Tamaulipas, Mexico. *Cotinga* 13:65, 67.

Gehlbach, F. R. 1995. Eastern Screech-Owl (*Megascops asio*). In *The birds of North America*, edited by A. Poole. Philadelphia: The Birds of North America.

Gill, F., and D. Donsker, eds. 2013. *IOC world bird list* (version 3.4). Available at http://www.worldbirdnames.org.

Gonzalez, J., H. Düttman, and M. Wink. 2009. Phylogenetic relationships based on two mitochondrial genes and hybridization patterns in Anatidae. *J. Zool.* 279:310–18.

Grieb, J. R. 1970. The shortgrass Canada Goose population. *Wildlife Soc. Monogr.* 22. Washington, D.C.

Grimes, S. A. 1953. Black-throated Oriole (*Icterus gularis*) nesting in Texas. *Auk* 70:207.

Griscom, L. 1920. Notes on the winter birds of San Antonio, Texas. *Auk* 37:49–55.

Griscom, L. L., and M. S. Crosby. 1925. Birds of the Brownsville region, southern Texas. *Auk* 42:432–40, 519–37.

———. 1926. Birds of the Brownsville region, southern Texas. *Auk* 43:18–36.

Groth, J. G. 1993. Call matching and positive assortative mating in Red Crossbill. *Auk* 110:398–401.

Grzybowski, J. A. 1982. Population structure in grassland bird communities during winter. *Condor* 84:137–51.

Grzybowski, J. A., J. W. Arterburn, W. A. Carter, J. S. Turner, and D. W. Verser. 1992. *Date guide to the birds of Oklahoma*. 2d ed. Tulsa: Oklahoma Ornithological Society.

Guthery, F. S., and J. C. Lewis. 1979. Sandhill Cranes in coastal counties of southern Texas: Taxonomy, distribution, and populations. In *Proceedings of the 1978 Crane Workshop*, edited by J. C. Lewis, 121–28. Fort Collins: Colorado State University Printing Service.

Hagar, C. N. 1944. Flamingo on the Texas Coast. *Auk* 61:301–2.

———. 1945. Harlequin Duck on the Texas coast. *Auk* 62:639–40.

Hagar, C. N., and F. M. Packard. 1952. *Check-list of the birds of the central coast of Texas*. Rockport, Tex.: C. N. Hagar and F. M. Packard.

Haller, K. W., and J. H. Beach III. 1984. A Glaucous Gull in Bryan County, Oklahoma and Grayson Co., Texas. *Bull. Okla. Ornith. Soc.* 17:27–28.

Hasbrouck, E. M. 1889. Summer birds of Eastland County, Texas. *Auk* :236–41.

Hatch, J. J. 1995. Changing populations of Double-crested Cormorants. *Colonial Waterbirds* 18:8–24.

Haucke, H. H., and W. H. Kiel Jr. 1973. Jabiru in South Texas. *Auk* 90:675–76.

Hawkins, A. S. 1945. Bird life of the Texas Panhandle. *Panhandle-Plains Hist. Rev.* 18:110–50.

Haynie, C. B. 1989. First photographic record of Common Redpoll in Texas. *Bull. Tex. Ornith. Soc.* 22:18–20.

Heidrich, P., C. König, and M. Wink. 1995. Bioacoustics, taxonomy, and molecular systematics in American Pygmy Owls. *Stuttgarter Beitraege zur Naturkunde Serie a (Biologie)* 534:1–47.

Heindel, M. 1996. Solitary Vireos. *Birding* 28:458–71.

Heiser, J. M. 1945. Eskimo Curlew in Texas. *Auk* 62:635.

Henderson, J. C. 1960. A Texas record of the Black Brant. *Auk* 77:227.

Hewetson, A., R. Kostecke, and B. Best. 2006. *Birds of the Texas South Plains.* Lubbock, Tex.: Llano Estacado Audubon Society.

Hoffman, J. C. 1999. Timing of migration of Short-billed Dowitchers and Long-billed Dowitchers in northeastern Oklahoma. *Bull. Okla. Ornith. Soc.* 32:21–29.

Hoffman, W., J. A. Wiens, and J. M. Scott. 1978. Hybridization between gulls (*Larus glaucescens* and *L. occidentalis*) in the Pacific Northwest. *Auk* 95:441–58.

Holdermann, D. A., S. Sorola Jr., and R. Skiles. 2007. Recent photo- and audio-documentation of Montezuma Quail from the Chisos Mountains, Big Bend National Park, Texas. *Bull. Tex. Ornith. Soc.* 40:62–67.

Holm, S. F., H. D. Irby, and J. M. Inglis. 1978. First nesting record of Double-crested Cormorant in Texas since 1939. *Bull. Tex. Ornith. Soc.* 11:50–51.

Horvath, E., and J. Karges. 2000. First Texas record of Buff-breasted Flycatcher. *Tex. Birds* 2 (1): 4–7.

Howell, S. N. G. 2012. *Petrels, albatrosses and storm-petrels of North America.* Princeton, N.J.: Princeton University Press.

Howell, S. N. G., J. Correa S., and J. Garcia B. 1993. First record of Kelp Gull in Mexico. *Euphonia* 2:71–80.

Howell, S. N. G., and S. Webb. 1995. *A guide to the birds of Mexico and northern Central America.* Oxford: Oxford University Press.

Hubbard, J. P. 1977. *The biological and taxonomic status of the Mexican Duck.* Bulletin of New Mexico Department of Game and Fish 16. Santa Fe: New Mexico Department of Game and Fish.

Hubbard, J. P., and D. M. Niles. 1975. Two specimen records of the Brown Jay from southern Texas. *Auk* 92:797–98.

Huckabee, J. W., T. Moore, and C. Dorn. 2010. Nesting of white sub-species of Great Blue Heron (*Ardea herodias occi-*

dentalis) in the Texas Coastal Bend 2006–2010. *Bull. Tex. Ornith. Soc.* 43:83–84.

Husak, M. S., and T. C. Maxwell. 2000. A review of 20th century range expansion and population trends of the Golden-fronted Woodpecker (*Melanerpes aurifrons*): Historical and ecological perspectives. *Tex. J. Sci.* 52 (4): 275–84.

Ingold, D. J. 1989. Nesting phenology and competition for nest sites among Red-headed and Red-bellied Woodpeckers and European Starlings. *Auk* 116:209–17.

Irwin, D. E., J. H. Irwin, and T. B. Smith. 2011. Genetic variation and seasonal migratory connectivity in Wilson's Warblers (*Wilsonia pusilla*): Species-level differences in nuclear DNA between western and eastern populations. *Mol. Ecol.* 20:3102–15.

James, P. 1960. Clay-colored Robin in Texas. *Auk* 77:475–76.

———. 1963a. Fork-tailed Flycatcher taken in Texas. *Auk* 80:85.

———. 1963b. Freeze loss in the Least Grebe (*Podiceps dominicus*) in lower Rio Grande delta of Texas. *Southwest. Nat.* 8:45–46.

James, P., and A. Hayes. 1963. Elf Owl rediscovered in lower Rio Grande delta of Texas. *Wilson Bull.* 75:179–82.

Jaramillo, A., J. D. Rising, J. L. Cooper, P. G. Ryan, and S. C. Madge. 2011. Emberizidae. In *Handbook of the birds of the world*. Barcelona: Lynx Ediciones.

Johnson, N. K., and J. A. Martin. 1992. Macrogeographic patterns of morphometric and genetic variation in the Sage Sparrow complex. *Condor* 94:1–19.

Johnson, W. P., and P. R. Garrettson. 2010. Band recovery and harvest data suggest additional American Black Duck records from Texas. *Bull. Tex. Ornith. Soc.* 43:34-40.

Johnson, W. P., and M. W. Lockwood. 2013. *Texas waterfowl*. College Station: Texas A&M University Press.

Jones, B. 1992. *A birder's guide to Aransas National Wildlife Refuge*. Albuquerque, N.Mex.: Southwest Natural and Cultural Heritage Association.

Jury, G. W. 1976. First record of White-winged Crossbill (*Loxia leucoptera*) for Texas. *Bull. Tex. Ornith. Soc.* 9:7.

Keefer, M. B. 1957. Varied Thrush in Texas. *Wilson Bull.* 69:114.

Kennedy, E. D., and D. W. White. 1997. Bewick's Wren (*Thryomanes bewickii*). In *The birds of North America*, No. 315, edited by A. Poole and F. Gill. Philadelphia: The Birds of North America.

Kincaid, E. B. 1956. Ringed Kingfisher at Austin, Texas. *Wilson Bull.* 68:324–25.

Kirn, A. J., and R. W. Quillin. 1927. *Birds of Bexar County, Texas*. San Antonio, Tex: Witte Memorial Museum.

Klicka, J., R. M. Zink, J. C. Barlow, W. B. McGillivray, and T. J. Doyle. 1999. Evidence supporting the recent origin and species status of the Timberline Sparrow. *Condor* 101 (3): 577–88.

Kostecke, R. M., D. Sperry, and D. A. Cimprich. 2006. Second record of an American Woodcock (*Scolopax minor*) breeding on the Edwards Plateau. *Bull. Tex. Ornith. Soc.* 39:1–2.

Kostecke, R. M., S. G. Summers, J. W. Bailey, and D. A. Cimprich. 2004. Confirmed nesting of a Lazuli Bunting with an Indigo Bunting on Fort Hood, Bell County. *Bull. Tex. Ornith. Soc.* 37:1–2.

Kratter, A. W., and D. W. Steadman. 2003. First Atlantic Ocean and Gulf of Mexico specimen of Short-tailed Shearwater. *N. Am. Birds* 57:277–79.

Kroodsma, D. E. 1989. Two North American song populations of the Marsh Wren reach distributional limits in the central Great Plains. *Condor* 91:332–40.

Kutac, E. A., and S. C. Caran. 1993. *Birds and other wildlife of south central Texas.* Austin: University of Texas Press.

Lacey, H. 1903. Notes on the Texas Jay. *Condor* 5:151–53.

———. 1911. The birds of Kerrville, Texas, and vicinity. *Auk* 28:200–219.

———. 1912. Additions to birds of Kerrville, Texas. *Auk* 29:254.

Lahrman, F. W. 1972. Eskimo Curlew in Texas. *Blue Jay* 30: 87–88.

Langham, J. M. 1980. Golden-crowned Warbler in Texas: A documented record for the ABA checklist area. *Birding* 12:8–9.

Lanyon, W. E. 1962. Specific limits and distribution of meadowlarks of the desert grassland. *Auk* 79:183–207.

Lasley, G. W. 1984. First Texas specimen of the White-collared Swift. *Am. Birds* 38:370–71.

Lasley, G. W., D. A. Easterla, C. W. Sexton, and D. A. Bartol. 1982. Documentation of the Red-faced Warbler in Texas and a review of its status in Texas and adjacent areas. *Bull. Tex. Ornith. Soc.* 15:8–14.

Lasley, G. W., and J. P. Gee. 1991. The first nesting record of the Hutton's Vireo (*Vireo huttoni*) east of the Pecos River, Texas. *Bull. Tex. Ornith. Soc.* 24:23–24.

Lasley, G. W., and M. Krzywonski. 1991. First United States record of the White-throated Robin. *Am. Birds* 45:230–31.

Lasley, G. W., and T. Pincelli. 1986. Gray Silky-flycatcher in Texas. *Birding* 18:34–36.

Lasley, G. W., C. W. Sexton, and D. Hillsman. 1988. First record of the Mottled Owl in the United States. *Am. Birds* 42:23–24.

Lawrence, R. B. 1927. Masked Duck (*Nomonyx dominicus*) in Texas. *Auk* 44:415.

Lee, C-T., and A. Birch. 2002. Notes on the distribution, vagrancy, and distribution of American Pipit and "Siberian Pipit." *N. Am. Birds* 56:388–98.

Lee, D. S., and S. W. Cardiff. 1993. Status of the Arctic Tern in the coastal and offshore waters of the southeastern United States. *J. Field Ornith.* 64:158–68.

Lehmann, V. W. 1941. *Attwater's Prairie Chicken: Its life history and management.* US Department of the Interior, North American Fauna 57.

Lethaby, N., and J. Bangma. 1998. Identifying Black-tailed Gull in North America. *Birding* 30:470–83.

Lieftinck, J. E. 1968. Report of an Eskimo Curlew from Texas coast. *Bull. Tex. Ornith. Soc.* 2:28.

Ligon, J. S. 1961. *New Mexico birds and where to find them.* Albuquerque: University of New Mexico Press.

Liguori, J. 2005. *Hawks from every angle.* Princeton, N.J.: Princeton University Press.

Lloyd, W. L. 1887. The birds of Tom Green and Concho Counties, Texas. *Auk* 4:181–93, 289–99.

Lockwood, M. W. 1992. First breeding record of *Aechmophorus* grebes in Texas. *Bull. Tex. Ornith. Soc.* 25:64–66.

———. 1995. A closer look: Varied Bunting. *Birding* 27 (2): 110–13.

———. 1997. A closer look: Masked Duck. *Birding* 29 (5): 386–90.

———. 1999. Possible anywhere: Fork-tailed Flycatcher. *Birding* 31 (2):126–39.

———. 2001. *Birds of the Texas Hill Country.* Austin: University of Texas Press.

———. 2005a. *Birds of the Edwards Plateau: A field checklist.* Austin: Texas Parks and Wildlife Department.

———. 2005b. Texas Bird Records Committee report for 2004. *Bull. Tex. Ornith. Soc.* 38:21–28.

———. 2006. Texas Bird Records Committee report for 2005. *Bull. Tex. Ornith. Soc.* 39:33–42.

———. 2007a. *Basic Texas birds.* Austin: University of Texas Press.

———. 2007b. Texas Bird Records Committee report for 2006. *Bull. Tex. Ornith. Soc.* 40:41–49.

———. 2008a. Occurrence of presumed Lucifer × Black-chinned Hummingbirds in Texas. *Bull. Tex. Ornith. Soc.* 41:1–4.

———. 2008b. Texas Bird Records Committee report for 2007. *Bull. Tex. Ornith. Soc.* 41:37–45.

———. 2009. Texas Bird Records Committee report for 2008. *Bull. Tex. Ornith. Soc.* 42:36-44.

———. 2010. Texas Bird Records Committee report for 2009. *Bull. Tex. Ornith. Soc.* 43:48–56.

————. 2011. Texas Bird Records Committee report for 2010. *Bull. Tex. Ornith. Soc.* 44:43–50.

————. 2012. Two recent additions to the breeding avifauna of the Guadalupe Mountains, Texas. *Bull. Tex. Ornith. Soc.* 45:43–45.

Lockwood, M. W., and R. Bates. 2005. First record for the United States of Black-headed Nightingale-Thrush. *N. Am. Birds* 59:350–51.

Lockwood, M. W., and T. W. Cooper. 1999. A Texas hybrid: Cinnamon × Green-winged Teal. *Tex. Birds* 1 (2): 38–40.

Lockwood, M. W., and B. Freeman. 2004. *The Texas Ornithological Society's handbook of Texas birds*. College Station: Texas A&M University Press.

Lockwood, M. W., and C. E. Shackelford. 1998. The occurrence of Red-breasted Sapsucker and suspected hybrids with Red-naped Sapsucker in Texas. *Bull. Tex. Ornith. Soc.* 31:2–6.

Loetscher, F. W., Jr. 1956. Masked Duck and Jacana at Brownsville, Texas. *Auk* 73:291.

Lorenz, S., C. Butler, and J. Paz. 2006. First nesting record of the Gray-crowned Yellowthroat (*Geothlypis poliocephala*) in the United States since 1884. *Wilson Bull.* 118:574–76.

Lyndon B. Johnson School of Public Affairs. 1978. *Preserving Texas' natural heritage*. Policy Research Project Report 31, 1–34. Austin: University of Texas at Austin.

MacInnes, C. D., and E. B. Chamberlain. 1963. The first record of Double-striped Thick-knee in the United States. *Auk* 80:79.

Marshall, J. T. 1967. Parallel variation in North and Middle American Screech-Owls. *Monogr. Western Found. Vert. Zool.*, No. 1.

Martinez-Morales, M. A., I. Zuria, L. Chapa-Vargas, I. MacGregor-Fors, R. Ortega-Alvarez, E. Romero-Aguila, and P. Carbo. 2010. Current distribution and predicted geographic expansion of the Rufous-backed Robin in Mexico: A fading endemism? *Divers. Distrib.* 16:786–97.

Maxwell, T. C. 1977. First record of Heermann's Gull for Texas. *Southwest. Nat.* 22:282–83.

————. 1979. Vireos (Aves: Vireonidae) in west-central Texas. *Southwest. Nat.* 24:223–29.

————. 1980. Significant nesting records of birds from western Texas. *Bull. Tex. Ornith. Soc.* 13:2–6.

————. 2013. *Wildlife of the Concho Valley*. College Station: Texas A&M University Press.

Maxwell, T. C., and M. S. Husak. 1999. Common Black-Hawk nesting in west-central Texas. *J. Raptor Res.* 33:270–71.

McAlister, W. H. 2002. *Birds of Matagorda Island: A field checklist*. Austin: Natural Resources Program, Texas Parks and Wildlife Department.

McCracken, K. G., W. P. Johnson, and F. H. Sheldon. 2001. Molecular population genetics, phylogeography, and conservation biology of the Mottled Duck (*Anas fulvigula*). *Conserv. Genet.* 2:87–102.

McGrew, A. D. 1971. Nesting of the Ringed Kingfisher in the United States. *Auk* 88:665–66.

McKenzie, P. M., and M. B. Robbins. 1999. Identification of adult male Rufous and Allen's Hummingbirds, with specific comments on dorsal coloration. *Western Birds* 30:86–93.

McKinney, B. 1998. *A checklist of lower Rio Grande birds*. Rancho Viejo, Tex.: B. McKinney.

McLaughlin, V. P. 1948. Birds of an army camp. *Auk* 65:180–88.

Meitzen, T. C. 1963. Additions to the known breeding ranges of several species in South Texas. *Auk* 80:368–69.

Merrill, J. C. 1878. Notes on the ornithology of South Texas, being a list of birds observed in the vicinity of Fort Brown, Texas, from February, 1876, to June, 1878. *Proc. US Nat. Mus.* 1:113–73.

Miller, F. W. 1955. Black Skimmer in north-central Texas. *Condor* 57:240.

———. 1959. The Barrow's Goldeneye in Texas. *Condor* 61:434.

Miller, W. D. 1906. Occurrence of *Progne chalybea* in Texas. *Auk* 23:226–27.

Moldenhauer, R. R. 1974. First Clay-colored Robin collected in the United States. *Auk* 91:839–40.

Moldenhauer, R. R., and K. B. Bryan. 1970. An interesting recovery of a Banded Evening Grosbeak during the 1968–69 winter incursion into East Texas. *J. Field Ornith.* 41:39.

Montgomery, T. H. 1905. Summer resident birds of Brewster County, Texas. *Auk* 22:12–15.

More, R. L., and J. K. Strecker. 1929. The summer birds of Wilbarger County, Texas. *Contr. Baylor Univ. Mus.*, No. 20.

Morgan, J. G., and T. L. Eubanks Jr. 1979. First documentation of Connecticut Warbler in Texas. *Bull. Tex. Ornith. Soc.* 12:21–22.

Morgan, J. G., T. L. Eubanks, V. Eubanks, and L. N. White. 1985. Yucatan Vireo appears in Texas. *Am. Birds* 39:244–46.

Morgan, J. G., and L. M. Feltner. 1985. A Neotropical bird flies north: The Greenish Elaenia. *Am. Birds* 39:242–44.

Neck, R. W. 1986. Expansion of Red-crowned Parrot, *Amazona viridigenalis*, into southern Texas and changes in agricultural practices in northern Mexico. *Bull. Tex. Ornith. Soc.* 19:6–12.

———. 1989. Winter Whooping Cranes in the Texas Hill Country in 1854. *Bull. Tex. Ornith. Soc.* 22:15–16.

Nehrling, H. 1882. List of birds observed at Houston, Harris County, Texas, and in the counties Montgomery, Galveston and Fort Bend. *Bull. Nutt. Ornith. Club* 7:6–13, 166–75, 222–25.

Neill, R. L. 1975. *The birds of the Buescher Division.* University of Texas Environmental Science Park at Smithville. Pub. No. 3.

Newfield, N. L. 1983. Records of Allen's Hummingbird in Louisiana and possible Rufous × Allen's Hummingbird hybrids. *Condor* 85:253–54.

Newman, G. A. 1974. Recent bird records from the Guadalupe Mountains, Texas. *Southwest. Nat.* 19:1–7.

Nirschl, R., and R. Snyder. 2010. First record of Bare-throated Tiger-Heron (*Tigrisoma mexicanum*) for the United States. *N. Am. Birds* 64:347–49.

Novy, F. O., and A. D. McGrew. 1972. Orange-breasted Bunting in southern Texas. *Auk* 91:178–79.

Oberholser, H. C. 1974. *The bird life of Texas.* Austin: University of Texas Press.

Ogilby, J. D. 1882. A catalogue of birds obtained in Navarro County, Texas. *Sci. Proc.*, Vol. III of the Royal Dublin Society.

Oring, L. W. 1964. Notes on the birds of Webb County, Texas. *Auk* 81:440.

Ortego, B., M. Ealy, G. Creacy, and L. LaBeau. 2011. Colonial waterbird survey. *Bull. Tex. Ornith. Soc.* 44:51–67.

Ortego, B., C. Gregory, D. Mabie, M. Mitchell, and D. Schmidt. 2009. Texas Bald Eagles. *Bull. Tex. Ornith. Soc.* 42:1–17.

Packard, F. M. 1946. California Gull on the coast of Texas. *Auk* 63:545–46.

———. 1947. Notes on the occurrence of birds in the Gulf of Mexico. *Auk* 64:130–31.

Pangburn, C. H., and J. M. Heiser Jr. 1945. The Phainopepla near San Antonio, Texas. *Auk* 62:146–47.

Papish, R., J. L. Mays, and D. Brewer. 1997. Orange-billed Nightingale-Thrush: First record for Texas and the U.S. *Birding* 29:128–30.

Parkes, K. C. 1948. Reddish Egret in Central Texas. *Auk* 65:308.

———. 1950. Further notes on the birds of Camp Barkeley, Texas. *Condor* 52:91–93.

Parkes, K. C., D. P. Kibbe, and E. L. Roth. 1978. First records of the Spotted Rail (*Pardirallus maculatus*) for the United States, Chile, Bolivia, and western Mexico. *Am. Birds* 32:295–99.

Partrikeev, M. 2009. First confirmed nesting of a Red-shouldered Hawk in Starr County. *Bull. Tex. Ornith. Soc.* 42:98–99.

Patten, M. A. 2000. Changing seasons, the winter season, 1999–2000: Warm weather and cross-continental wonders. *N. Am. Birds* 54:146–49.

Patten, M. A., and G. W. Lasley. 2000. Range expansion of the Glossy Ibis in North America. *N. Am. Birds* 54:241–47.

Patten, M. A., and C. L Pruett. 2009. The Song Sparrow, *Melospiza melodia*, as a ring species: Patterns of geographic variation, a revision of subspecies, and implications for speciation. *Syst. Biodivers.* 7:33–62.

Pemberton, J. R. 1922. A large tern colony in Texas. *Condor* 24:37–42.

Peters, H. S. 1931. Abert's Towhee, a new bird for Texas. *Auk* 48:274–75.

Peterson, J. J., G. W. Lasley, K. B. Bryan, and M. Lockwood. 1991. Additions to the breeding avifauna of the Davis Mountains. *Bull. Tex. Ornith. Soc.* 24:39–48.

Peterson, J. J., and B. R. Zimmer. 1998. *Birds of the Trans Pecos.* Austin: University of Texas Press.

Peterson, R. T. 1960. *A field guide to the birds of Texas.* Boston: Houghton Mifflin.

Petrides, G. A., and W. B. Davis. 1951. Notes on the birds of Brazos County, Texas. *Condor* 53:153–54.

Petrovic, C. A., and J. King Jr. 1972. Common Eider and King Rail from the Dry Tortugas, Florida. *Auk* 89:660.

Pettingell, N. 1967. Eskimo Curlew: Valid records since 1945. *Bull. Tex. Ornith. Soc.* 1 (3/4): 14, 21.

Phillips, A. R. 1950. The Great-tailed Grackles of the Southwest. *Condor* 52:78–81.

———. 1986. *The known birds of North and Middle America, part 1.* Published privately.

———. 1991. *The known birds of North and Middle America, part 2.* Published privately.

Phillips, H. W., and W. A. Thornton. 1949. The summer resident birds of the Sierra Vieja range in southwestern Texas. *Tex. J. Sci.* 1:101–31.

Phillips, J. N. 1998. A survey of wintering Rufous Hummingbirds in Texas. *Bull. Tex. Ornith. Soc.* 31:65–67.

Pranty, B., J. L. Dunn, S. Heinl, A. W. Kratter, P. Lehman, M. W. Lockwood, B. Mactavish, and K. J. Zimmer. 2007. Annual report of the ABA Checklist Committee: 2007. *Birding* 39 (6): 24–31.

———. 2008. Annual report of the ABA Checklist Committee: 2007–2008. *Birding* 39 (6): 32–38.

Pruett, C. L., S. E. Henke, S. M. Tanksley, M. F. Small, K. M. Hogan, and J. Roberson. 2000. Mitochondrial DNA and morphological variation in White-winged Doves in Texas. *Condor* 102:871–80.

Pulich, W. M. 1955. A record of the Mexican Crossbill (*Loxia curvirostra stricklandi*) from Fort Worth, Texas. *Auk* 72:299.

———. 1961a. *Birds of Tarrant County*. Fort Worth, Tex.: Allen Press.

———. 1961b. A record of the Yellow Rail from Dallas County, Texas. *Auk* 78:639–40.

———. 1966. A specimen of the Little Gull from Dallas County, Texas. *Auk* 83:482.

———. 1968. The occurrence of the Crested Hummingbird in the United States. *Auk* 85:322.

———. 1971. Some fringillid records for Texas. *Condor* 73:111.

———. 1976. *The Golden-cheeked Warbler, a bioecological study*. Austin: Texas Parks and Wildlife Department.

———. 1979. *The birds of Tarrant County*. 2d ed. Fort Worth, Tex: Branch Smith.

———. 1980. A Thayer's Gull specimen from Texas: A problem in identification. *Southwest. Nat.* 25:257–58.

———. 1988. *The birds of north central Texas*. College Station: Texas A&M University Press.

Pulich, W. M., Jr. 1982. Documentation and status of Cory's Shearwater in the western Gulf of Mexico. *Wilson Bull.* 94:381–85.

Pulich, W. M., and J. E. Parrot. 1977. The occurrence of the Gray Vireo east of the Pecos River. *Southwest. Nat.* 21:551–52.

Pulich, W. M., and W. M. Pulich Jr. 1973. First Brown Booby specimen for Texas. *Auk* 90:683–84.

Pyle, P. 1997. *Identification guide to North American birds, part I*. Bolinas, CA: Slate Creek Press.

———. 2008. *Identification guide to North American birds, part II*. Bolinas, CA: Slate Creek Press.

Quillin, R. W. 1935. New bird records from Texas. *Auk* 52:324–25.

Quillin, R. W., and R. Holleman. 1918. The breeding birds of Bexar County, Texas. *Condor* 20:37–44.

Quinn, T. W., G. F. Shields, and A. C. Wilson. 1991. Affinities of the Hawaiian Goose based on two types of mitochondrial DNA data. *Auk* 108:585–93.

Ragsdale, G. H. 1879. Lewis' Woodpecker in middle Texas. *Sci. News* 1:208.

———. 1881. *Larus glaucus* in Texas. *Bull. Nutt. Ornith. Club* 6:187.

———. 1885. Lewis' Woodpecker in Texas. *Ornithologist and Oologist* 10:79.

Rappole, J. H., and G. W. Blacklock. 1985. *Birds of the Texas Coastal Bend: Abundance and distribution*. College Station: Texas A&M University Press.

Reid, M., and D. Jones. 2009. First North American record of White-crested Elaenia (*Elaenia albiceps chilensis*) at South Padre Island, Texas. *N. Am. Birds* 63:10–14.

Remsen, J. V., Jr. 2001. True winter range of the Veery (*Catharus fuscescens*): Lessons for determining winter ranges of species that winter in the tropics. *Auk* 118 (4): 838–48.

Risser, A. 1932. Unusual summer birds from the vicinity of Brownsville, Texas. *Auk* 49:106–7.

Robbins, M. B., D. L. Dittmann, J. L. Dunn, K. L. Garrett, S. Heinl, A. W. Kratter, G. Lasley, and B. Mactavish. 2003. ABA Checklist Committee 2002 annual report. *Birding* 34:138–44.

Robinson, J. A., and G. Aumaun. 1997. An American Woodcock nest in Galveston County, Texas. *Bull. Tex. Ornith. Soc.* 30:20–23.

Rojas-Soto, O. R. 2003. Geographic variation in the Curve-billed Thrasher (*Toxostoma curvirostre*) complex. *Auk* 120:311–22.

Ruegg, K. 2007. Divergence between subspecies groups of Swainson's Thrush (*Catharus ustulatus ustulatus* and *C. u. swainsoni*). In *Festschrift for Ned K. Johnson: Geographic Variation and Evolution in Birds*, edited by Carla Cicero and J. V. Remsen Jr., 67–77. Ornithological Monographs No. 63. Berkeley: University of California Press.

Runnels, S. R. 1975. Rose-throated Becard in Jeff Davis County, Texas. *Condor* 77:221.

———. 1980. Louisiana Heron (*Hydranassa tricolor*) breeding in north central Texas. *Bull. Tex. Ornith. Soc.* 13:23.

Rupert, J. R., and T. Brush. 1996. Red-breasted Mergansers, *Mergus serrator*, nesting in southern Texas. *Southwest. Nat.* 41:199–200.

Rylander, K. 2002. *The behavior of Texas birds*. Austin: University of Texas Press.

Sauer, J. R., J. E. Hines, J. E. Fallon, K. L. Pardieck, D. J. Ziolkowski Jr., and W. A. Link. 2011. *The North American Breeding Bird Survey, results and analysis 1966–2010* (version 12.07.2011). Laurel, Md.: USGS Patuxent Wildlife Research Center.

Schaefer, R. R. 1998. First county records of Red Crossbill in the Pineywoods region of eastern Texas. *Bull. Tex. Ornith. Soc.* 31:63–64.

Schmidt, J. R. 1976. First nesting record of Anna's Hummingbird in Texas. *Bull. Tex. Ornith. Soc.* 9:6–7.

Schorger, A. W. 1955. *The Passenger Pigeon: Its natural history and extinction*. Caldwell, N.J.: Blackburn Press.

Seidel, S. 2010. Genetic structure and diversity of the Eastern Wild Turkey (*Meleagris gallopavo silvestris*) in East

Texas. Master's thesis, Stephen F. Austin University, Nacogdoches, Texas.

Sell, R. A. 1918. The Scarlet Ibis in Texas. *Condor* 20:78–82.

Sexton, C. W. 1999. The Vermilion Flycatcher in Texas. *Tex. Birds* 1 (2): 41–45.

Seyffert, K. D. 1984. Wintering White-throated Swifts in the Texas Panhandle. *Bull. Okla. Ornith. Soc.* 17:31.

———. 1985a. The breeding birds of the Texas Panhandle. *Bull. Tex. Ornith. Soc.* 18:7–20.

———. 1985b. A first nesting of the Wilson's Phalarope in Texas. *Bull. Tex. Ornith. Soc.* 18:27–29.

———. 1988. Breeding status of the Eared Grebe in the Texas Panhandle. *Bull. Okla. Ornith. Soc.* 21:5–6.

———. 1991. Does the Cedar Waxwing nest in the Texas Panhandle? *Bull. Tex. Ornith. Soc.* 24:54–56.

———. 1993. Nesting of the Yellow-headed Blackbird in the Panhandle of Texas. *Bull. Okla. Ornith. Soc.* 26:1–4.

———. 2001a. *Birds of the High Plains and Rolling Plains of Texas: A field checklist.* Austin: Texas Parks and Wildlife Department.

———. 2001b. *Birds of the Texas Panhandle.* College Station: Texas A&M University Press.

Shackelford, C. E. 1998. Compilation of published records of the Ivory-billed Woodpecker in Texas: Voucher specimens versus sight records. *Bull. Tex. Ornith. Soc.* 31:35–41.

Shackelford, C. E., D. Saenz, and R. R. Schaefer. 1996. Sharp-shinned Hawks nesting in the Pineywoods of eastern Texas and western Louisiana. *Bull. Tex. Ornith. Soc.* 29:23–25.

Shackelford, C. E., and G. G. Simons. 1999. *An annual report of the Swallow-tailed Kite in Texas: A survey and monitoring project for 1998.* Austin: Texas Parks and Wildlife Department.

Shifflett, W. A. 1975a. First photographic record of Brown Jay in the United States. *Auk* 92:797.

———. 1975b. Ruddy Ground Dove in South Texas. *Auk* 92:604.

Sibley, C. G., and B. L. Monroe Jr. 1990. *Distribution and taxonomy of birds of the world.* New Haven, Conn.: Yale University Press.

Simmons, G. F. 1914. Spring migration (1914) at Houston, Texas. *Wilson Bull.* 26:128–40.

———. 1915. On the nesting of certain birds in Texas. *Auk* 32:317–31.

———. 1925. *Birds of the Austin region.* Austin: University of Texas.

Singley, J. A. 1887. Observations on eggs collected in Lee County, Texas, etc. *Ornithologist and Oologist* 12:163–65.

Slack, D. R., and K. A. Arnold. 1985. Nesting of the Magnificent Hummingbird in Jeff Davis County, Texas. *Bull. Tex. Ornith. Soc.* 18:27.

Smith, A. P. 1916. Additions to the avifauna of Kerr Co., Texas. *Auk* 33:187–93.

———. 1917. Some birds of the Davis Mountains, Texas. *Condor* 19:161–65.

Smith, P. W. 1987. The Eurasian Collared-Dove arrives in the Americas. *Am. Birds* 41:1370–79.

Snell, R. R. 2002. Iceland Gull (*Larus glaucoides*) and Thayer's Gull (*Larus thayeri*). *In The birds of North America*, No. 699, edited by A. Poole and F. Gill. Philadelphia: The Birds of North America.

Stevenson, J. O. 1937a. The Alaska Longspur and Oregon Horned Lark in Texas. *Condor* 39:44.

———. 1937b. The Red Phalarope in Texas. *Condor* 39:92.

———. 1942a. Birds of the central Panhandle of Texas. *Condor* 44:108–15.

———. 1942b. Whooping Cranes in Texas in summer. *Condor* 44:40–41.

Stevenson, J. O., and T. F. Smith. 1938. Additions to the Brewster County, Texas bird list. *Condor* 40:184.

Stewart, R. E., and J. W. Aldrich. 1956. Distinction of maritime and prairie populations of Blue-winged Teal. *Proc. Biol. Soc. Wash.* 69:29–36.

Stiles, F. G. 1972. Age and sex determination in Rufous and Allen's Hummingbirds. *Condor* 74:25–32.

Stone, W. 1894. Capture of *Ceryle torquata* (Linn.) at Laredo, Texas: A species new to the United States. *Auk* 11:177.

Strecker, J. K., Jr. 1912. *The birds of Texas: An annotated check list*. Baylor University Bulletin 25. Waco, TX: Baylor University Press.

———. 1927. *Notes on the ornithology of McLennan County Texas*. Baylor University Press Bulletin No. 1. Waco, TX: Baylor University Press.

Stringham, E. 1948. *Kerrville, Texas, and its birds*. Kerrville: Pacot Publications.

Stutzenbaker, C. D. 1988. *The Mottled Duck, its life history, ecology and management*. Austin: Texas Parks and Wildlife Department.

Sutton, G. M. 1949. The Rose-throated Becard in the Lower Rio Grande Valley of Texas. *Auk* 66:365–66.

———. 1960. Flammulated Owl in Lubbock County, Texas. *Southwest. Nat.* 5:173–74.

————. 1967. *Oklahoma birds*. Norman: University of Oklahoma Press.

Tacha, R. C., D. C. Martin, and C. T. Patterson. 1981. Common Crane (*Grus grus*) sighted in West Texas. *Southwest. Nat.* 25:569.

Telfair, R. C., II. 1980. Additional inland nesting records in Texas of four species of colonial waterbirds. *Bull. Tex. Ornith. Soc.* 13:11–13.

————. 1995. Neotropic Cormorant (*Phalacrocorax brasilianus*) population trends and dynamics in Texas. *Bull. Tex. Ornith. Soc.* 28:1, 7–16.

Texas Ornithological Society. 1995. *Checklist of the birds of Texas*. 3d ed. Austin: Capital Printing.

Thompson, B. C., M. E. Schmidt, S. W. Calhoun, D. C. Morizot, and R. D. Slack. 1992. Subspecific status of Least Tern populations in Texas: North American implications. *Wilson Bull.* 104:244–62.

Thompson, W. L. 1952. Summer birds of the Canadian "breaks" in Hutchinson County, Texas. *Tex. J. Sci.* 4:220–29.

————. 1953. The ecological distribution of the birds of the Black Gap area, Brewster County, Texas. *Tex. J. Sci.* 5:158–77.

Thornton, W. A. 1951. Ecological distribution of the birds of the Stockton Plateau in northern Terrell County, Texas. *Tex. J. Sci.* 3:413–30.

Tipton, H. C., P. F. Doherty Jr., and V. J. Dreitz. 2009. Abundance and density of Mountain Plover (*Charadrius montanus*) and Burrowing Owl (*Athene cunicularia*) in eastern Colorado. *Auk* 126:493–99.

Tomlinson, R. E., D. D. Dolton, R. R. George, and R. E. Mirarchi. 1994. Mourning Dove. In *Migratory shore and upland game bird management in North America*, edited by T. C. Tacha and C. E. Braun. Washington, D.C.: International Association of Fish and Wildlife Agencies.

Trautman, M. B. 1964. A specimen of the Roadside Hawk, *Buteo magnirostris griseocauda*, from Texas. *Auk* 81:435.

Traweek, M. S. 1978. *Texas waterfowl production survey*. Fed. Aid Proj. No. W-106-R-5. Final report. Austin: Texas Parks and Wildlife Department.

Van Tyne, J. 1929. Notes on some birds of the Chisos Mountains of Texas. *Auk* 46:204–6.

————. 1936a. The discovery of the nest of the Colima Warbler (*Vermivora crissalis*). *Univ. Mich. Mus. Zool.*, Misc. Publ. No. 33.

————. 1936b. *Spizella breweri taverneri* in Texas. *Auk* 53:92.

Van Tyne, J., and G. M. Sutton. 1937. The birds of Brewster County, Texas. *Mich. Mus. Zool.*, Misc. Publ. No. 37.

Wauer, R. H. 1967. First Thick-billed Kingbird record for Texas. *Southwest. Nat.* 12:485–86.

———. 1969. Winter bird records for Chisos Mountains and vicinity. *Southwest. Nat.* 14:252–54.

———. 1970. The occurrence of the Black-vented Oriole, *Icterus wagler*, in the United States. *Auk* 87:811–12.

———. 1971. Ecological distribution of birds in the Chisos Mountains. *Southwest. Nat.* 11:1–29.

———. 1973. Status of certain parulids in West Texas. *Southwest. Nat.* 18:105–10.

———. 1985. *A field guide to birds of the Big Bend*. Austin: Texas Monthly Press.

———. 2001. Breeding avifaunal baseline for Big Bend National Park. *Occas. Publ. Tex. Ornith. Soc.* No. 3.

Wauer, R. H., and D. G. Davis. 1972. Cave Swallows in Big Bend National Park, Texas. *Condor* 74:482.

Wauer, R. H., and J. D. Ligon. 1974. Distributional relations of breeding avifauna of four southwestern mountain ranges. In *Transactions of the Symposium of the biological resources of the Chihuahuan Desert region*, edited by R. H. Wauer and D. H. Riskind, 567–78. Washington, D.C.: National Park Service.

Wauer, R. H., P. C. Palmer, and A. Windham. 1994. The Ferruginous Pygmy-Owl in South Texas. *Am. Birds* 47:1071–76.

Wauer, R. H., and M. K. Rylander. 1968. Anna's Hummingbird in West Texas. *Auk* 85:501.

Weeks, R., and M. A. Patten. 2000. First Texas record of Yellow-footed Gull. *Tex. Birds* 2 (1): 25–33.

Weske, J. S. 1974. White-winged Junco in Texas. *Condor* 76:119.

West, S. 1976. First Presidio County and Texas winter record of the Olivaceous Flycatcher. *Bull. Tex. Ornith. Soc.* 9:8.

Wetmore, A., and H. Friedmann. 1933. The California Condor in Texas. *Auk* 35:37–38.

Wheeler, B. K. 2003. *Raptors of western North America*. Princeton, N.J.: Princeton University Press.

Wheeler, B. K., and W. S. Clark. 1995. *A photographic guide to North American raptors*. San Diego, Calif.: Academic Press.

Whitbeck, M. 2004. Discovery of a fledgling American Bittern (*Botaurus lenginosus*) in Chambers County, Texas. *Bull. Tex. Ornith. Soc.* 37:14–15.

White, M. 1999. Inland occurrences of Nelson's Sharp-tailed Sparrow. *Tex. Birds* 1 (1): 34–39.

———. 2000. Range expansion of Fish Crow in northeast Texas. *Bull. Tex. Ornith. Soc.* 33:6–9.

———. 2002. *Birds of northeast Texas*. College Station: Texas A&M University Press.

Wiedenfeld, C. C. 1983. Lark Buntings (*Calamospiza melano-corys*) breeding in the Edwards Plateau of Texas. *Bull. Tex. Ornith. Soc.* 16:32–33.

Williams, G. G. 1959. Probable Eskimo Curlew on Galveston Island, Texas. *Auk* 76:539–41.

Williams, S. O. 1987. A Northern Jacana in Trans-Pecos Texas. *Western Birds* 18:123–24.

Wolf, D. E. 1978. First record of an Aztec Thrush in the United States. *Am. Birds* 32:156–57.

Wolf, D. E, C. E. Shackelford, G. G. Luneau, and C. D. Fisher. 2001. *Birds of the Pineywoods of eastern Texas: A field checklist.* Austin: Texas Parks and Wildlife Department.

Wolf, L. L. 1961. Specimen of Yellow-green Vireo from Texas. *Auk* 78:258–59.

Wolfe, L. R. 1956. *Check-list of the birds of Texas.* Lancaster, Penn.: Intelligencer Printing.

———. 1965. *Check list of the birds of Kerr County, Texas.* Kerrville, Tex.: L. R. Wolfe.

Wormington, A., and R. M. Epstein. 2010. Amazon Kingfisher (*Chloroceryle amazona*): New to Texas and to North America north of Mexico. *N. Am. Birds* 64:208–11.

Wright, J. S., and P. C. Wright. 1997. Stygian Owl in Texas. *Field Notes* 51:950–52.

Yovanovich, G. D. L. 1995. Collared Plover in Uvalde, Texas. *Birding* 27:102–4.

Zimmer, B., and K. Bryan. 1993. First United States record of the Tufted Flycatcher. *Am. Birds* 47:48–50.

Zink, R. M. 1994. The geography of mitochondrial DNA variation, population structure, hybridization, and species limits in the Fox Sparrow (*Passerella iliaca*). *Evolution* 48:96–111.

Zink, R. M., and R. C. Blackwell-Rago. 2000. Species limits and recent population history in the Curve-billed Thrasher. *Condor* 102:881–86.

Zinn, K. S. 1977. Olivaceous Cormorants nesting in north central Texas. *Southwest. Nat.* 21:556–57.

INDEX